Lincoln Christian College
P9-DEO-985

GENES, MIND, AND CULTURE

Genes, Mind, and Culture

The Coevolutionary Process

Charles J. Lumsden
and
Edward O. Wilson

HARVARD UNIVERSITY PRESS
Cambridge, Massachusetts, and London, England 1981

Printed in the United States of America

Library of Congress Cataloging in Publication Data

Lumsden, Charles J 1949–
Genes, mind, and culture.

Bibliography: p.
Includes index.
1. Sociobiology. 2. Developmental psychology.
3. Social evolution. 4. Cognition and culture.
5. Nature and nurture. I. Wilson, Edward Osborne,
1929– joint author. II. Title.
GN365.9.L85 306′.4 80-26543
ISBN 0-674-34475-8

For Wei-yee Lumsden and Irene Wilson

Contents

Preface

This book contains the first attempt to trace development all the way from genes through the mind to culture. Many have sought the grail of a unifying theory of biology and the social sciences. In recent years the present authors have come to appreciate the probable existence of some form of coupling between genetic and cultural evolution, and we have undertaken our effort with the conviction that the time is ripe for the discovery of its nature. The key, we feel, lies in the ontogenetic development of mental activity and behavior and particularly the form of epigenetic rules, which can be treated as "molecular units" that assemble the mind midway along the developmental path between genes and culture.

Why has gene-culture coevolution been so poorly explored? The principal reason is the remarkable fact that sociobiology has not taken into proper account either the human mind or the diversity of cultures. Thus in the great circuit that runs from the DNA blueprint through all the steps of epigenesis to culture and back again, the central piece—the development of the individual mind—has been largely ignored. This omission, and not intrinsic epistemological difficulties or imagined political dangers, is the root cause of the confusion and controversy that have swirled around human sociobiology.

In attempting to develop a theory of gene-culture coevolution, we have taken care to examine the steps that lead from genes through mind to culture, and to develop explicit models that connect individual mental development to culture and culture to genetic evolution. Of necessity we have drawn on ideas and data from several previously disparate fields, such as population genetics, cultural anthropology, and mathematical physics. Our own primary research interests cover several of these disciplines. One of us (CJL) is a physicist who has extended his research into theoretical biology. The other (EOW) is a biologist with a special interest in evolution and social systems. In the course of our study we have both developed a renewed concern and abiding respect for social theory, the neurosciences, and psychology.

We have also not hesitated to call on colleagues in other disciplines for help. The following persons read parts of the manuscript: David P. Barash, Gary Beauchamp, Daniel Bell, John T. Bonner, Marc H. Bornstein, Irven DeVore, Mildred Dickemann, Robert M. Fagen, James G. Greeno, Paul Harvey, Richard J. Herrnstein, Bert Hölldobler, David H. Hubel, Melvin J. Konner, John Pfeiffer, J. M. Rendel, and Lynn E. H. Trainor. Other forms of assistance, particularly instruction on special topics, was given us by Gary Beauchamp, Scott A. Boorman, Joel Cohen, John E. Dowling, Patricia Draper, Robin Fox, David M. Green, Douglas R. Hofstadter, Robert J. Kiely, Thomas C. Schelling, Joseph Shepher, John Terrell, Richard F. Thomas, Pierre van den Berghe, Harrison C. White, and Johannes Wilbert. Kathleen M. Horton assisted in the bibliographic research and typed the manuscript through several difficult versions. William Minty drafted the original figures. We are grateful to all of these associates and friends but of course in no way hold them responsible for any errors that persist in the final text. Part of the original research was conducted under National Science Foundation Grant No. DEB 77-27515 held by EOW. During the collaboration CJL was supported by a Postdoctoral Fellowship and a NATO Postdoctoral Fellowship, both from the Natural Sciences and Engineering Research Council of Canada.

The contents of *Genes, Mind, and Culture* can be summarized very briefly as follows. We begin with an examination of the possible forms of socialization and show that one of them, gene-culture transmission, is the most likely in any species that attains the advanced form of culture (*euculture*) found in man. A persistent tabula rasa state is unlikely; one can estimate the average time species spend in such a state and specify the conditions that hasten their departure during the course of their evolution. In order to proceed with our analyses, we define a *culturgen,* the basic unit of inheritance in cultural evolution.

In Chapters 2 and 3 we review the epigenetic rules thus far discovered during studies of human behavioral development. These constraints bias individuals toward the assimilation of one culturgen or set of culturgens as opposed to another. A distinction is made between primary epigenetic rules, expressed mostly through sensory screening and perception, and secondary epigenetic rules, which operate during the later stages of memory storage and recall, valuation, and decision making.

In Chapter 4 we introduce the concept of gene-culture translation, which is the transformation of individual cognition and choices of culturgens into cultural patterns. Cognition is constrained by the epigenetic rules, which in turn are genetically determined. The rules can be dissected into innate bias and such context-dependent parameters as the pattern of response to choices made by other members of the society. Using both of these factors in models, we demonstrate that cultural patterns are remarkably sensitive to small changes in the epigenetic rules. Data from developmental and social psychology are employed to analyze actual cases of gene-culture translation.

Chapter 5 begins our evolutionary modeling. We deal with the particularly tractable case of epigenetic rules that depend only on innate programming and are not influenced by cultural patterns. Such models approximate important ethnographic conditions, such as the phenomena of brother-sister incest avoidance and color classification dealt with in earlier chapters. For such systems we investigate the magnitudes of the innate learning biases that prevail in evolutionary time, given specific environmental conditions. We demonstrate the existence of a novel gene-culture adaptive landscape when conditions are steady, and extend the model to cover heterogeneity in space, time, and developmental phenotype.

In Chapter 6 we move to a new conception, in which the epigenetic rules assemble the mind and channel information processing and decision making. In order to model the mechanisms correctly, we extract the key relevant traits of human cognition from current experimental research. Culturgens are defined anew with respect to knowledge structures in long term memory, and genetic fitness is correlated more directly with cognitive process. Using these newly conceived relationships, we construct and analyze the first such complete models of coupled genetic and cultural evolution.

In Chapter 7 we use models similar to those employed in biogeography to examine the accumulation of information within societies. Culturgens are treated as entities that "colonize" minds and become extinct through disuse and memory loss, the net effect of which is to create standing cultures of various sizes that at any given moment may be growing, declining, or constant. The possible constraints on the mind's occupancy of the "civilization niche" are investigated. We use the resulting theory to speculate on why the capacity for euculture has originated so rarely—in fact, only once—thus far in evolution.

Finally, in Chapter 8, the relations of the theory of gene-culture coevolution to the remainder of sociobiology and the social sciences are explored. We discuss the possibility that these subjects will eventually permit a deeper understanding of history.

We are aware that a complete reading of the book may prove taxing to some, because of both the mathematical flavor and the unavoidably multidisciplinary nature of the subject. We have labored to keep the mathematics on a tight leash; it is used where its power can promote insight into the mechanisms that link genes with mind and culture and illuminate the content of our theory. For readers not wishing to pursue the details, we have provided paragraphs, set off in italic type, that summarize the essential content and results of the mathematical sections. These paragraphs appear at the start of each such major section. We recommend that for rapidity of overall comprehension the summaries of the chapters be read first, in sequence, then Chapter 1 in its entirety, and finally the other chapters in part or whole as the reader's interest may suggest.

CJL
EOW

GENES, MIND, AND CULTURE

CHAPTER ONE

Introduction

The subject of this book is what we have come to call gene-culture coevolution. At first glance such an expression might seem to imply a coupling of processes that is unlikely and perhaps impossible. But this is not the case. The linkage between biological and cultural evolution is a logical possibility, the exploration of which has become an increasingly clear major intellectual challenge. Many philosophers and scientists still consider the gap between the biological and social sciences to be a permanent discontinuity, grounded in epistemology and reinforced by a fundamental difference in goals on the part of specialists. We view it instead as a largely unknown evolutionary process—a complicated, fascinating interaction in which culture is generated and shaped by biological imperatives while biological traits are simultaneously altered by genetic evolution in response to cultural innovation.

With a few exceptions[1] evolutionary biologists have hesitated to extend the concepts of biological causation and natural selection to the study of culture. They have been inhibited by the prevalence of what may be called the promethean-gene hypothesis: that genetic evolution produced culture, but only in the sense of creating the capacity to evolve by culture; thus a group of promethean genes has freed the human mind from the other genes. For their part social scientists, again with notable exceptions,[2] have concurred in this view and affirmed the autonomy of the so-

1. In particular, Alexander (1979a,b); Bonner (1980); Boyd and Richerson (1976); Cloninger et al. (1979a,b); Durham (1976,1979); Feldman and Cavalli-Sforza (1976,1979); Pulliam and Dunford (1979); Rice et al. (1978).

2. Reviewed in Campbell (1975); Chagnon and Irons (1979); Fishbein (1976); Freedman (1979); Harrison (1977); Tiger and Fox (1971); van den Berghe (1979).

cial sciences. Theirs is a biologically nondimensional view of social evolution deduced from a second key hypothesis, the psychic unity of mankind. This opinion holds that human culture evolved during too short a time for genetic evolution to have occurred, and that in any case it depends largely upon a single promethean genotype.

In spite of the immediate plausibility of the promethean-gene and psychic-unity hypotheses, the conditions they postulate are special, extreme cases that still await evaluation within a broader theory of coupled genetic and cultural evolution. Such a "comparative social theory" would treat mankind as one in an array of species, both real and conceivably evolvable (Wilson, 1980a). The experience of the natural sciences teaches that the strongest theory is created when the real world is visualized within a matrix of possible worlds. The trajectory of human history can be plotted when the arrays of genetic-cultural properties are deduced across many imaginable species and beyond the arbitrary limits of human variation and are used to define the mathematical elements needed to describe the evolutionary process.

Ethologists and sociobiologists have by and large been sympathetic to this position. They characterize man as one species among a variety of primate species, each adapted in idiosyncratic ways to particular environments. In evaluating the peculiarities of human behavior, they tend to speak of patterns of behavior having a human-specific genetic basis, hence of genes that prescribe behavior. When learning is treated at all, it is viewed largely as a process by which packets of information that encode explicit behaviors leap between generations and colonize the brain—in the same way that pathogens invade hosts.

For mankind at least, these postulates are radically incorrect. Behavior is not explicit in the genes, and mind cannot be treated as a mere replica of behavioral traits. In this book we propose a very different view in which the genes prescribe a set of biological processes, which we call epigenetic rules, that direct the assembly of the mind. This assembly is context dependent, with the epigenetic rules feeding on information derived from culture and physical environment. Such information is forged into cognitive schemata that are the raw materials of thought and decision. Emitted behavior is just one product of the dynamics of the mind, and culture is the translation of the epigenetic rules into mass patterns of mental activity and behavior. In contrast to the approaches of traditional ethology and sociobiology, including previous approaches to gene-culture coevolution, we take account of the free-ranging activities of the mind and of the diversity of cultures created by them. Genes are indeed linked to culture, but in a deep and subtle manner.

We shall embark on our enterprise with a simple evolutionary classification. Consider a series of species that vary in ability to teach and learn. They can be arranged into sets that possess various combinations of four components basic to the evolution of culture: simple learning, imitation, teaching, and reification. The last term means the construction of symbols and other abstract representations of the environment, a function we shall discuss in more detail shortly. Each of the components might conceivably evolve independently of all the others, although in most if not all phylogenetic lines of animal evolution they have appeared in the order in which we gave them: simple learning → imitation → teaching → reification. In Table 1–1 combinations of the components are used to define five evolutionary grades that roughly parallel the gradual emergence of cultural behavior. Each species known to us (see reviews by Wilson, 1975; Alcock, 1979; Beck, 1980; and Bonner, 1980) falls into one or another of the evolutionary grades. In the animal kingdom as a whole, the proportion of species declines precipitously at each step, from the acultural (I + II) to the eucultural grades (see Figure 1–1). Man alone has attained the eucultural grade, that is, the most advanced or "true" cultural state.

We define culture in the broad sense, to include the sum total of mental constructs and behaviors, including the construction and employment of artifacts, transmitted from one generation to the next by social learning. Although recognizing the preeminence of symbols in human culture, we do not agree with Geertz (1966; 1973:89), Schneider (1980), and some other social scientists in considering them to be exclusively diagnostic of

Table 1–1 The components of learning and teaching that define grades of cultural evolution. The most advanced grade (eucultural) can be reached by one or another of several evolutionary routes consisting in a stepwise accumulation of components.

	Components			
Grades	Learning	Imitation	Teaching	Reification (including symbolization)
Acultural I				
Acultural II	•			
Protocultural I	•	•		
Protocultural II	•	•	•	
Eucultural	•	•	•	•

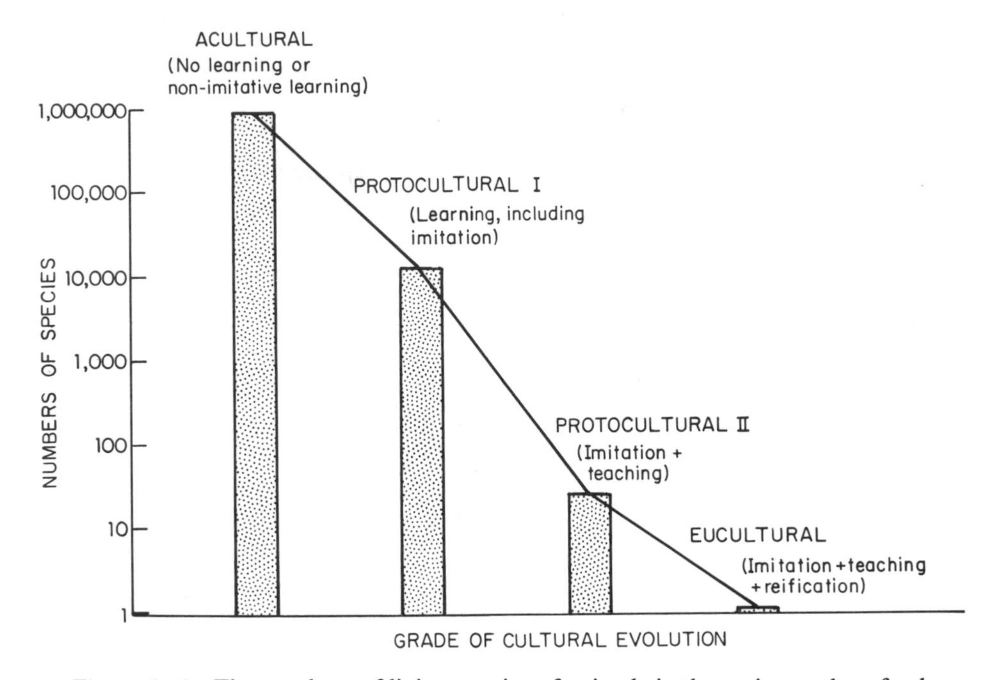

Figure 1–1 The numbers of living species of animals in the major grades of cultural behavior. The acultural group was taken to be composed of all the invertebrates and cold-blooded vertebrates; the 1,000,000 species in the group is an order-of-magnitude estimate (see for example Wilson et al., 1978). The protocultural I group is taken to include roughly the 8,600 species of birds and 3,200 species of mammals exclusive of those in the protocultural II group, in which are tentatively placed the 7 species of wolves and dogs (*Canis*), the single species of African wild dog (*Lycaon*), the one species of dhole (*Cuon*), the one species of lion, both species of elephant, and all 11 species of anthropoid apes. Man is the single eucultural species.

culture. This semantic restriction arbitrarily sets apart a substantial class of imitative behaviors, some of which grade imperceptibly into symbolization. It underestimates the complexity of the cognitive process, much of which cannot be analytically subsumed under the meaning normally given to symbolization. Most importantly, the narrower definition excludes the key word, culture, from the theoretical analysis of animal and ancestral human species.

Mankind has attained euculture in part by means of teaching conducted during the socialization of the young. Programs of intentional instruction, consisting in at least informal questioning and answering, are employed by all human societies (Williams, 1972a,b; Davidson, 1977; Hansen, 1979;

Patricia Draper, personal communication). Adults are strongly prone to provide complex instructions, and the young are predisposed to follow them. As Waddington (1960) aptly expressed this relationship, man is the authority-bearing species.

It is true that teaching is not unique to human beings. In a broad sense honeybees can be said to instruct nestmates during the waggle dance. The direction and duration of the straight run, the central segment of the figure-eight dance pattern, symbolize the location of newly discovered nectar sources or some other target. Other bees automatically follow the dances, memorize the coded instructions, and act on the information to find the target. The mothers of chimpanzees and other primates go further, using signals and physical force to guide the play and imitative movements of their offspring. When infant chimpanzees at the Gombe Stream National Park in Tanzania climbed too high, for example, their mothers tapped on the tree trunks, causing them to come down immediately (Goodall, 1965). As Bonner (1980) notes, teaching is diversified into a continuum across the small number of animal species practicing it, ranging from actions that are performed with the apparent function of inducing imitation to step-by-step reinforcement of imitative and explorative behavior. But even the most elaborate forms of animal teaching fall far short of the intensive and complicated programs of instruction practiced by human societies.

In short, human beings differ quantitatively from animals in the magnitude of the enculturation process. There is in addition a unique activity that fully separates mankind from the most advanced protocultural animal species and makes it the only known eucultural species. This is the process we have called reification. The operations of the human mind incorporate (1) the production of concepts and (2) the continuously shifting reclassification of the world. Insects, cold-blooded vertebrates, and other relatively small-brained animals filter out most signals at the level of peripheral sensory cells and lower associative centers and then respond principally to a very restricted set of "sign stimuli" among the signals remaining. The human mind, in contrast, absorbs vastly greater numbers of chaotically timed stimuli, most of which lack immediate relevance, and constructs an internal reality from them. Gradually varying configurations are broken into categories by a complex but specific set of operations that are beginning to be studied in depth by cognitive psychologists (Posner, 1973; Rosch et al., 1976; Getty et al., 1979). The categories are often two in number—for example, ingroup/outgroup, child/adult, sacred/profane—while their boundaries are enhanced by ritual and taboo. And wholly new productions, "mentifacts" (Huxley, 1958), are created to encompass processes that are emotionally potent but only weakly comprehended by the rational portions of the mind. Thus gods, spirits, and totems can be interpreted as the outward representations of sanctifying, group-binding activities of the mind (Lenski and Lenski, 1970; Rappaport, 1971). Metaphors are created to link more directly perceived physical phenomena with those less easily grasped. One pictures the rush of thought across this page, for example, or the eagle that represents the hereditary unity of the tribe, or the exact differential, dx/dt, signifying change in an arbitrarily chosen entity.

The enabling device of reification is symbolization, which at once serves to aid memory, trigger emotion, classify the environment, and transmit information and feelings to others (Needham, 1979). Human language is largely the manipulation of symbols to convey the reified constructs of the mind. One result of the language-training experiments on chimpanzees has been a partial devaluation of the pure mechanics of language. These animals can be taught over a hundred words by means of sign language, as well as elementary sentence-like combinations of three or more words that convey emotion or present requests and commands (Savage-Rumbaugh and Rumbaugh, 1978, and Terrace et al., 1979). What they evidently lack is the human capacity and sheer drive to reify their experience into new concepts that can be transmitted to others by any means whatever, including single symbols or sentences. Although grammatical rules and phonetics are essential transmittal devices, abstraction and symbolization appear to have been the primary achievements in the evolution of euculture.

The advance of men over chimpanzees and other higher primates was attained during an exceptionally rapid burst of evolution. Over a period of approximately three million years, from the time of the beginnings of a materials-based culture in the ancestral *Australopithecus* to the Upper Paleolithic era of modern *Homo sapiens,* the brain tripled in size. The cranial capacity of *Australopithecus* was 400–500 cc, comparable to that of the chimpanzee and gorilla. Two million years later *Homo erectus* had attained a capacity of 1,000 cc. Another million years saw an increase to 1,400–1,700 cc in Neanderthal man and 900–2,000 cc in modern *Homo sapiens*. The young of modern *Homo sapiens* also undergo a much longer period of socialization, a phase that evidently originated while the human neocortex was expanding. The juvenile period of the chimpanzee, orangutan, and gorilla is seven to eight years. In modern man it is about fourteen and accompanied by a relatively slow, programmed unfolding of cognitive and physiological development. Moreover, the development has incorporated wholly new elements that lead to advanced levels of cognition, thinking, and learning (see Figure 1–2).

In short, the attainment of euculture by the single species *Homo sapiens* was a unique event in more than one respect. It was achieved through an acceleration of neuroanatomical and behavioral evolution unprecedented in the history of life. One can visualize the process in almost physical terms: the crossing of a eucultural threshold by the species, followed—perhaps inevitably—by a sustained autocatalytic reaction in which genetic and cultural evolution drove each other forward.

In order to examine this remarkable process more closely, it is necessary to consider the ways in which genetic and cultural evolution can interact through programs of individual development. As illustrated in Figure 1–3, three classes of programs are possible in a social species. Imagine for the moment an array of transmissible behaviors, mentifacts, and artifacts, which we propose to call *culturgens* (from L. *cultur(a)*, culture, + L. *gen(o)*, produce; and pronounced by us "kul′ tur jens"). The unit is the equivalent of the artifact type employed in archaeology (Clarke, 1978), and it is similar in variable degree to the mnemotype of Blum (1963), idea of Huxley (1962) and Cavalli-Sforza (1971), idene of H. A. Murray (in Hoagland, 1964), sociogene of Swanson (1973), instruction of Cloak (1975), culture type of Boyd and Richerson (1976), meme of Dawkins (1976a), and concept of Hill (1978). A justification for using this neologism, and a more precise definition of it, are given in Appendix 1–1. The culturgens of the simplified scheme in Figure 1–3 are equally accessible for both teaching and learning. They are processed through a sequence of *epigenetic rules,* which are the genetically determined procedures that direct the assembly of the mind, including the screening of stimuli by peripheral sensory filters, the internuncial cellular organizing processes, and the deeper processes of directed cognition. The rules comprise the restraints that the genes place on development (hence the expression "epigenetic"), and they affect the probability of using one culturgen as opposed to another. The probability distributions themselves are appropriately termed *usage bias curves* or, more simply, just *bias curves*.

Suppose that a naive population or a naive individual of a given genotype within a species was confronted with a set of culturgens. These elements might be an assortment of food items, an array of carpenter's tools, a variety of alternative marriage customs to be adopted or discarded, or any comparable array of choices. If the development of individual

SENSORIMOTOR-INTELLIGENCE SERIES

STAGES	TACTILE/ KINESTHETIC			VISUAL/ BODY			VISUAL/ FACIAL			VISUAL/ GESTURAL			VOCAL			AUDITORY			EXAMPLE OF BEHAVIOR
	M	G	H	M	G	H	M	G	H	M	G	H	M	G	H	M	G	H	
1. REFLEX	X	X	X	X	X	X	X	?	X				X	?	X				Roots and sucks
2. PRIMARY CIRCULAR REACTION	X	X	X				•	?	X				•		X				Repetitive hand-hand clasping
3. SECONDARY CIRCULAR REACTION	X	X	X	X	X	X	•	•	X	•	X	X	•	•	X		X	X	Swings object and observes motion
4. COORDINATION OF SECONDARY BEHAVIORS	X	X	X	X	X	X		?	X		X	X			X		X	X	Sets aside one object to get to another
5. TERTIARY CIRCULAR REACTIONS (EXPERIMENTATIONS)	X	X	X	X	X	X		X	X		X	X			X			X	Experimentally learns to use stick to obtain another object
6. INVENTION OF NEW MEANS THROUGH MENTAL COMBINATIONS (INSIGHT)	X	X	X	X	X	X		X	X		X	X			X			X	Mentally discovers how one object can be used to obtain another

IMITATION SERIES

STAGES	VISUAL/ BODY			VISUAL/ FACIAL			VISUAL/ GESTURAL			VOCAL			AUDITORY			DESCRIPTION OF BEHAVIOR
	M	G	H	M	G	H	M	G	H	M	G	H	M	G	H	
1. REFLEXIVE CONTAGIOUS IMITATION												X				Crying or other reflexive behavior stimulated by model
2. SPORADIC SELF-IMITATION				X								X				Imitation by model of infant's motor pattern stimulates vocal activity
3. PURPOSEFUL SELF-IMITATION (SOCIAL FACILITATION)	X	X	X	X	X	X	X	X	X	X	X	X		X	X	As above, but matching becomes more precise
4. IMITATION USING UNSEEN BODY PARTS	?	X	X		X	X		X	X			X		X	X	As above, but infant's body parts can be unseen
5. IMITATION OF NEW BEHAVIOR PATTERNS ('TRUE IMITATION')	?	X	X		X	X		X	X			X			X	Imitation of model by repeated attempts at matching
6. DEFERRED IMITATION	?	X	X		X	X		X	X			X			X	Imitation symbolic or deferred

M Stumptail macaques
G Great apes: gorillas and chimpanzees
H Homo sapiens

☒ Observed behavior
• Parallel stages qualitatively different from human behavior
? Abilities suspected but not observed
☐ Behavior not observed

Figure 1–2 A comparison of the developmental programs of young monkeys, anthropoid apes, and human beings. The classifications used are derived from those of Piaget and follow series of stages in the development of sensorimotor intelligence and imitative abilities. (Based on Chevalier-Skolnikoff, 1977.)

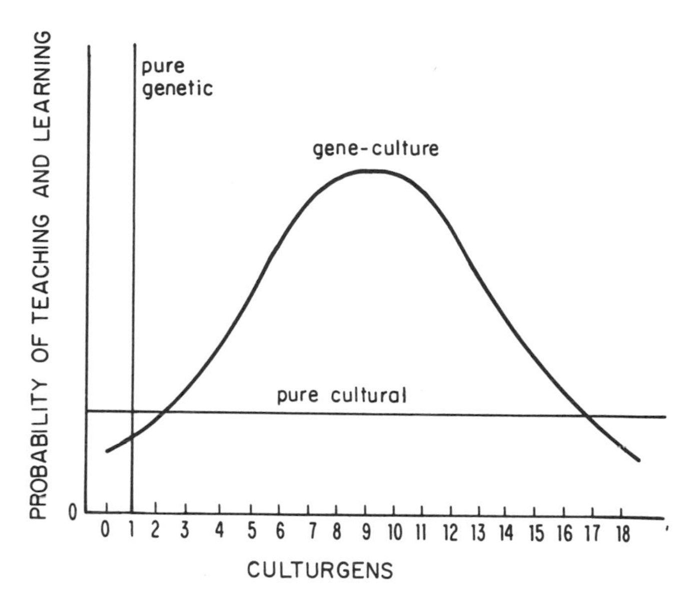

Figure 1–3 The three conceivable classes of programs of informational transmission in a social species. According to the properties of the epigenetic rules (the genetically determined processes of cellular peripheral screening and directed cognition), the probability of transmission of culturgens will be limited to a single choice (pure genetic transmission), all choices equiprobable (pure cultural transmission), or multiple choices that are not equiprobable (gene-culture transmission). These curves are the products of the epigenetic rules possessed by individual organisms of given genotypes with reference to particular categories of culturgens. (Modified from Lumsden and Wilson, 1980a.)

members of the society is genetically constrained in such a way that the same culturgen is selected each time, the transmission is said to be *pure genetic transmission*. The epigenetic field, which is the time course of all developmental choices, is a single narrow channel stretching from birth to maturity. In noncultural species, behavioral choices that are similarly constrained are called fixed-action patterns by ethologists. Depending on the species and behavioral category, such responses can be entirely programmed within the central nervous system, requiring no learning whatever. Or, if learning occurs, it is so tightly directed that only the one genotype-specific behavior is attained. Interference with the developmental process results in a rudimentary version of the behavior or no behavior at all, instead of some alternative fully developed pattern (Alcock, 1979). Thus, the young of many species of birds must learn the species-specific song because they are incapable of learning any other.

Pure genetic transmission has hitherto been of primary interest to biologists, especially neurobiologists and ethologists, because it is typically identified with "instinct," the polar opposite of culture. But note that a highly social species might conceivably evolve that would possess an advanced language and culture all of which is learned, and yet be able to transmit only one set of behaviors and therefore one culture. In other words, a pure genetic culture is a logical possibility! The shift from one preferred culturgen to another because of a change in the epigenetic rules in a totally constrained species, whether it is programmed for fixed-action patterns or for fully channeled learning, is pure genetic evolution.

At the opposite extreme, also illustrated in Figure 1–3, all of the available culturgens are equally likely to be utilized. Epigenetic rules have evolved so as to remove all forms of bias in individual development that could result from peripheral sensory screening, internuncial cellular organization, or innate predispositions in deeper processes of cognition. This *pure cultural transmission* is what many social scientists have in mind when they interpret human evolution:

> Humanness is socio-culturally variable. In other words, there is no human nature in the sense of a biologically fixed substratum determining the variability of socio-cultural formations. There is only human nature in the sense of anthropological constants (for example, world-openness and plasticity of instinctual structure) that delimit and permit man's socio-cultural formations (Berger and Luckmann, 1966:46–47).

This hypothesis appears to be the prevailing view. In an analysis of twenty-four introductory sociology textbooks, Petryszak (1979) found the following assumptions to be basic: "that any consideration of biologic factors believed to be innate to the human species is completely irrelevant in understanding the nature of human behavior and society . . . that human culture is comprised solely of the ideational and technological aspects of society and excludes any consideration of having a biological basis . . . that man's ability to learn and to be susceptible to the processes of socialization and the opinion of others is due solely to the absence of all instincts." Shifts in culturgen preference in such a wholly unconstrained species would be pure cultural evolution. At first thought such an arrangement might appear to constitute a relatively simple, generalized, and stable state, since social behavior has been "liberated" from the genes. But the opposite is more likely to be the case. Physiological equalization of preferences among culturgens requires a genotype that encodes multiple complex fine-tuning mechanisms. Pure cultural transmission can be sustained only with precise controls that depend upon continuing genetic evolution.

The intermediate case is *gene-culture transmission,* defined as transmission in which more than one culturgen is accessible and at least two culturgens differ in the likelihood of adoption because of the innate epigenetic rules. *Gene-culture coevolution* is correspondingly defined as any change in the epigenetic rules due to shifts in gene frequency, or in culturgen frequencies due to the epigenetic rules, or in both jointly. As we shall demonstrate, both kinds of shifts inevitably occur, given enough time, and they exert a reciprocal influence. The formal conceptualization and analysis of this interaction may be called gene-culture coevolution theory or, more concisely, *gene-culture theory.*[3]

On a priori grounds gene-culture transmission appears to be the most likely mode of inheritance for all categories of culturgens, in the human species and in all other imaginable species in which the capacity for culture evolves. Pure cultural inheritance is an improbable alternative outcome of genetic evolution, for the following reasons.

(1) In each of the human sensory modalities (vision, hearing, taste, smell, touch, humidity perception, heat perception) sensitivity varies in degree over the array of accessible stimuli. In the case of vision and hearing, narrow upper and lower limits bracket the wave frequencies that can be discriminated or even perceived. It would be difficult or impossible for an evolving biological system, dependent on transducer cells and chains of neurons, to attain sufficient quality control so as to approach uniform sensitivity and at the same time to abolish limits on the frequency spectrum. The same considerations apply to the reading of pattern and complexity by coding devices among interneurons. In short, central nervous systems are inherently unprepared to perceive and classify equally all ranges of primary stimuli from culturgens. There is certain to be some bias built *ab initio* into their screening and coding apparatus.

(2) *Homo sapiens,* and any other theoretically conceivable species placing strong reliance on cultural inheritance, is unlikely to begin its evolution from an undifferentiated behavioral development. As in all animal

3. Coevolution is a well-known phenomenon in biology. Theoretical and experimental studies have been conducted on the interaction of competing species, character displacement in premating isolating mechanisms between hybridizing species, predators and prey, hosts and parasites, and partners in mutualism. Excellent reviews of these forms of interaction are given by Roughgarden (1979), Slatkin and Maynard Smith (1979), and D. S. Wilson (1980).

species, the immediate ancestral populations must have been dependent on automatic peripheral screening and pattern coding, as well as forms of prepared learning such as imprinting and inhibition resulting from an adaptive, biased preference for certain stimuli.

(3) Even if a species could somehow start with uniform epigenetic rules, the strategy would generally be unstable, leading in the course of gene-culture coevolution to the reappearance of nonuniform epigenetic rules.

Consider a population of diploid organisms with individual genotypes G_iG_j, where each paired i and j denote the allelomorphs on each of the loci. In a cultural species the genetic fitness w of an organism is determined not only by its genotype G_iG_j but also by its cultural heritage as expressed by its set of assimilated culturgens, $\mathbf{c}$. The genetic fitness w is subject to changes due to learning, innovation, and other time-dependent cultural processes that alter the culturgen content:

$$w = w(i,j,\mathbf{c}) = w[i,j,\mathbf{c}(t)].$$

As indicated by Figure 1–4, genetic fitness is enhanced when the probability per unit time $v_{0k}(t)$ for the change of culturgen set from $\mathbf{c}_0$ to a set $\mathbf{c}_k$ with equivalent or higher relative fitness is greater than the probability per unit time $v_{0m}(t)$ for a change to some other set $\mathbf{c}_m$ with lower relative

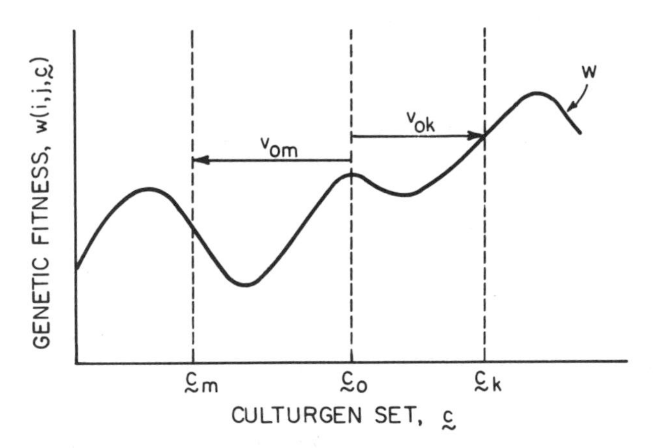

Figure 1–4 The evolution of epigenetic rules. Natural selection favors organisms that consistently make the probability of shift v_{0k} from culturgen set $\mathbf{c}_0$ to culturgen set $\mathbf{c}_k$ greater than the probability of shift v_{0m} from $\mathbf{c}_0$ to $\mathbf{c}_m$, where the genetic fitness conferred by $\mathbf{c}_0$ is greater than that conferred by $\mathbf{c}_m$ but less than or equal to that conferred by $\mathbf{c}_k$.

fitness. The epigenetic rules provide this capability, shaping the organism's pattern of culturgen use in such a way that transitions to relatively advantageous sets such as $\mathbf{c}_k$ occur with greater frequency than transitions to relatively deleterious sets such as $\mathbf{c}_m$.

Now consider a population of tabula rasa organisms, which change their culturgen sets and alter the degree of control over behavior exercised by specific culturgens without reference to the effects on genetic fitness. In other words, the developmental field is flat. The population is exposed to an environment that contains both adaptive and deleterious culturgens, but it is unable to distinguish them. At the same time it is open to cultural programming that could at whim set $v_{0m} > v_{0k}$. Over a period of generations the population is unstable against invasion by genetic mutants that program epigenetic rules biasing individuals toward assimilation of relatively adaptive sets. The epigenetic rules will then tend to channel cognitive development toward certain culturgens as opposed to others. We refer to this relation informally as the "leash principle" in order to make it metaphorically more vivid: genetic natural selection operates in such a way as to keep culture on a leash.

The leash symbolizes genetically prescribed tendencies to use culturgens bearing certain key features that contribute to genetic fitness. These tendencies are distinct from the hardwired algorithms that lead with certainty to the assimilation of specific culturgens (and hence permit no leash at all). Several other factors will however conspire to favor a lengthening of the leash in cultural species. In terms of pure physiology, the benefits accruing from a neural subsystem able to identify and classify the adaptive features of culturgens will eventually be offset as its precision increases by the costs of the ontogenetic pathways required to create the system and by the metabolic costs of maintaining it in a functional state (see Figure 1–5). This limitation of the resolution capacity produces additional flexibility in culturgen assimilation capacity and increases the length of the leash. In the case of eucultural and some protocultural organisms we expect the culturgen set to be more than just a passive receptacle for assimilated cultural information. It is also a site where new culturgens can be invented. For the innovating cultural organism the adaptive value of a culturgen is partly dependent on the potential of the organisms to "seed" innovations that possess adaptive features and to promote their assimilation. Even when novel culturgens carry high risk, they can bear a correspondingly high probability of assimilation under epigenetic rules previously favored by genetic natural selection. Such culturgens may or may not bear short term fruit before their risk is realized.

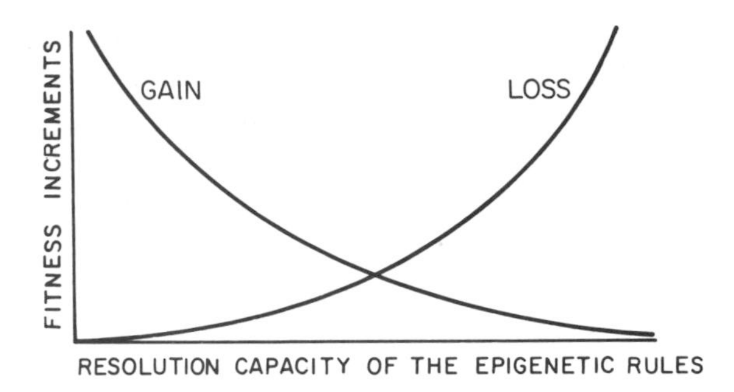

Figure 1–5 The resolution capacity of epigenetic rules as constrained by an eventual reduction in the net gain of genetic fitness. This reduction results from both the decline in the increment that accrues from improved resolution (labeled "gain" in the figure) and the rising cost of the more precise physiological mechanisms required for the improvement in resolution (labeled "loss"). In the imaginary case depicted here the optimum is reached when the curves cross, or at a level of intermediate resolution capacity.

The transition probabilities can be perturbed from the tabula rasa state by either the incorporation of appropriate new genotypes or the discovery of new culturgens. The former alters the effect on genetic fitness of previously existing culturgens by changing the epigenetic rules that affect the probability of culturgen usage, while the latter alters the fitness of previously existing genotypes. The important point is that both kinds of events move the population from the tabula rasa state.

We can go so far as to estimate the waiting times before culturgen innovation will have this perturbing effect. Such information makes it possible to evaluate in a preliminary manner the relative importance of the contributing factors. In the case of culturgen innovation, where the distinctness of the culturgens from one another is a random variable and culturgens are created at a constant rate, the time to the introduction of a fitness-altering culturgen is

$$t = [\nu MI(\pi^{1/2}\alpha)^N]^{-1}\,\Gamma\left(\frac{N}{2}+1\right)$$

where ν is the rate of culturgen innovation in culturgens per person in unit time, I is the number of persons in the population, M is the number of fitness-altering culturgens accessible, N is the number of qualities by which a culturgen can be distinguished from others, α is a measure of the difficulty of distinguishing individual culturgens, and Γ denotes the gamma function. As illustrated by the curves in Figure 1–6, systems appear to be most sensitive to variation in the ability to discriminate cul-

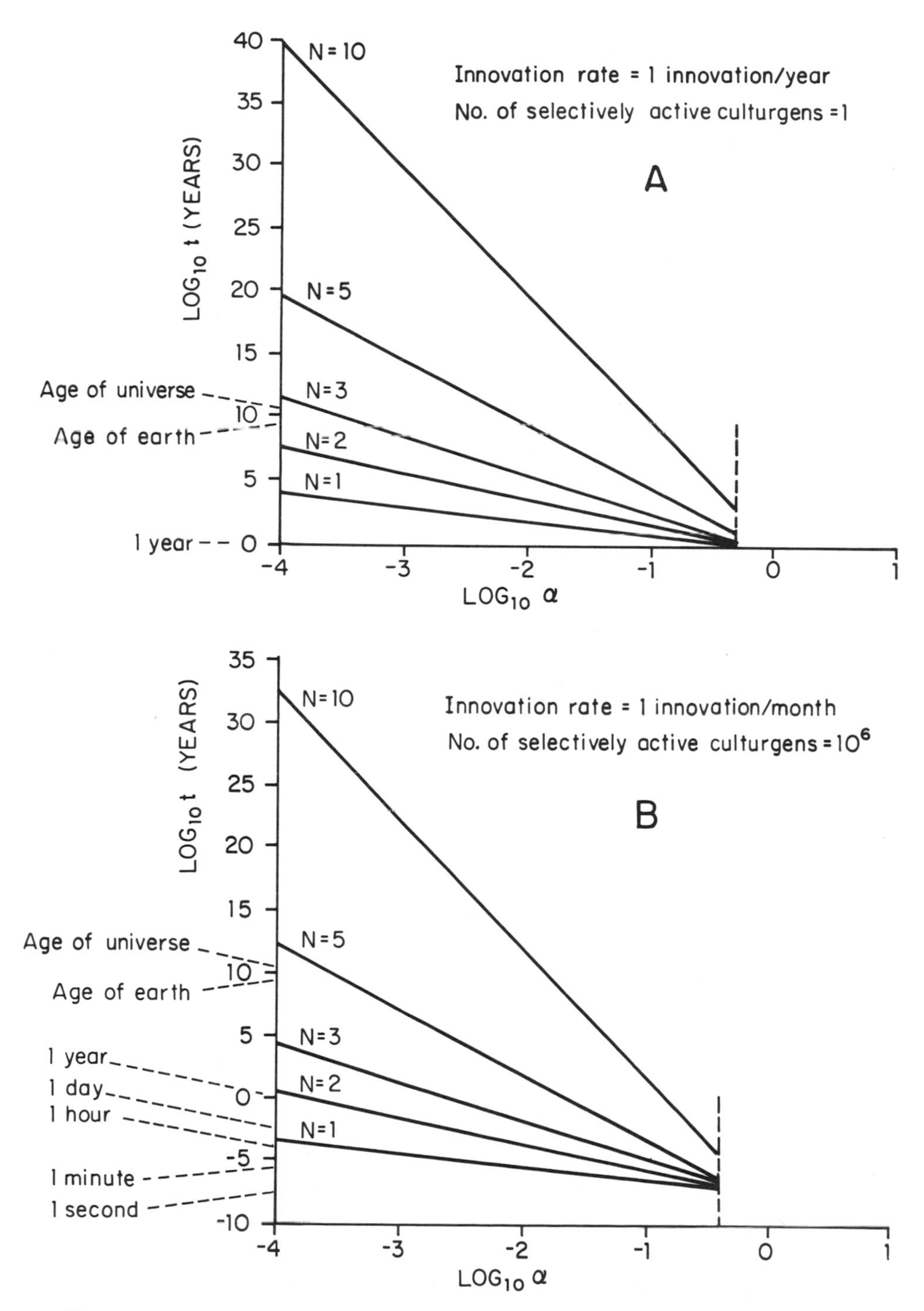

Figure 1–6 Waiting times to the departure of a society from a tabula rasa state, as a function of the number N of attributes used to discriminate culturgens and of the resolving power α of the cognitive system. The innovation rates shown in the diagram are those of the whole population, in other words νI.

turgens. As the number of dimensions in the culturgen space (N) is increased and the ease of recognition in the separate dimensions is enhanced (by reduction of α), the rate at which fitness-altering culturgens are added falls very steeply. The same principle holds for the assimilation of "deadly culturgens," extreme fitness-altering innovations that endanger the welfare of the entire society. Such improvements in culturgen discrimination would slow the departure from the tabula rasa state, but it could not avoid it altogether. (For a derivation of the waiting-time estimate formula, see Appendix 1–2.)

Is the human species engaged in gene-culture coevolution, as suggested by the foregoing argument? Four classes of evidence seem to us to be minimally necessary for such a conclusion. First, it must be shown that nonuniform epigenetic rules exist, that they are commonplace if not universal, and that they can be analyzed in such a way as to test the details of gene-culture coevolutionary theory. We shall demonstrate in Chapters 2 and 3 that this first requirement is fully met.

The second requirement is that genetic variance in epigenetic rules must exist within human populations. Because the study of epigenetic rules is still in an early stage, the extent and causes of their variation have seldom been explicitly considered—and certainly not in the context of gene-culture theory. But pedigree analysis and standard comparisons of fraternal and identical twins, in some instances strengthened by longitudinal studies of development, have yielded evidence of genetic variance in virtually every category of cognition and behavior investigated by this means, including some that either constitute epigenetic rules or share components with them. These categories include color vision, hearing acuity, odor and taste discrimination, number ability, word fluency, spatial ability, memory, timing of language acquisition, spelling, sentence construction, perceptual skill, psychomotor skill, extroversion/introversion, homosexuality, proneness to alcoholism, age of first sexual activity, timing of Piagetian developmental stages, some phobias, certain forms of neurosis and psychosis, including manic-depressive behavior and schizophrenia, and others (Heston and Shields, 1968; McClearn and DeFries, 1973; Ehrman and Parsons, 1976; Farley, 1976; Loehlin and Nichols, 1976; Martin et al., 1977; Bohman, 1978; R. S. Wilson, 1978; Ashton et al., 1979; Comings, 1979; Rainer, 1979; Vandenberg and Wilson, 1979).

The third condition for the establishment of gene-culture coevolution is the verification of a link between cultural practice and genetic fitness within human populations. In fact, such a relation has already been documented in a wide array of behavioral categories, often entailing fine differences among culturgens. For example, certain practices in tattooing and other modes of body marking, as well as in circumcision, treatment of menstrual and afterbirth blood, and diet, are known to transmit viruses and other infectious agents that profoundly affect mortality, birth rate, and even sex ratio (Blumberg and Hesser, 1975; Gajdusek, 1977; Drew et al., 1978). Documentation also exists for the direct effects of sexual practice, marital customs, early mother-infant attachment, differential infanticide, formalized techniques of aggression, and economic organization on genetic fitness (Daly and Wilson, 1978; Chagnon and Irons, 1979; D. G. Freedman, 1979; Kennell and Klaus, 1979). However, the long-term effects of such practices have not been measured.

Relatively detailed connections have been made between culture and genetic fitness in the case of cuisine. Humans are genetically unable to biosynthesize the amino acid lysine, which must be obtained from a balanced diet. Maize, the only cultivated cereal of the New World, possesses substantial quantities of lysine, but two-thirds of the compound is locked up in the indigestible glutelin fraction of the endosperm and germ. The simplest method for releasing the sequestered portions is alkali cooking. Several variations of this technique have been invented and then spread by cultural diffusion. Among the fifty-one societies in North, Central, and South America reviewed by Katz and associates (1974), there exists a strong positive correlation between the intensity of maize cultivation, the use of alkali cooking, population density, and the complexity of social organization. Archaeological data and observations of contemporary societies have shown that population density increased when heavy reliance was placed on maize as a food staple. But maize is not a nutritious food unless processed for lysine release. It seems unlikely that the many New World societies adopting alkali cooking could have directly perceived this procedure as the solution to their biochemical shortfall and hence as a requisite for further population growth and social evolution.

A different kind of gene-culture relation is exemplified in the use of the fava bean, one of the most popular and easily cultivated crops of the Mediterranean region. The gene $G6PD^-$ is a sex-linked recessive, which when unaccompanied by the $G6PD^+$ allele results in a deficiency of the red-blood-cell enzyme glucose-6-phosphate dehydrogenase. Among Mediterranean populations the incidence of $G6PD^-$ ranges from 0.05 to 0.30. Its relatively high frequency is generally attributed to the heightened resistance to malaria that it confers. However, the purely biochemical deficiency caused by the gene is worsened by the consumption of fava beans, causing illness and sometimes death. There appears to exist a

$G6PD^-/G6PD^+$ polymorphism, with the $G6PD^-$ frequency almost certainly kept lower by the consumption of fava beans and the resulting disorder of favism. Not surprisingly, fava beans have been the object of special rituals in cooking, selective taboos, and legends in Mediterranean societies throughout most of recorded history. Yet there is little evidence that the peoples affected made any direct, rational connection between their beliefs and what is now perceived to be the true nature and cause of favism (Katz, 1980).

The fourth and final condition necessary for gene-culture coevolution is the existence of molecular and cellular mechanisms that directly connect genes to cognitive development. Intermediate links have been extensively documented in sensory reception and behavior in animal species (McClearn and DeFries, 1973; Ehrman and Parsons, 1976; Hall and Greenspan, 1979). While the evidence for a linkage in human cognition is fragmentary, some of the essential steps are known. Even the most complex epigenetic rules can be altered by genetic changes that affect molecular and cellular mechanisms. At the level of peripheral sensory screening, for example, single mutations altering color vision and sensitivity to phenylthiocarbamide are based on changes in molecular structure in the primary receptor cells. Similarly elementary genetic controls affect the ability to detect certain odorants such as pentadecalactone and might underlie sex differences in perception of the musk-like substance exaltolide.

At the next level of organization, some of the twenty or more known neurotransmitters, including the monoamines serotonin, noradrenaline, and dopamine, exercise profound effects on mood, concentration, sleeping habits, and social behavior. They operate at specific receptor sites, and their effects can be enhanced by the use of substances that inhibit their specific oxidases or promote persistence at the synapses by other means. Any mutation that to any significant extent alters the production of the behaviorally potent neurotransmitters or the properties of their receptor sites is likely to alter epigenetic rules as well. For example, the most severe clinically definable forms of schizophrenia are evidently under partial genetic control (but see the reservations of Taylor and Abrams, 1977). They are associated with higher densities of dopamine receptors or higher concentrations of dopamine (Greenberg, 1978). Similar possibilities exist in the adjustment of the melanocyte-stimulating hormone, a peptide that affects memory and anxiety levels, and the endorphins, which mediate pain reception, anxiety, and depression (Arehart-Treichel, 1978). Neurobiologists and physiological psychologists are optimistic that the cellular mechanisms responsible for behavioral phenomena in man will eventually be understood (see for example Boddy, 1978, and Schmitt and Worden, 1979). Farley (1976) and other biological psychiatrists have gone so far as to conclude that neuroses and psychoses are mental conditions caused by physiological events often just beyond the ordinary genetically mediated range of activity. Sufficient perturbations can be produced by unusual genotypes, by early developmental accidents, by environmental stress, or by interactions among these processes. Thus fundamental changes in behavior can sometimes follow relatively small amounts of genetic change that affect key cellular activities.

At a still higher level of organization, that of cytoarchitecture, the pleasure centers of the brains of rats, rhesus monkeys, and human beings are now known to consist not of local cell clusters but of entire fiber tracts that extend from the limbic area into the frontal cortex. Research has shown that activity in this "brain reward system" is affected differentially by the monoamine neurotransmitters (Routtenberg, 1978; Stevens, 1979). We suggest that a genetic alteration of the system would redirect cognitive development and thus constitute a modification of the epigenetic rules. At the highest level, there is no shortage of models to explain the neuronal basis of consciousness and the mind. Indeed, this is currently one of the most active areas of theoretical neurobiology, and its range is exemplified in the writings of Pribram (1971), Buser and Rougeul-Buser (1978), Colby (1978), Edelman and Mountcastle (1978), Grossberg (1978), Roederer (1978), Wassermann (1978), and Simon (1979). Some of the models are remarkably subtle and draw inventively on a large amount of experimental data on brain structure and action. While the body-mind connection remains to be worked out in detail, its feasibility has been established in concrete reconstructions based on neurobiological information.

To summarize briefly to this point, we have argued that gene-culture transmission was a virtually inevitable property of the human species as it entered the most advanced, "eucultural" grade of social organization, because the alternative, pure cultural transmission, is inherently a rare and unstable condition in any cultural species. We have further shown that the genetically based epigenetic rules expected in gene-culture coevolution do occur in human beings and that all of the conditions necessary for their empirical analysis also exist.

We come now to the central empirical problem of gene-culture coevolutionary theory: the degrees of rigidity and specificity of the epigenetic rules created by the interaction of genetic and cultural evolution in human populations. It is an important consideration that only a slight difference, on the order of 1 percent or less, in the selection rate of two phenotypes based on epigenetically constrained culturgens can produce genetic evolution through the population as a whole sufficient to bias individual development toward the favored culturgen. Furthermore, the genetic shift can become detectable in as few as ten generations. This inference is based on both the diallelic and polygenic models of classical population genetics (Crow and Kimura, 1970; Roughgarden, 1979), and we shall later confirm it directly in the models of gene-culture coevolution. It has been amply documented also in selection experiments on orientation and other elementary forms of animal behavior (Dobzhansky et al., 1972). Its particular implications for gene-culture coevolution will be explored in Chapter 6.

Conversely, only a small bias in the selection of culturgens is sufficient to yield a striking new pattern of social organization. In *Micromotives and Macrobehavior* (1978), Schelling provided some impressive examples of large effects on pattern based on relatively small personal decisions, including traffic jams caused by ten-second slowdowns by individual drivers, housing segregation resulting from nothing more than a mild desire to join a majority, the dying of institutions through a small decline in interest by the members to below a threshold level, and so forth. It seems equally likely that weak innate predispositions toward the adoption of one subset of culturgens out of a larger set available (for example the tendency to prefer certain food items, dress, or marriage arrangements) can be amplified under certain conditions into much larger events in cultural evolution. For example, the relatively minor innate differences in early temperament in young boys and girls (Blurton Jones and Konner, 1973; Maccoby and Jacklin, 1974) are magnified into consistent role differences in all societies, including extreme male dominance in a few (Rohner, 1975; Draper, 1976).

An important consequence of this amplification is the ease with which the pattern alteration can be initiated by different causes and reached along separate pathways. Hence micromotives are relatively difficult to deduce from a knowledge of macrobehavior alone. Rapid historical change is often cited as proof of the absence of genetic constraint in human social behavior (Allen et al., 1976; M. Harris, 1979). However, it is not the amount of cultural evolution that is relevant, but the consistency of its direction. Also important are the obedience of the pattern to detailed predictions from sociobiological theory and, above all, the cause-effect relation between the change and epigenetic rules in individual development. The significant theoretical questions of the social sciences will become more tractable when precise information from developmental psychology is incorporated into the models of gene-culture coevolution.

The study of culture should therefore benefit from a body of work that focuses on the relation between genetic and cultural evolution as mediated by the epigenetic rules. We suggest that the evolution of the rules entails two complementary and reciprocating modes of selection. The first is *genetic assimilation,* which consists in the following sequence of events: the frequency distribution of phenotypes produced by the interaction of genotype and the ordinary environment of the species is extended by a change in the environment, creating a novel phenotype; the tendency to produce the phenotype in the new environment varies among the preexisting genotypes; the novel phenotype is at a genetically selective advantage in the new environment, causing an increase in the frequency of the genotypes predisposing development to it; after a sufficient number of generations, depending on the intensity of the selection and the mode of inheritance, the population as a whole becomes more prone to develop the phenotype and the trait may even persist as part of the norm of reaction when the species returns to the original environment. Although genetic assimilation has hitherto been documented only in anatomical and physiological traits (Waddington, 1962; Futuyma, 1979; Milkman, 1979), it should be equally attainable in behavioral phenotypes, including cultural innovations which themselves partly constitute the altered environment. The genetic assimilation of cultural capacity is expected to occur through a shift in epigenetic rules, making transmission of the novel culturgens more likely (see Figure 1–7).

The inverse process can be called *culturgen assimilation:* if developmental flexibility is great enough, certain novel culturgens are likely to be invented and spread. In other words, culture tends to fill out the space permitted by the genetically determined epigenetic rules. In addition, culture is expected to be richest in those categories of behavior where the rules most favor it. We would expect culture to pile up as nodes around the conventions most affected by biased epigenetic rules, such as incest avoidance, courtship, and discrimination between in-groups and out-groups. The most ritualized forms of culture will not tend to replace the epigenetic rules, as many social scientists have thought, but to reinforce them.

Epigenetic rules can evolve by genetic assimilation under the influence of a few selectively advantageous novel culturgens, an alteration that admits still other culturgens. In other words, culturgen assimilation follows genetic assimilation. Alternatively, culturgen assimilation can

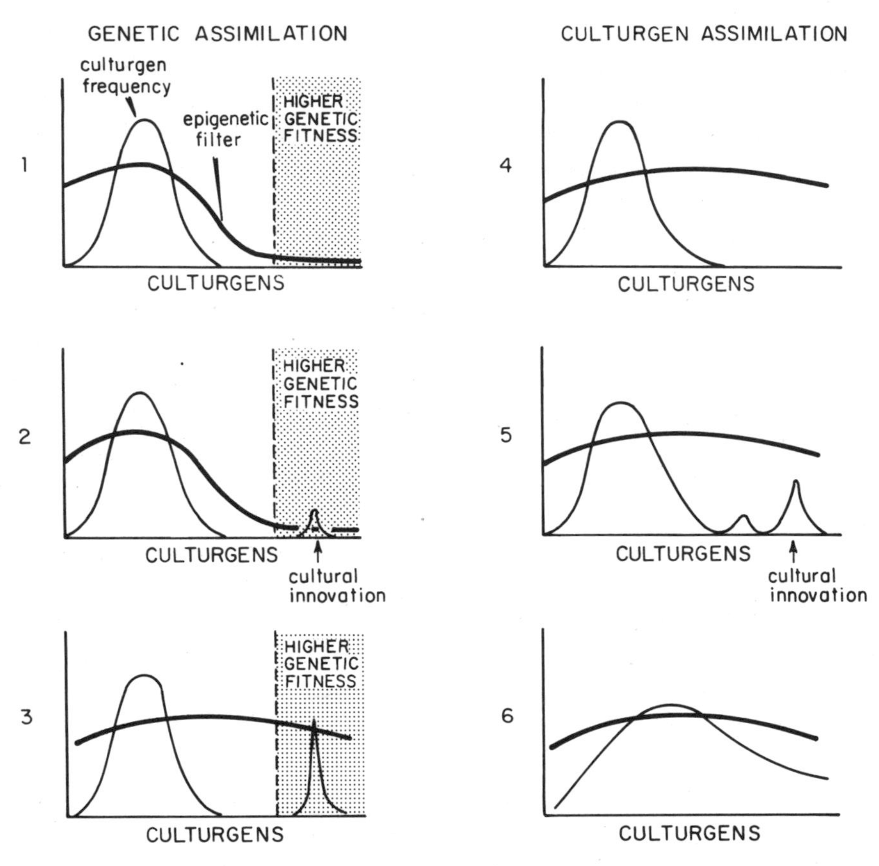

Figure 1–7 A simplified representation of the hypothesized reciprocal effects of genetic and cultural transmission during gene-culture coevolution. The thin line denotes the frequency of alternative culturgens in the society, while the thicker line represents the propensities to learn, use, and teach certain culturgens. *Genetic assimilation:* a culturgen is invented that conveys greater genetic fitness (1, 2); subsequent natural selection alters the epigenetic curve to make transmission more likely (3). *Culturgen assimilation:* the epigenetic rules are already generally permissive, with the result that culturgens are more easily invented and spread (4–6). Although the two modes of assimilation are displayed in sequence in this particular scheme, they can also occur simultaneously. Furthermore, the direction of assimilation can be reversed, resulting in a narrowing of the epigenetic rules and the impoverishment of culture.

follow cultural impoverishment, entailing no genetic alteration in the epigenetic rules.

The elementary properties of gene-culture coevolution can be perceived more clearly if they are expressed in the abstract imagery of a single life cycle (see Figure 1–8). In the course of each generation the culturgens of a society are swept through "filters" consisting of the epigenetic rules. If the culture is composed of elements easily taught and learned and is not innovative, it will remain stable from one generation to the

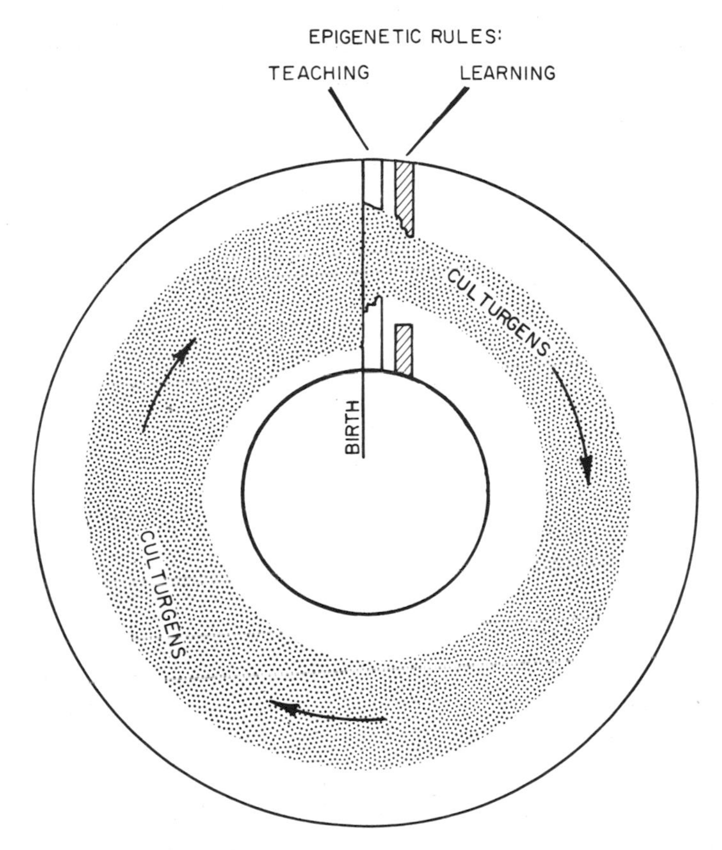

Figure 1–8 An abstract representation of gene-culture coevolution. The torus contains the life cycle of a member of the society. Culturgen assimilation occurs when culturgens are introduced to the society during the individual's life span; in this particular example, an exceptionally large number are displayed. The epigenetic rules that affect both the proneness of individuals to teach particular culturgens and of others to learn them are shown in the diagram as filters. These rules, which affect the rate and extent of cultural evolution, are placed here as a single occurrence in early childhood, but they can occur in multiples and throughout life. Genetic assimilation occurs when the filters are altered over many generations by natural selection operating on the epigenetic rules. When the novel culturgens consistently confer higher genetic fitness, the filters are eroded. When the culturgens lower genetic fitness, the filters grow inward.

next. The culturgens "feel right" to the members of the society, and over many generations they result at most in a relatively weak intensity of stabilizing genetic natural selection, which in turn holds the epigenetic rules approximately constant. If on the other hand new culturgens are introduced by innovation or diffusion, they will be incorporated with relative ease, causing the culture to expand or shift, until they are opposed by the epigenetic rules. This is the process we have called culturgen assimilation. Culturgens disfavored by the epigenetic rules will be eliminated more quickly and come to occur in lower frequencies across many societies. The process is expected to continue indefinitely unless the novel culturgens confer higher genetic fitness, in which case the epigenetic rules themselves can be altered by genetic evolution over a period of generations in a direction more permissive to the novel culturgens. This form of genetic assimilation is represented in the figure as an eroding of the epigenetic filters. But the opposite process can also occur: if new culturgens appear that consistently lower fitness, the epigenetic rules will be tightened. In the imagery of the metaphor, the filters now grow inward.

We can now restate the goal of this book in a more precise form: it is to formulate a theoretical framework that will generate a full spectrum of models from all-genetic to all-cultural transmission, with reference to either the entire repertory of species or particular categories of behavior within the repertory. This program is based on our postulate that human cultural transmission is ultimately gene-culture transmission, and human cultural evolution represents only a small set of traces within a far larger array of possible transmission histories. The transmission space can be made comprehensible through the development of a comparative social theory, in which a wide range of inferred nonhuman solutions is added to the conventional social sciences.

The theory of gene-culture coevolution will greatly extend the scope of sociobiology as well. Zoological studies have hitherto focused on a relatively limited set of phenomena, such as kin selection, territoriality, and caste systems, that are most likely to incorporate cooperation into the instinctual repertories of animals. Sociobiology has not coped with learning and cognition or the consequences of the deep, thorough socialization that characterizes the behavioral development of human beings. The theory of gene-culture coevolution is designed to address these issues and permit the entry of evolutionary theory into the study of the mechanisms that produce the human mind and behavior. As we will show in the coming chapters, it derives patterns of cultural diversity from biological ground rules. And because human socialization and cognition affect virtually every behavioral trait short of the reflexive and autonomic, gene-culture theory extends sociobiology far beyond the conventional topics of the animal-based studies. It indicates the existence of a large realm of distinctly human epigenetic rules, most of which remain unstudied, whose consequences in the channeling of cultural evolution are profound and still largely unappreciated.

Summary

Within the framework of the newly proposed comparative social theory, mankind is classified as a eucultural species, in which mental activity is based to a large extent on reification and symbolization and the young are socialized through purposive programs of teaching.

During socialization an array of behaviors and artifacts, which we have termed culturgens, are processed through a sequence of epigenetic rules. These rules are the genetically determined peripheral sensory filters, interneuron coding processes, and more centrally located cognitive procedures of perception, learning, and decision making. They affect the probability of transmitting one culturgen as opposed to another. The probability distributions themselves constitute bias curves, which will be used in later chapters to link human cognition to patterns of social behavior.

Enculturation can theoretically be composed of pure genetic transmission, in which all members are genetically constrained to learn one culturgen within a given category; or pure cultural transmission, in which no innate predisposition exists favoring one culturgen over another; or gene-culture transmission, in which one or more culturgens are favored because of bias from the innate epigenetic rules (see Figure 1–3). Gene-culture coevolution is correspondingly defined as any change in the epigenetic rules due to shifts in the frequencies of the prescribing genes, in culturgen frequencies due to the influence of the epigenetic rules, or in both jointly. On the basis of theoretical considerations and evidence from genetics and neurobiology, it appears that eucultural species will always tend to evolve toward gene-culture transmission. In the simplest circumstances it is possible to estimate the time of departure from a tabula rasa state to nonuniform epigenetic rules. We show that such a shift should occur most rapidly when the number of dimensions used to distinguish culturgens is small. It will also be enhanced when the ability to distinguish culturgens within any one dimension is small, and when the number of discoverable culturgens that are advantageous or disadvantageous in genetic natural selection is large (see Appendix 1–2).

Four conditions are necessary for the analysis of gene-culture coevolution: nonuniform epigenetic rules exist and can be studied in such a way as to test gene-culture coevolutionary models; some variance in the expression of the epigenetic rules is heritable; cultural practice affects genetic fitness; and causal links exist between genetically controlled processes at the molecular and cellular level and the epigenetic rules. All of these processes have been documented in contemporary human populations.

Gene-culture coevolution includes both genetic assimilation, in which epigenetic rules predisposing individuals toward advantageous culturgens are strengthened by natural selection, and culturgen assimilation, in which cultural innovation is speeded by the preexistence of permissive epigenetic rules. The two processes are envisaged as acting in a reciprocating and often nonequilibrium manner (see Figures 1–7 and 1–8).

The theory of gene-culture coevolution extends sociobiology well beyond the conventional topics of animal-based studies, such as kin selection, territoriality, and caste systems. It indicates the existence of a large realm of distinctively human epigenetic rules that affect socialization and cognition in virtually every category of behavior. Most of these rules remain unstudied, and their role in the channeling of cultural evolution is only beginning to be understood.

Appendix 1–1

The Definition of Culturgen

New terms should be put forward reluctantly, for jargon is the anesthetic of scholarship. But we recommend an exception in the case of the culturgen, for the reason that a neologism can be defined more precisely. No word burdened with a history, as Nietzsche said, can be defined perfectly. Culturgen does not share exactly the same meaning as the other expressions—idene, instruction, and meme—that have been used in various ways for approximately the same category, and it can therefore be incorporated without ambiguity into this first comprehensive gene-culture theory. Other advantages that the term has over the alternatives is its proper (albeit hybrid) hellenistic derivation, the availability of a reasonably graceful adjectival form (culturgenic), and the fact that the word is easily recognizable and the correct operational meaning implicit (culturgens generate culture).

Whatever scholars choose to call the unit, they will find it necessary to devise a consistent, practical definition. This has in fact been accomplished to a reasonable level of satisfaction by Clarke (1978) and other archaeologists who, perhaps more than other social scientists, have seen the need for a rigorous definition of the operational unit of culture. The unit of archaeology is the artifact type, which can be regarded as merely a special kind of culturgen. Its definition can be modified as follows for the more general category: *a culturgen is a relatively homogeneous set of artifacts, behaviors, or mentifacts (mental constructs having little or no direct correspondence with reality) that either share without exception one or more attribute states selected for their functional importance or at least share a consistently recurrent range of such attribute states within a given polythetic set.* Later, in Chapter 6, we shall show how culturgens act via relational networks in long term memory, and in many instances can be identified with them.

The concept of the polythetic set is derived from numerical taxonomy, a methodology that attempts to quantify the degrees of relationship among organisms, species, and other conceivable objects for which classifications are needed (Jardine and Sibson, 1971; Sneath and Sokal, 1973; Doran and Hodson, 1975). A polythetic group is any set of entities, such as an array of swords or marriage ceremonies, in which each entity possesses a large number of the attributes of the group, where the attributes might be size, geometric shape, duration of a process, and so forth. Furthermore, each attribute is shared by large numbers of entities, while no single attribute is both sufficient and necessary for group membership. As illustrated in Table 1–2, polythetic groups are distinguished from monothetic groups, which are defined by the fact that they all possess—without exception—one or more diagnostic attributes. The proposed definition of a culturgen embraces such polythetic or monothetic groups of particular artifacts, behaviors, and mentifacts, where it is further understood that one or some combination of the attribute states are the functions served by the culturgen.

Clarke has suggested that among artifact sets there exist tightly correlated core clusters of closely covarying attributes, some of whose links are as high as 90–100 percent. Most of the attributes might be connected at a level of perhaps 60 percent, while a minor "penumbra" of sporadic attributes will be linked at less than 10 percent with any other. Each artifact type can be represented as a core of attributes within an outer group

Table 1–2 Definition of a culturgen as a group of particular usages that can be distinguished either as a monothetic set or polythetic set. +, attribute present; −, attribute absent.

		Artifacts or behaviors							
		A	B	C		D	E	F	G
Attributes	1	+	+	+					
	2	+	+	+	monothetic				
	3	+	+	+	group				
	4	+	+	+					
	5	+	+	+					
	6					+	+	−	+
	7				polythetic	−	+	+	−
	8				group	+	−	+	+
	9					−	+	−	+
	10					+	−	+	−

of attributes of decreasing correlation. Each type can also be represented as a more or less discrete cluster of specimens in a space of discriminant functions. An example of the latter display, from the analysis of Hallstatt swords by Doran and Hodson, is given in Figure 1–9.

There remains the problem of the clustering of clusters of artifacts, behaviors, and mentifacts into larger taxa of culturgens to form a higher classification. In this example, which is the culturgen: all bronze swords, all bronze flange-tanged swords, or all bronze flange-tanged Erbenheim

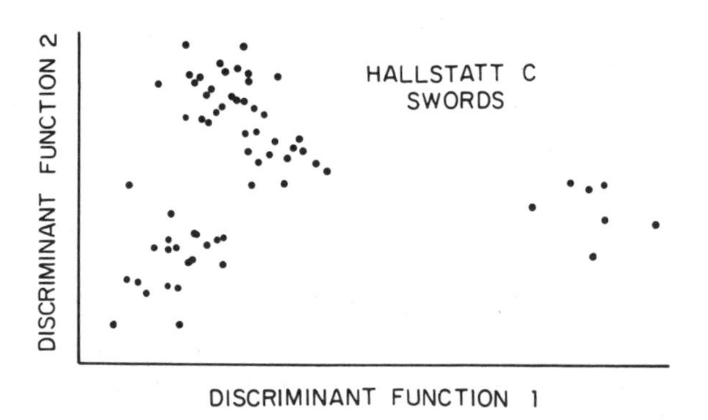

Figure 1–9 The differentiation of culturgens based on a formal definition of the culturgen as a discrete unit. The scatter diagram is derived from a *k*-cluster analysis of sixty-five Hallstatt C swords described by nineteen attributes. (Modified from Doran and Hodson, 1975.)

swords? It is no less the case in the classification of culturgens than in the classification of organisms that the choice of the level of discrimination used to define the taxonomic category must be arbitrary. However, as demonstrated by numerical taxonomists, the level need not be subjective. Entities such as artifacts, behaviors, and mentifacts can be examined by phenetic analysis, which establishes the degrees of their similarity. Those at the 90-percent level of similarity can be arbitrarily placed in the same culturgen; those at some other level, such as 70 percent or 45 percent, can be equally well clustered. Other techniques developed by psychologists elevate sets of units (definable as culturgens) compared pairwise by experimental subjects and the data matrix simplified through multiple dimensional scaling (Shepard and Arabie, 1979; Davis, 1979). The point is that similarity can be measured objectively and classifications made replicable even in cases where natural processes do not operate to divide entities in any predictable or intuitively clear manner.

It is possible to watch culturgens evolve over time as shifting patterns of artifact or behavior clusters defined by multiple attributes and mapped onto discriminant planes (see Figure 1–10). Clarke (1978) has described the process vividly as a perception of archaeological data: "Gradually, within the polythetic constellation of the ancestral artefact-type a new and growing nucleus appears, steadily increasing in content and intercorrelation level—a consistent variant is developing . . . Eventually the extended set will divide on functional lines as the emergent properties of the new artefact format are increasingly appreciated, leading to the deliberate formalization, increasing differentiation and the specialized development of divergent artefact usage patterns." The flint chopper, for example, may

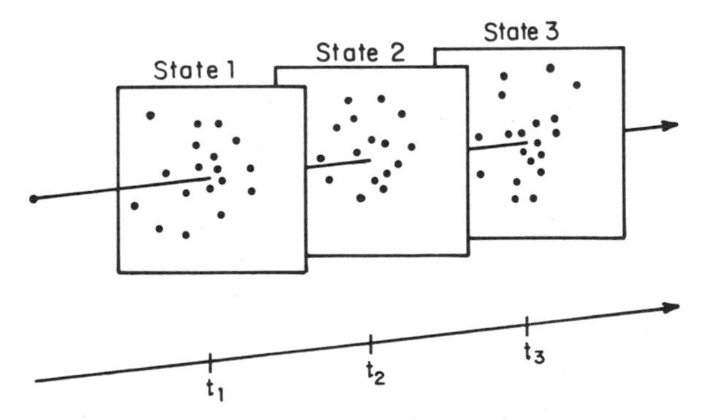

Figure 1–10 The evolution of a culturgen through time. Each plane is a representation of discriminant functions of the kind illustrated in Figure 1–9. (Modified from Clarke, 1978.)

be laterally flattened to accommodate the tool more readily to the hand, then in some specimens the secondary working around the radial edge will be accentuated, at the same time that other specimens are given a more pointed shape, and so on. Then the chopper/chopping-tool type continues as one variant, but a pointed and increasingly preferred piriform variant is shaped in response to functional requirements, with the result that the hand ax, a new category of artifact, is born. There are many striking analogs of culturgen evolution to the process of speciation in animal and plant populations.

Such a detailed procedure for culturgen discrimination seems justifiable in only a few exceptionally difficult or pivotal cases, such as the diffusion of Mayan serving vessels as analyzed by Fry (1979). The more formal approach has been explained here in order to address the general question of classification of complex, continuously varying arrays and to demonstrate the feasibility of precise and replicable analysis within gene-culture theory.

It will be noted, and the book will illustrate, that culturgens vary greatly in the clarity with which they can be distinguished. Culturgen categories range all the way from those in which the variation consists in two obvious states, such as the acceptance or rejection of incest, to much more subtle and complex phenomena that must be subdivided in an arbitrary manner. A similar variety can be found in biological units that have nevertheless proved their worth in theory. Although cells are most often discrete units centered around a single diploid nucleus, biologists must also contend with multinucleate syncytia and anucleate "cells." In classical Mendelian genetics the elementary segregating units, the genes, are well defined; but in modern molecular genetics the meaning of the term "gene" has become far more ambiguous. The segments of DNA that segregate are not necessarily the ones that mutate or prescribe polypeptide chains. Furthermore, "genes" grade insensibly into large and even less well defined segregating units, such as the supergene complexes held together by linkage disequilibrium and chromosome aberrations. The species can be objectively defined where reproductively isolated populations coexist in the same place and time, but the unit becomes increasingly arbitrary when populations are separated. It becomes fully arbitrary in the case of the asexually reproducing "agamospecies." The best research strategy for gene-culture coevolutionary theory, and the one we have employed in this book, would seem to be the same as that employed in biology and ethnography: start with examples in which the units are most sharply and readily definable, establish them as paradigms, and then proceed into more complex phenomena entailing less easily defined units.

Appendix 1–2

Waiting Time for Departure from Tabula Rasa

Consider an N-dimensional space in which the axes correspond to the properties or attributes of culturgens. Owing to the finite capacity of sensory and cognitive processing systems, the feasible culturgens are contained in a finite volume of size V in the attribute space. For realism we assume that the information processing capacity of the organisms is finite, so that two culturgens are perceived as essentially the same if the distance between them is r_0 or less.

We suppose that culturgens appear uniformly at random throughout V. In other words, if c denotes a newly innovated culturgen and v is any volume within V, then the probability that c lies in v is equal to v/V. This assumption models the extreme environmentalist view of culture as an imposition of arbitrary form on the environment and of people as entirely malleable with reference to culturgen usage. But within the attribute space there exist some points corresponding to culturgens that are selectively neutral and other points corresponding to culturgens that are selectively active with reference to the genes affecting the epigenetic rules of the population. We have already demonstrated that this is a realistic premise for most or all categories of culturgens. When a newly innovated culturgen falls within r_0 of a selectively active point, the two are equated and thus a selectively active culturgen enters the population.

By definition, our tabula rasa population possesses only selectively neutral culturgens at the outset. As soon as it discovers and begins to transmit selectively active culturgens, its epigenetic rules are subject to alteration by natural selection and it departs from the tabula rasa state. A certain fraction of the potential new culturgens are selectively active. The waiting time to the first appearance of a selectively active culturgen is a function of both the size of this fraction and the innovation rate of the population.

Specifically, if during some time interval δt an innovation occurs, the probability that its culturgen is selectively active is equal to the summed volume of all radius r_0 spheres centered on selectively active points, divided by the total volume of the attribute space, V. To make the relation precise, we suppose that the N-dimensional signal space is equipped with Euclidean geometry; the volume of an N-dimensional sphere of radius r_0 is then

$$V_0 = \frac{\pi^{N/2} r_0^N}{\Gamma\left(\frac{N}{2} + 1\right)} \tag{1-A1}$$

where $\Gamma\left(\frac{N}{2} + 1\right)$ is the gamma function; for any N, Γ is a constant.

There exists a natural unit of length $V^{1/N}$ associated with the space such that r_0 can be expressed as

$$r_0 = \alpha V^{1/N} \tag{1-A2}$$

where α is a constant. If M selectively active points exist in the space, the probability that an innovated culturgen will be identified with one of them is

$$p = \frac{M}{V} \frac{\pi^{N/2}}{\Gamma\left(\frac{N}{2} + 1\right)} (\alpha V^{1/N})^N = \frac{M(\pi^{1/2}\alpha)^N}{\Gamma\left(\frac{N}{2} + 1\right)}. \tag{1-A3}$$

It is useful to model the dynamics of culturgen invention as a walk on a one-dimensional lattice. If n is the number of selectively active culturgens innovated by time t, then the society starts with $n = 0$ at $t = 0$ and there is a constant probability per unit time, λ, that such a culturgen will be invented and thus increase n by one. We allow innovations to remain in circulation once introduced, so that the extinction rate is $\mu = 0$. Hence in the walk process,

$$\begin{aligned} &\text{Prob}\ \{n \longrightarrow n + 1 \text{ in } (t, t + \delta t)\} = \lambda \delta t, \\ &\text{Prob}\ \{n \longrightarrow n - 1 \text{ in } (t, t + \delta t)\} = \mu \delta t = 0. \end{aligned} \tag{1-A4}$$

In terms of the previously defined quantities, λ is the probability of an innovation occurring per unit time multiplied by the probability that the innovation is a selectively active culturgen. Let $P(n,t)$ be the time-dependent probability density for the number n of selectively active culturgens at time t, given that no such culturgens exist at time $t = 0$. The course of innovation is a Poisson process with density

$$P(n,t) = \begin{cases} e^{-\lambda t}(\lambda t)^n/n!, & n \geq 0 \quad \text{(1-A5a)} \\ 0, & n < 0 \quad \text{(1-A5b)} \end{cases}$$

and expected value

$$\langle n \rangle = \lambda t. \tag{1-A5c}$$

The quantity λ acts as a characteristic growth rate for the increase in the expected number of selectively active culturgens and gives this process a characteristic time scale of

$$\tau = \lambda^{-1}. \tag{1-A6}$$

From Eq. (1-A5c) the time required to accumulate a mean level of $\langle n \rangle$ selectively active culturgens is

$$t = \langle n \rangle \lambda^{-1} = \langle n \rangle \tau \tag{1-A7}$$

and the first such culturgen is expected within time τ. If ν is the average innovation rate per innovator, and I is the number of such individuals in the population, then

$$\lambda = \nu I p \tag{1-A8}$$

and we can write the *mean time to departure from the tabula rasa* as

$$t = [\nu MI(\pi^{1/2}\alpha)^N]^{-1}\, \Gamma\left(\frac{N}{2} + 1\right). \tag{1-A9}$$

This relation can be rewritten in the following, more transparent linear form:

$$\log_{10} t = \log_{10}\left[\frac{\Gamma\left(\frac{N}{2} + 1\right)}{\nu MI}\right] - N \log_{10}(\pi^{1/2}\alpha), \tag{1-A10}$$

where, to recapitulate,

I = number of innovators in the society,
ν = individual innovation rate,
M = number of selectively active points that are accessible in V,
N = dimension of the signal space (number of qualities used to distinguish culturgens),
α = ratio of r_0, the distance within which culturgens are indistinguishable, to $V^{1/N}$, the volume of the total signal space raised to the power $1/N$.

In Figure 1–6 we have presented two families of curves that illustrate the sensitivity of the active-culturgen waiting time to the parameters of the model.

The most realistic conditions under which a short waiting time is expected are the following:

(1) *A low signal-space dimension, N.* Although archaeologists have distinguished twenty or more dimensions in statistical studies of artifacts, it is likely that most artifacts are classified in daily life through the process of "feature extraction" demonstrated in studies of cognitive psychology, whereby perceptually important traits are extracted from the pattern while other information is lost. Many vocalizations and paralinguistic signals appear intuitively to us to be separated on the basis of one or two dimensions. Relatively few studies have been reported in the literature. In one instance naive observers were given the relatively difficult task of classifying visual sonograms of complex sounds; when a multidimensional scaling procedure was applied to their similarity judgments, it yielded a three-dimensional perceptual space (Getty et al., 1979). However, the designs were still simple compared to many encountered in everyday life. The estimate of the number of dimensions of perceptual space is an important problem in cognitive psychology that will prove of importance to gene-culture theory when applied to a wide range of culturgen categories.

(2) *A large* α, which means a large r_0 and hence few distinguishable culturgens along any one axis.

(3) *A very high M,* or number of culturgens active in natural selection, the result expected if the environment is exceptionally heterogeneous.

The elementary model presented here can of course be made more complex and realistic. We have presented this elementary version to provide a first notion of the parameters that we believe to be most potent in the determination of waiting times and to suggest qualities in cognition that might thereby prove of greatest significance in future studies of cultural evolution.

CHAPTER TWO

The Primary Epigenetic Rules

The key element in the theory of gene-culture coevolution is the role of the epigenetic rules in culturgen choice. These rules and the bias curves they generate can serve as the molecular units in postulational-deductive models of the social sciences. In spite of the relative sophistication of the field of developmental psychology, the epigenetic rules have never been systematically described, and the data concerning them have remained scattered and unconnected to evolutionary theory. Perhaps the principal reason is that students of cognitive development have largely neglected the study of preference—especially innate preference—among multiple competing stimuli and schemata. Instead, they have focused on reinforcement and learning with reference to a single stimulus or at most paired stimuli chosen for operational convenience. Although Bruner, Skinner, and other leading investigators perceive learning capacity as a complex process unfolding with age, a view basically congenial with genetic theory, they have paid relatively little attention to the possibility of innate constraints that channel learning toward certain choices as opposed to others within the same stimulus category. They have concentrated on attempting to explain development with the most generalized stages and learning rules, as stressed in the recent review by Logue (1979).

Almost the opposite approach has been taken by psycholinguists, psychologists, and anthropologists belonging to the "structuralism" school. Piaget, Lévi-Strauss, Chomsky, and their coworkers postulate the existence of innate constraints of a kind more or less consistent with gene-culture coevolution theory. Although their approach has been fruitful and stimulating, their methods are largely nonexperimental, yielding few data adequate for the construction of a true postulational-deductive theory of epigenesis (see, for example, the critique by Brainerd, 1978). In fact, the very notion of linking structuralist conceptions to neurobiology and genetics to create a "biogenetic structuralism" is relatively new (Laughlin and d'Aquili, 1974), and it has not been transformed into either working theories or specific models that summon information from developmental psychology into the service of evolutionary theory.

Epigenesis is defined as the total process of interaction between genes and the environment during development, with the genes being expressed through epigenetic rules. The imagery invoked by Waddington (1957) and other biologists in their original conception of the epigenetic field, or "landscape," is the influence of the epigenetic rules viewed through successive time transects over the entire course of development. Each epigenetic rule affecting behavior comprises one or more elements of a complex sequence of events that occur at sites distributed throughout the nervous system. They range from the selective filtering of arrays of stimuli by the retina, cochlea, and other primary receptors, through the integration of visual and auditory information by interneurons responsive to patterns of activity among the feeder neurons, to innate predisposition to use certain culturgens as opposed to others. Epigenetic rules are the outcome of specificity in cell structure, neuron circuitry, and the timing of hormone release, which properties are themselves more fundamental products of epigenesis at the cellular level.

Existing information on cognition is most efficiently organized with reference to gene-culture theory by classifying the epigenetic rules into two classes that occur sequentially within the nervous system. *Primary epigenetic rules* are the more automatic processes that lead from sensory filtering to perception. Their consequences are the least subject to variation due to learning and other higher cortical processes. For example, the cones of the retina and the internuncial neurons of the lateral geniculate nucleus are constructed so as to facilitate a perception of four basic colors. The *secondary epigenetic rules* act on color and all other information displayed in the perceptual fields. They include the evaluation of perception through the processes of memory, emotional response, and decision making through which individuals are predisposed to use certain culturgens in preference to others. Thus fear of strangers represents a form of prepared learning in which human infants from six or eight months to eighteen months old display an aversion toward adults to whom they are not accustomed, and the response is automatically intensified if the stranger stares (Argyle and Cook, 1976; Eibl-Eibesfeldt, 1979).

This rough two-part classification is consistent with synopses of information processing employed independently by some cognitive psychologists. In their analysis of visual illusions, Girgus and associates (1975) found important differences between the "structural properties" of the optical and nervous systems and the cognitive components of the processing that occurs in the higher centers of the central nervous system. Similarly, in a study of complex visual figures Getty and colleagues (1979) distinguished the automatic perception of key features from the decision processes by which the similarities of the figures are judged. Massaro's model (1975) of auditory information processing separates preperceptual auditory storage, which depends on properties of the receptor system and is not subject to change by learning, from later phases of feature evaluation and decision making that are to some extent modifiable by experience.

We now present an account of the best-understood primary epigenetic rules, noting some of their consequences in cultural evolution. It appears generally true that filters consisting of peripheral sensory screening and internuncial coding operate through rigidly canalized developmental sequences under a minimum of influence from cortex-mediated learning. Thus myelin deposition in the optic radiations of the brain is complete at approximately five months postpartum in children with normal sight, within a few weeks of the time that deviant behavior becomes marked in congenitally blind children. Similarly, myelinization in the acoustic radiations is completed at the third to fourth year, at about the time parents of congenitally deaf children begin to notice significantly different behavior (Yakovlev and Lecours, 1967; D. A. Freedman, 1979). Lateral organization of vision occurs within the first few weeks of life in infant human beings and rhesus monkeys. In monkeys the right and left afferent neurons entering the cortex are intermingled through the prenatal period, but separation of the bands proceeds quickly after birth and is completed within about three weeks. If one eye is occluded, the afferents of the remaining eye form synapses with neurons that would normally receive input from the damaged eye (Hubel et al., 1977; Goldman and Rakic, 1979). At the opposite extreme from these almost purely cellular events are some of the apparently most complex secondary epigenetic rules, such as automatic brother-sister incest avoidance and early language development, which entail learning and conscious assessment of the results (Lenneberg, 1967; Shepher, 1971; Brown, 1973).

The actual case histories to be examined next illustrate the diversity of the primary epigenetic rules and their effects on the development of social behavior. These examples also reveal that this domain of psychology, though experimentally tractable, is still in a very early stage of exploration. Its further pursuit can be expected to produce additional results important to social theory. The development of many, perhaps most, of the categories of behavior will prove to be channeled by combinations of primary and secondary rules. The first example to be given, taste and smell, shows how such a combination can lead to relatively complex specific behavior.

Taste and Smell

The chemical senses are more poorly developed in human beings than in most kinds of animals, and chemosensory learning is subject to several peculiar constraints. When given a single chance to identify substances by smell, individuals are able to identify only six to twenty-two items. This repertory can be increased to a hundred or more when all of the following three conditions are met: the substances have long been familiar to the person, there has been a long-standing connection between the odors and their names, and some aid is provided in recalling the names (Cain, 1979). Thus odor memory, and probably taste memory as well, can be notably enhanced when coupled to word memory. The inferiority of chemosensory information processing is reflected in human languages, which contain far fewer words denoting qualities of odor and taste than words designating qualities of sight and sound. In Table 2–1 we have recorded counts of all the words applying to both transmission and reception in eight sensory modalities, taken from randomly selected ten-page sections of dictionaries in each of the following independently evolved languages: English, Japanese, Zulu, San ("Bushman"), and Dakota. Only terms signifying distinct qualities of intensity, frequency, or pattern in ordinary usage are included. Technical expressions such as anechoic, pheromone, and paraxial have been omitted. By inspection it can be seen that there exists a correspondence across independently evolved cultures between human discriminatory ability within a sensory modality and the number of words applied to the modality. It is further true and of no little significance for evolutionary studies that the greatest sensitivity and capacity to discriminate odors lies in the class of substances associated with human food and mammalian body smell (Amoore, 1977). Thus the relative powers of the sensory system and the general linguistic expression of this sensory hierarchy are correlated species-specific traits, and in the case of chemoreception they appear to have direct biological adaptive value.

The properties of odor memory are strikingly different from those of audiovisual memory. Engen and Ross (1973) found that the visual details

Table 2–1 Distribution of words that describe qualities of eight sensory modalities in five independently evolved languages. The numbers given are percentages of the total number of words encountered that denote sensory qualities. The two chemoreceptive modalities, smell and taste, are combined into a single category.

Language	Source	Total no. words encountered	Vision	Hearing	Smell and taste	Touch, including surface and density	Temperature	Humidity	Electric field
English	*Random House Dictionary*, unabridged ed. (1966)	100	0.49	0.32	0.10	0.06	0.01	0.01	0.01
Japanese	*Kenkyusha's New Japanese-English Dictionary*, American ed. (1942)	141	0.45	0.30	0.06	0.11	0.04	0.02	0.02
Zulu	*Zulu-English Dictionary*, C. M. Doke and B. W. Vilakazi (1953)	137	0.36	0.32	0.07	0.16	0.04	0.04	0.01
San (''Bushman''), mixed dialects	*Bushman Dictionary*, Dorothea F. Bleek (1956)	117	0.25	0.37	0.08	0.13	0.12	0.05	0
Teton Dakota Sioux	*Dictionary of the Teton Dakota Sioux Language*, E. Buechel (1970)	86	0.28	0.36	0.07	0.14	0.08	0.07	0

of pictures can be memorized quickly and accurately for short periods of time but are largely forgotten within three months. In contrast, odors are memorized with greater difficulty and less short-term accuracy, but the memory remains undiminished for three months or longer. It is a commonplace observation that simple cues from odor and taste are exceptionally potent in evoking vivid, detailed recollections of people and places. Furthermore, the memories are typically emotional and nonverbal, as poets have noted. Proust's inspiration from a madeleine comes to mind unavoidably, while Baudelaire has provided another striking literary example in *Flowers of Evil:*[1]

> Some wardrobe, in a house long uninhabited,
> Full of the powdery odours of moments that are dead—
> At time, distinct as ever, an old flask will emit
> Its perfume; and a soul comes back to live in it.

In *Mary, a Novel* (1970), Nabokov noted that "memory can restore to life everything except smells, although nothing revives the past so completely as a smell that was once associated with it."

Hence it is clear that the role of chemosensory information in cultural evolution is constrained in a severe and peculiar manner. Furthermore, although the finer details of the ontogeny of taste and smell are in an early stage of investigation and still poorly known, the following principal steps in the development of dietary preference have been established. These elaborations entail postperceptual valuation and hence are influenced by what we have called secondary epigenetic rules.

(1) Infants innately prefer sugar, a bias that persists into childhood and beyond, and they show distinctive aversive responses to acid, salty, and bitter flavors.

(2) If at weaning children are allowed to experiment with foods provided ad libitum, they are relatively undiscriminating but nevertheless come to select a nutritionally balanced diet.

(3) Thereafter food habits become exceptionally conservative, remaining stable even after other cultural elements have been changed. The conservatism is based on the long chemosensory memory of individuals, a general correlation between the degree of familiarity with odors and pleasure received from them, and a general aversion to new odors and tastes.

1. Translation from the French by Edna St. Vincent Millay of Charles Baudelaire's *Flowers of Evil,* Harper & Row. Copyright 1936, 1963 by Edna St. Vincent Millay and Norma Millay Ellis.

The documentation supporting these generalizations has been accumulated piecemeal by many investigators. Using ingestion experiments, Maller and Desor (1974) found that newborn infants prefer a variety of sugar solutions (sucrose, fructose, lactose, and glucose, in that order) over plain water through at least part of the range of concentrations detectable by adults. This selectivity continues into later life. When high-caloric protein food is supplemented with sucrose, children increase their intake substantially, whereas they are indifferent to the addition of the taste of almonds, the most preferred of seven flavors separately tested (Grewal et al., 1973). The trait is reasonably postulated to be species specific. G. K. Beauchamp (personal communication) has predicted that herbivores and omnivores such as rodents and human infants should innately prefer sweets, which generally signal a high caloric yield in digestible food items, but carnivores should not. In a pilot study Beauchamp and associates (1977) found that four species of cats tested (lion, tiger, leopard, jaguar) are in fact indifferent to sugars, whereas they have a strong preference for protein and fat supplements.

Newborn infants not only prefer sweet solutions but they discriminate further among tastes that are acid, salty, or bitter, showing distinctive facial reflexes in response to each that resemble various adult aversive expressions to strong and unpleasant tastes (Chiva, 1979; Steiner, 1979). Russell (1976) found that by six weeks of age breast-fed infants prefer the odors left by their mothers on breast pads to those left on breast pads by other lactating women—a remarkable feat of discrimination, yet one that is matched by the newborn young of other mammals (Rosenblatt, 1972). The Maller-Desor experiment nevertheless does demonstrate a much lower level of selectivity toward dietary flavors in infants than exists in adults. Its results are consistent with an earlier finding that children three to four years of age report amyl acetate, synthetic feces odor, and synthetic sweat odor to be equally pleasant. By the age of six to eight they regard the odors of feces and sweat as unpleasant (Stein et al., 1958). In the opinion of Engen (1974) this conclusion is compromised by the tendency of young children to say "I like it" to almost any stimulus; but the reaction may in fact simply reflect greater tolerance for chemosensory stimuli rather than crudity of expression.

At the time of weaning, children are less discriminating and more likely to experiment with foods than at older ages. This trial-and-error method nevertheless does not result in a random diet. In a classic experiment Davis (1928) permitted three newly weaned children to select their own diet from a wide variety of foods offered ad libitum. Each soon arrived at a nutritious and balanced diet that included milk, cereal, vegetables, fruit, eggs, and other animal products. Although significant differences occurred among the children in the proportions of these categories, the degree of convergence within the array of all possible diets was remarkably close. Should this result be confirmed, we can only speculate on the physiological and psychological mechanisms that guide the choices (Engen, 1979). The behavior is closely comparable to self-selection of diet by rats placed on a "cafeteria" regime. As Richter and his associates showed (for example, Richter and Rice, 1945), the animals automatically seek foods rich in those components which they lack at the time. The rats treat foods deficient in the essential nutrients they most need as though the foods were slow poisons. They have the capacity to learn aversive taste stimuli by delayed action, the "Garcia effect" found in many studies (reviewed by Logue, 1979). Later diets containing the stimuli are avoided in favor of those lacking them, with the eventual result that a balanced diet is achieved (Rozin, 1976). Delayed learning also results when more direct negative effects, including gastrointestinal distress and systemic poisoning, follow the ingestion of particular food items. There may also be innate properties of attractiveness in the constituent chemicals of the food items. In a multiple-dimensional scaling analysis of odorants, Davis (1979) found that human subjects automatically classify alcohols and rank them according to pleasantness in a manner correlated with increasing molecular chain length.

In summary, a set of very general learning rules guides the further narrowing of dietary preference beyond innate preference for sugars during the weaning period. The homeostasis achieved can be precise, and we suggest that it is subject to genetic alterations that shift the major pattern of dietary choice to new steady states. The evidence for this conclusion is indirect. An autosomal "obesity gene" has been demonstrated in mice, which in the homozygous state (*ob*/*ob*) induces heavier eating, the choice of a higher percentage of fat (52 percent compared to 29 percent chosen by their nonobese siblings), and somewhat less sensitivity to variation in sweetness (Mayer et al., 1951; Ramirez and Sprott, 1979). A similar phenomenon in human genetics is the well-known polymorphism affecting ability to taste phenylthiocarbamide (PTC). The frequency of nontasters, who are homozygous for the opposing, recessive allele, varies geographically as follows: 30 percent in Europeans, 10.6 percent in Chinese, 6.4 percent in Afro-Americans, and 1.9 percent in the Amerindians of highland Peru. It is generally believed that the inability to taste PTC is maladaptive in parts of the world where endemic goiter is highest—as in the

Peruvian Andes—and where natural goitrogens, the taste of which is bitter like that of PTC, are apt to occur in the diet. The high frequencies of both alleles might be explained by the greater tendency of tasters to develop hyperthyroidism, which when countered by the vulnerability of nontasters to goitrogens leads to a state of genetic polymorphism (Greene, 1974).

Specific anosmias, the lower abilities of individuals to smell certain classes of substances but without reduction in the abilities to smell other classes of substances, are diverse and very widely distributed in human beings (Beets, 1979). The pattern of their occurrence has led Amoore (1977) to suggest the existence of primary odors in human beings comparable to the innate primary colors. At least one of the anosmias, the loss of sensitivity to the musk-like substance pentadecalactone, is based upon a recessive gene (Whissell-Buechy and Amoore, 1973). Some inconclusive evidence exists that insensitivity to the odor of skunk, *n*-butyl-mercaptan, and the scent of freesia flowers is based on autosomal recessives. On the other hand, variation in sensitivity to acetic acid, isobutyric acid, and 2-*sec*-butyl-cyclohexanone does not appear to have a genetic component (Hubert et al., 1980).

Time and practice steadily reinforce the idiosyncracies of cuisine. To a point, the more familiar an odor associated with food, the more pleasant it is rated; the less familiar, the more likely it is to be actively avoided, a phenomenon paralleling the well-documented feeding neophobia of rats (Engen, 1974; Rozin, 1976).

Food and the customs surrounding mealtime have been thoroughly ritualized to communicate and enforce correct behavior in virtually every other facet of social life, including greeting, conciliation, alliance formation, courtship, dominance, hygiene, and religion (Lévi-Strauss, 1969; Douglas, 1979). No evidence exists that specific ritual forms have ever been genetically assimilated. Nevertheless, the forms are clearly enmeshed in the epigenetic rules that guide chemosensory learning, and they are constrained by the rules affecting other primary social functions served by culinary ritualization.

Color Classification

The brain processes two fundamental components of visual information. The first is contour, which is perceived as the spatial rate of change in luminance and conveys the shapes of objects. The second component is color, which conveys information about the surface of objects.

It is a remarkable fact that the brain perceives variation in luminance along a continuum but divides color into categories (Figure 2–1). It is further true that all cultures employ language categories to describe color. Many social scientists (for example, Whorf, 1956) used to believe that the divisions into red, green, and so forth are arbitrary, but linguistic and cross-cultural studies have shown that they are in fact closely tied to natural color perception (Berlin and Kay, 1969; Rosch, 1973; Ratliff, 1976).

Research by Bornstein and his associates (Bornstein et al., 1976; Born-

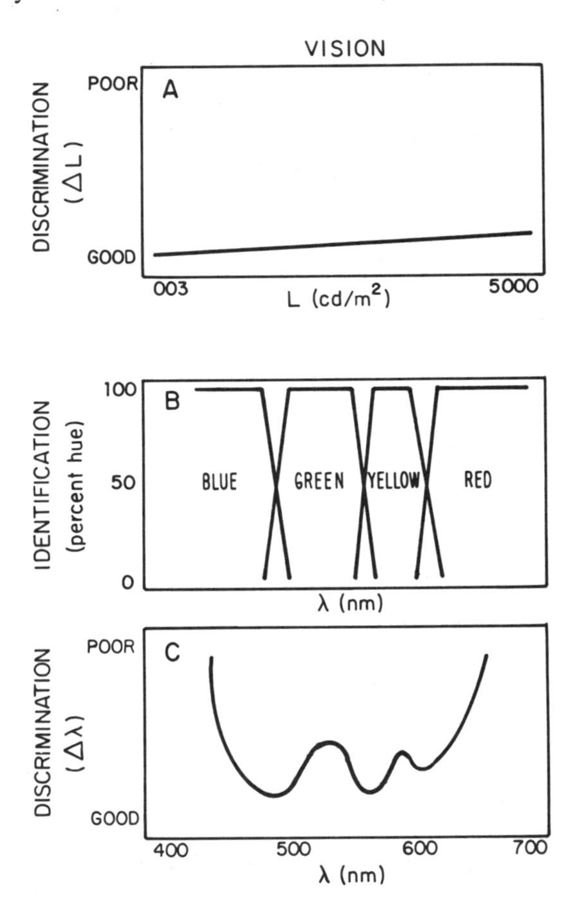

Figure 2–1 Epigenetic rules in vision. *A*: Luminance discrimination is a continuum, varies little as luminance is increased, and is overall quite accurate. *B*: During infancy light wavelength is perceived psychologically as though it were broken into four principal categories—blue, green, yellow, and red. *C*: Wavelength discriminability varies abruptly between good and poor along the visible spectrum, with peaks roughly corresponding to the boundaries between the four major hue categories. (Modified from Bornstein, 1979.)

stein, 1979) has provided valuable new insights into the developmental origin of color classification. Four-month-old infants were found to respond to variation in wavelength as though the brain discriminates four basic categories of hue corresponding to the adult categories of blue, green, yellow, and red. This discretization was detected by measuring the span of the infants' attention to various monochromatic lights distributed along the visible spectrum. After a short time the infants became accustomed to the presentation of one light and looked away. Recovery from habituation was greater when the wavelength to which the infant was next exposed was located in a hue category adjacent to the habituation light than when it came from the same category, even when the two stimuli tested were equidistant (in nanometers) from the original, habituating wavelength. The categorization scheme is indicated in Figure 2–1B.

The innate human color classification starts from the differentiation of the retinal cones into three types (Wald, 1969), whose respective maximal sensitivities correspond to blue (440 nm), green (535 nm), and yellow-green (565 nm) (Mollon, 1980). The mechanism that reorganizes these sensitivities into perceptions of the four principal colors is still under debate.

Psychologists have made some progress in mapping the color-perceptual space. When human subjects judge various pairs of colors according to degree of similarity and the data are transformed onto a plane by multidimensional scaling procedures, the successive wavelength categories do not form a straight line. Rather, their path curves back onto itself, a form known as the circumplex (Guttman, 1954; Shepard, 1978). As a result, the opposite ends of the visible spectrum, red and violet, are judged to resemble each other almost as closely as adjacent primary colors on the wavelength scale (see Figure 2–2). Circumplexes are a widespread phenomenon in human perceptual fields. Figures resembling the color circumplex have also been discovered in the perception of phonemes and musical sounds (Shepard, 1978; Shepard and Arabie, 1979), as well as in the detection of odor of alcohols through a series of increasing molecular chain length (Davis, 1979).

The epigenetic constraints in color perception are reflected in the verbal color classifications employed in the languages of all cultures thus far studied. In an important study by Berlin and Kay (1969), native speakers of twenty languages around the world (which included Arabic, Bulgarian, Cantonese, Catalan, Hebrew, Ibidio, Japanese, Thai, Tzeltal, Urdu, and others) were shown arrays of chips classified by color and brightness in the Munsell system. They were asked to place each of the principal color

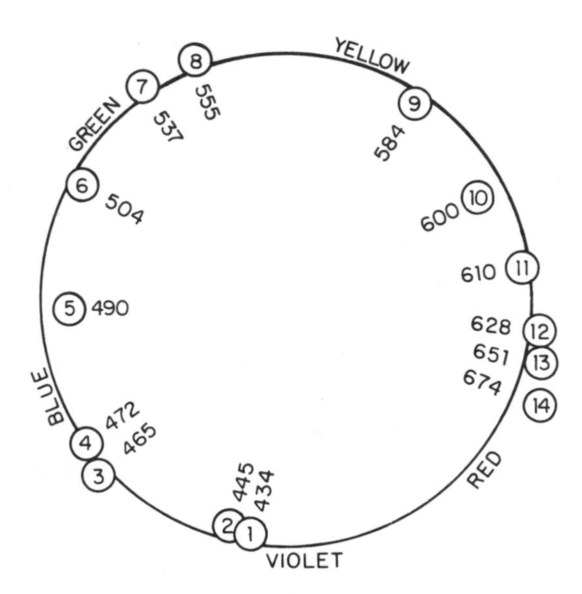

Figure 2–2 The circumplex form of the perceptual field of color. Red and violet, located at the opposite ends of the visible spectrum, are judged to resemble each other almost as closely as adjacent primary colors. The numbers given adjacent to each of the fourteen positions are wavelengths in nanometers. (Modified from Shepard, 1978; based on data from Ekman, 1954.)

terms of their language within this two-dimensional array. The results, given in Figure 2–3, show clearly that the languages have evolved in a way that conforms closely to the epigenetic rules of color discrimination. The terms fall into largely discrete clusters that correspond, at least in an approximate manner, to the principal colors that appear to be innately distinguished by infants. This central result has been subsequently confirmed by many investigators, as noted in the reviews by Kay (1975), Bolton (1978), and von Wattenwyl and Zollinger (1979). Rosch (1973) found that Dani men of New Guinea, whose color classification is rudimentary, learned a "natural" classification based on the clusters of the Berlin-Kay data more rapidly than a competing scheme based on other, arbitrarily selected clusters. Thus Rosch was able to demonstrate a learning rule that operates in adults.

It is further true that when color classifications are compared across cultures, those of most cultures fall into one or the other of the following categories considered to represent an evolutionary sequence: black, white; black, white, red; black, white, red, green *or* yellow; black, white, red, green, *and* yellow; the previous colors plus blue (Berlin and Kay,

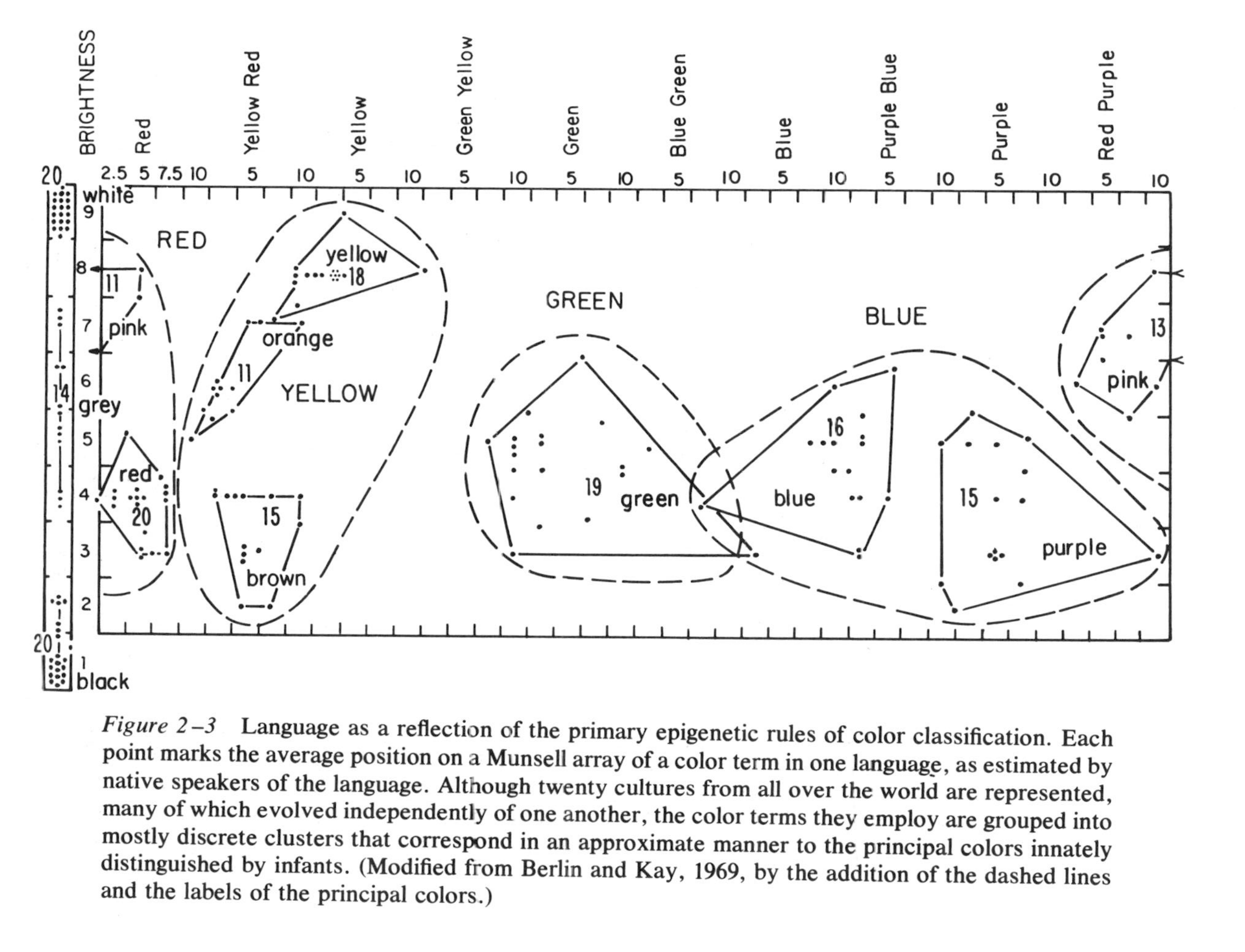

Figure 2–3 Language as a reflection of the primary epigenetic rules of color classification. Each point marks the average position on a Munsell array of a color term in one language, as estimated by native speakers of the language. Although twenty cultures from all over the world are represented, many of which evolved independently of one another, the color terms they employ are grouped into mostly discrete clusters that correspond in an approximate manner to the principal colors innately distinguished by infants. (Modified from Berlin and Kay, 1969, by the addition of the dashed lines and the labels of the principal colors.)

1969; Bolton and Crisp, 1979). This sequence matches the order of acquisition of color terms by children (Harkness, 1973; Johnson, 1977).

Categories at the blue end of the spectrum are collapsed in some languages, so that green and blue, or blue and black, or green, blue, and black are not distinguished by names; this tendency toward conflation increases with proximity to the equator (Bornstein, 1973). According to von Wattenwyl and Zollinger, these and other cumulative data are consistent with existing neurophysiological information. They favor the "opponent model," in which color vision is based on three pairs of opponent processes, for the perception of dark (no reflection) and light, red and green, and blue and yellow. The primary tristimuli photochemical reaction is then encoded by the cells of the lateral geniculate nucleus and possibly other interneurons.

The structure of hue perception can influence other functions of visual perception as well. Pollack (1972), for example, found that the magnitude of the Müller-Lyer illusion is influenced by color. In this effect lines placed between outward pointing chevrons (↔) appear to be shorter than lines of identical length placed between inward pointing chevrons (>—<). The illusion is greatest when the lines are blue or red, corresponding to the zones of minimum wavelength discrimination as shown in Figure 2–1C, and less by as much as 30 percent when the lines are green, yellow, or black.

Finally, a paradigm of the genetic evolution of color vision has been provided by recent experimental studies of Kovach (1980) on the quail *Coturnix coturnix japonica*. At the age of one day, individual chicks kept in the dark since hatching are allowed to orient visually for the first time as they find their way through a series of compartments. Under these conditions they differ slightly among themselves in their preference for red as opposed to blue stimuli. Using standard biselection procedures, Kovach was able to separate two lines of chicks possessing much more marked preferences for red or blue respectively. Complete divergence, with no overlap in choice scores, was achieved within five generations. The differences between the two lines were estimated to be based on four to eight segregating units of inheritance. This result is important because it demonstrates two phenomena anticipated by theory, namely that the genetic basis of epigenetic rules can be altered within ten generations or less by a sufficiently rigid selection regime, and that complex behavior can be affected by relatively small ensembles of genes. We shall return to the fuller meaning of these principles in Chapter 6.

Hearing

The infant begins life with built-in acoustic behavior that helps to shape its subsequent communication and social existence. Its response to a sudden loud sound or jarring of the crib is the Moro reflex: "The infant lying on his back extends his arms forward, stiffens the lower extremities and contorts his face into a grimace; after a second or two he brings the arms slowly together into a sort of embrace, emits a cry and then gradually relaxes" (Holt and Howland, 1939:32). In four to six weeks the Moro reflex is replaced by the startle response, perhaps the most complicated true reflex displayed by children and adults. Within a fraction of a second after an unexpected loud noise is heard, the eyes close, the mouth opens, the head drops, the shoulders and arms sag, and the knees buckle slightly. In short, the individual is prepared as well as he can be to absorb a blow to the body. The natural startle response occurs more quickly than any imitation of its movements performed deliberately by volunteer human subjects.

Newborn infants can further discriminate between noise and tone. During tests conducted by Levarie and Rudolph (1978), neonates were disturbed by 85 dB sounds of objects struck together but not by an 85 dB tone pulse at 293 Hz.

Infants also possess innate rules of speech perception that are adult-like and facilitate the development of language (Liberman et al., 1967; Eimas et al., 1971). As noted in the case of luminance, variation in pitch is perceived as a continuum. But distinctions of voicing, like distinctions of hue, are automatically classified into categories, in this case into phonemes. For example, sounds ranging between */ba/* and */ga/* and */s/* and */v/* are not heard as continua but as one or the other of these paired units. A principal component of phoneme discrimination is voice onset time (VOT), which is the timing of the formants or energy bands relative to one another (see Figure 2–4).

Recent investigations have revealed that the development of phoneme discrimination is channeled, proceeding through more than one stage during the first year of life or longer (Eilers et al., 1977). Voice onset time alone is not sufficient to explain the discrimination data. Context is used also: the recognition of stop and fricative consonants depends on the extent of the first formant and the direction of the second formant. Moreover, the infant's phoneme perception cannot be mapped readily onto adult phoneme perception (Eilers and Minifie, 1975; Eilers, 1977). But the

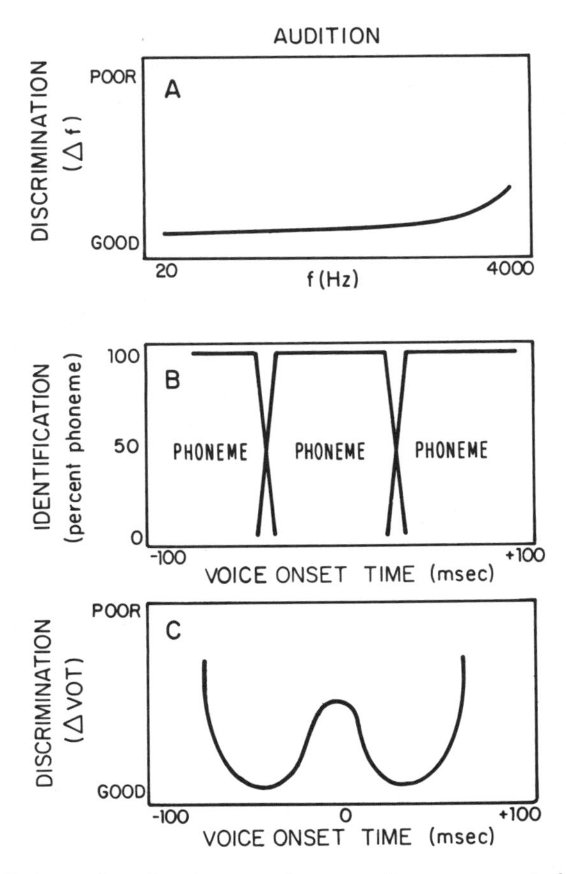

Figure 2–4 Epigenetic rules in speech perception represented in schematic form. Discrimination of pitch (*A*) from 20 to 4,000 Hz is continuous, varies only slightly across the spectrum, and is generally acute. In contrast, phoneme identification (*B*) is based on strongly varying discrimination (*C*) between successive energy bands, according to the time separating them (− 100 to + 100 msec). Where discrimination is poor, the formants appear to belong to the same category; where it is acute, phonemes can be distinguished. (Modified from Bornstein, 1979.)

correspondence is strong enough to yield consistent patterns in the temporal values of voice onset time across many languages. Studies of eleven languages by Lisker and Abramson (1964) revealed that total variability is not restricted solely by limitations in anatomical capability. There is in addition a fundamental and restrictive strategy: in every language one or two values along each VOT continuum serve as points of reference and segregate the continuum into two or three phonetic categories. The total repertory is twenty to sixty phonemes, the precise number varying according to culture. Moreover, the phonemes fall into spontaneous categories based on the intuitive perception of differing degrees of similarity among them (see Figure 2–5).

Because study of the epigenetic rules of speech perception is in an early stage, the relation of these rules to the diversity of word formation within and between languages has not yet been effectively addressed. The same is true of grammatical rules, including the "deep grammar" considered by many psycholinguists to contain innate components and hence to unfold according to hereditary programs. It is apparent that the epigenetic rules have affected the short-term evolution of languages. It is reasonable to speculate that over longer periods of time the relentless cultural pressure

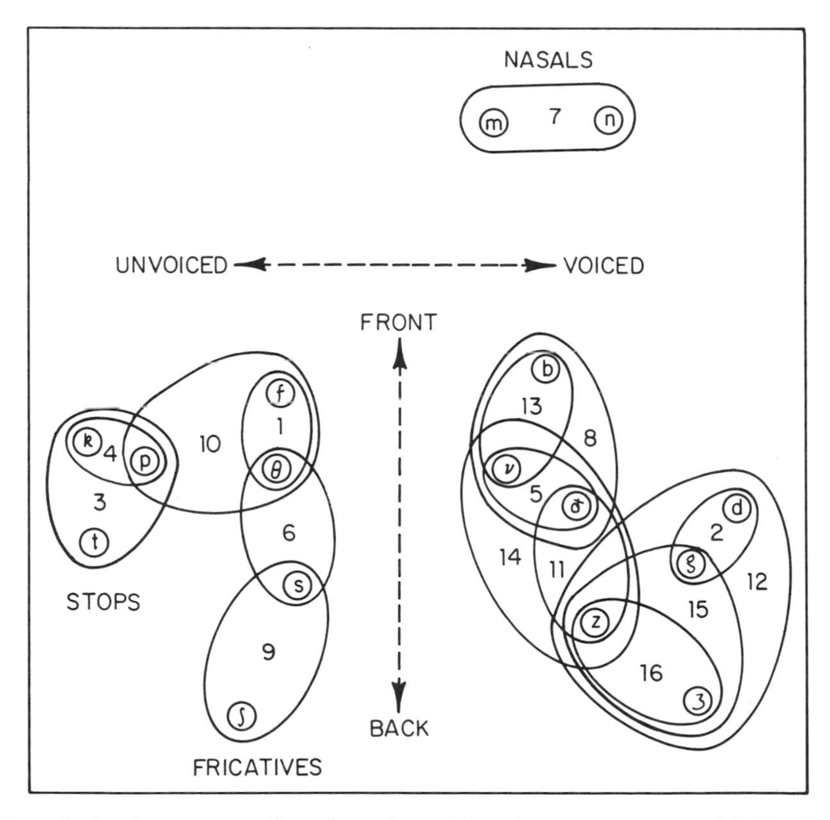

Figure 2–5 A representation of portions of the phoneme perceptual field. Clusters of sixteen consonant phonemes based on the similarity perceived by experimental subjects have been embedded in a two-dimensional space. (Analysis by Shepard and Arabie, 1979; based on the data of Miller and Nicely, 1955).

to expand language has shaped the genetic evolution of the epigenetic rules.

Summary

Epigenesis is the total process of interaction between genes and the environment during the course of development, with the genes being expressed through epigenetic rules. Each epigenetic rule affecting behavior comprises one or more elements of a complex sequence of events occurring at various sites throughout the nervous system. It is useful to divide these elements into two classes: primary epigenetic rules, which range from sensory filtering to perception; and secondary epigenetic rules, which include the procedures of feature evaluation and decision making through which individuals are predisposed to transmit certain culturgens in preference to others. Many, perhaps most, categories of cognition and overt behavior are channeled by combinations of the two classes of rules.

The primary rules are the more genetically restricted and inflexible. Examples have been provided from taste and smell, color classification, and hearing. Each exercises important constraining effects on the operation of the mind and has resulted in parallel or convergent evolution in independently derived cultures.

CHAPTER THREE

The Secondary Epigenetic Rules

We have seen how the primary epigenetic rules play an important role in determining perceptual spaces and thereby influence cultural evolution. But this form of constraint is only the beginning of the relationship between gene, mind, and culture. As the present chapter will show, many additional steps lead to the secondary epigenetic rules and the final ontogeny of social behavior.

Each stimulus configuration processed by the brain is dissected and distributed into what can be abstractly characterized as points or envelopes in the perceptual spaces of the various sensory modalities. The perception can be as simple as a flash of a single color or the scent of an aromatic compound, each occupying a single point on the perceptual space. Or it can be as complicated as the nursing of an infant, which creates envelopes of sound, vision, touch, and odor. The pathways leading from the sensory receptors to the cortical regions in which perception is transformed into conscious images have been the subject of substantial research, reviewed for example by Bullock and colleagues (1977) and by Boddy (1978).

Most culturgens are relatively complex and engage multiple sensory modalities. The brain does not act upon all of the information the configurations of these culturgens carry inward. The mind's decisions are based on certain features extracted from the final representations of the culturgens within the perceptual spaces. The inner physical basis of feature extraction remains largely unknown, unlike the relatively well studied neuron circuitry that carries sensory information to the higher centers. The outer qualities of the extraction process identified by both human ethologists (Eibl-Eibesfeldt, 1975, 1979) and cognitive psychologists (Neisser, 1976) are rapidity, precision, and comparative simplicity.

An interesting physical analog of the subjectively inferred process is provided by Hofstadter's "trip-let" (Figure 3–1). The central figure in the photograph can be used to represent a culturgen, or more precisely the configuration of a culturgen within the perceptual space. The mind does not act on all of the information present in the structure. It selects certain features which on the basis of a separate set of neural mechanisms are the subjects of decision making. These features are represented by the letters silhouetted in the beams of light. For lower animals the letters are roughly equivalent to sign stimuli or, more precisely, those patterns of neuron firing evoked by the sign stimuli. In the case of human behavior they are the neuronal activity patterns corresponding to schemata, which in turn are constituted from varying proportions of inherited circuitry and cellular modification due to learning and thinking.

Figure 3–1 Hofstadter's "trip-let," an abstract physical analog of feature extraction of complex stimulus configurations and hence of the initial interpretation of culturgens by the mind. (From Hofstadter, 1979.)

We shall return shortly to some of the most significant properties of human cognition that imply the existence of secondary epigenetic rules. For purposes of clarity, however, let us proceed first to one of the key manifestations of behavioral development, namely the *bias curves*. These curves are the probability distributions of the usage of various culturgens, a form that can be translated into the models of gene-culture coevolution. Usage is defined broadly as one or more links in a complex chain of decisions by individuals, which include the initial learning or failure to learn certain culturgens, the preference for one culturgen over another upon reflection, and the actual employment of particular culturgens. The bias curves can pertain to any one of these events. Consider the two curves displayed in Figure 3–2. The left-hand one plots the probabilities of use, u, of various culturgens in a very large array; it is therefore given a continuous form. The right-hand curve depicts the extreme opposite condition: members of the society can choose between only two culturgens.

Some bias curves are relatively rigid. This is especially the case of those derived from the primary epigenetic rules. Others vary greatly, to the extent that the disposition to adopt one culturgen over another can be reversed, in ways that are predictable yet still dependent upon context. The ethnographic literature contains a great many examples of reversals that are correlated with differences in habitat, mode of production, and nutritional status. Cultural anthropologists, especially those who have lived for long periods in societies markedly different from their own, speak of the "contextual" or "holistic" nature of culture, which means the correlation of particular culturgens with many others in a functional manner. They employ contextual analysis, "a methodological procedure and a theoretical assumption . . . which notes that each aspect of cul-

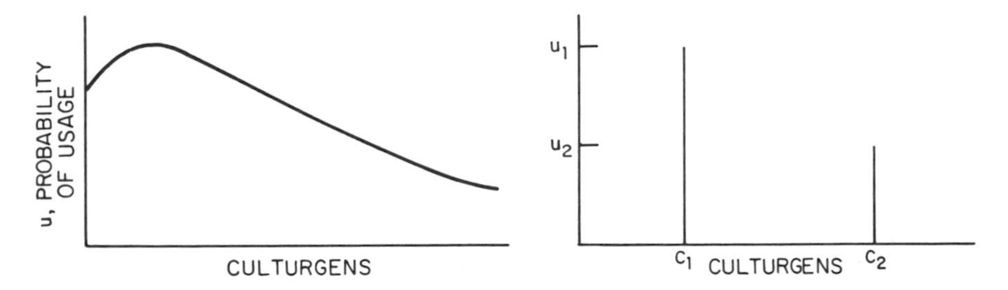

Figure 3–2 The two extremes of bias curves according to numbers of culturgens available. The curve on the left gives the probabilities of usage, u, across a very large array of culturgens; that on the right gives the probabilities of usage involving only two culturgens, c_1 and c_2.

ture must be viewed and studied in terms of its interdependence with other aspects of culture'' (Williams, 1972a:288). Thus female puberty initiation rites originate most frequently in societies where a young girl continues to reside at least half the time in her mother's household (in other words, where uxorilocal or bilocal residence rules are in force), as well as in societies that depend heavily upon the work of girls and young women to make a major contribution to sustenance activities.

Recognizing that some culturgens are more derivative and communicative in function than others, it is possible to delineate what appear to be chains of causation connecting the principal habitat, economic strategy, and both primary and derivative cultural choices. A very reduced example is the following:

Habitat and economy		Principal culturgens		Derivative culturgens
Desert herders	⟶	Polygyny, patrilineality, patrilocality	⟶	Absence of formal female initiation ceremony

The divisions cited in this particular case correspond roughly to the infrastructure and superstructure of Marxian anthropologists (Terray, 1975; Godelier, 1975, 1977) and more closely to the infrastructure, structure, and superstructure distinguished by Marvin Harris (1979) in his exposition of cultural functionalism.

Cultural anthropologists usually perceive this kind of sequence as the outcome of conscious choices made according to culturally acquired rules of inference and valuation. The existence of different routes through the causal network suggests great flexibility. Hence man is judged to be ''culturally determined,'' not genetically determined to any meaningful degree. Human beings are thought to pursue their own interest and that of their society on the basis of a very few simply structured biological needs by means of numerous, arbitrary, and often elaborate culturally acquired behaviors. In contrast to this conventional view, our interpretation of the evidence from cognitive and developmental psychology indicates the presence of epigenetic rules that have sufficiently great specificity to channel the acquisition of rules of inference and decision to a substantial degree. This process of mental canalization in turn shapes the trajectories of cultural evolution. Furthermore, the habitat and economic strategy are not necessarily the prime movers. They represent boundary conditions whose selection is influenced by the epigenetic rules and which constrain rather than direct the choices made by individual members of the society.

The observed variability in outcome of cultural evolution does not by itself suggest the absence of such structure in the epigenetic rules. To make this point more explicit, define ϕ as the difference between the probability of usage of culturgens in a two-culturgen state, so that

$$\phi = u_2 - u_1, \qquad -1 \leq \phi \leq 1.$$

As illustrated in Figure 3–3, ϕ can remain constant across a given range of environments, modes of production, or other patterns of the ambient culture. Or, in contrast, ϕ can change through any one of many patterns as the environmental or cultural context is altered. The important point is that both forms of response can be based upon equally rigid, genetically based rules of perception and decision making. The rules remain the same from one culture to the next; only the starting points and context differ.

It has often been argued, for example very cogently by Geertz (1966) and Marvin Harris (1979), that social anthropology should focus on the differences among cultures and the causation of the differences rather than upon the common properties of human cognition. But the history of the natural sciences suggests that such an approach, while yielding natural history vital to any discipline, never results in much more than surface description and correlation. Both the fundamental rules and the context are necessary for a full, analytic account. Consider bodies in motion. An

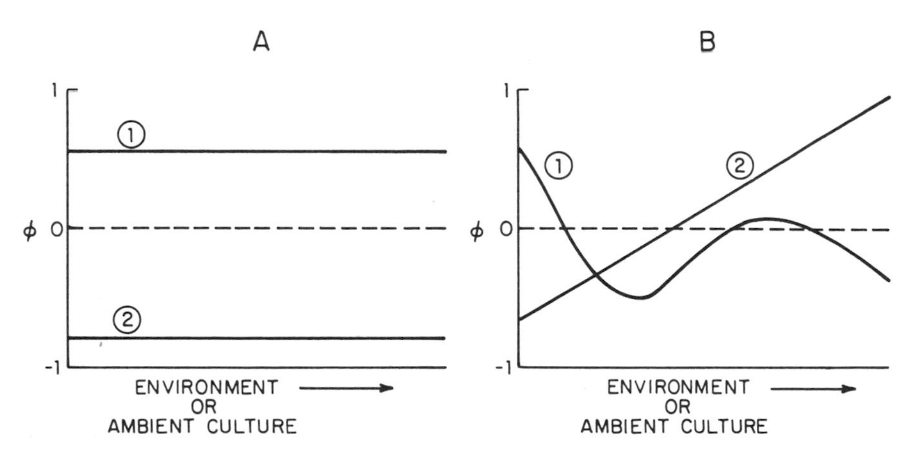

Figure 3–3 The difference ϕ in predisposition to acquire one or the other of two culturgens. As illustrated in *A*, this difference can remain constant as changes occur in the environment or culture. Or, as represented in *B*, the difference can change through any one of many patterns as the context is altered. Both types of response can be under the control of genetically determined core rules of cognition and decision.

orbiting satellite, an airplane, and a falling meteor have radically different trajectories. But the variation among them cannot be understood without the application of Newton's laws of motion, even though these regularities may at first seem irrelevant or even counterintuitive. Closer to the subject at hand, phenylketonuria and normal phenylalanine metabolism represent a phenotypic variation that can be described nicely as an outcome of nutrition. It can be individually prescribed by diet and its effects observed down to the biochemical level. Yet the origin and distribution of the variation within human populations are comprehensible only by reference to the laws of Mendelian genetics to which they are covertly obedient.

The question of the fundamental, biologically grounded rules in human cultural evolution is empirical, to be settled by the analysis of epigenesis within the context of cognitive and developmental psychology aided by neurobiology and genetics. The existence of primary epigenetic rules, which impose constraints in perception, was documented in Chapter 2. We now examine some of the studies that have addressed the procedures of higher information processing and decision making in human beings. The goal of our analysis will be to demonstrate striking regularities that run through the diversity of facts known about human cognition—regularities that we interpret as evidence for the existence of secondary epigenetic rules predicted by our theory.

Elements of Human Information Processing

Information processing is composed of at least five procedures: feature discrimination, storage, interpretation, recall, and computation. In recent years cognitive psychologists have made important advances in identifying the human form of these activities. A summary of the information-processing segment of human cognition is presented in Figure 3–4. Although this formulation cannot be regarded as definitive, it does incorporate a diversity of experimental data. It perceives that in the

Figure 3–4, opposite A model summarizing human information processing. The size of short term memory represents the number of symbols that the conscious mind is able to manage simultaneously. Processes that would correspond to the subconscious are not at all understood in the framework shown here and are not displayed explicitly. The modification times per chunk shown for long term memory refer to the intervals per old chunk required to form a new chunk. See the discussion of the 10*J* Rule in the text. (Based on data and models in Newell and Simon, 1972; Massaro, 1975; Norman and Rumelhart, 1975; Lindsay and Norman, 1977; Oden and Massaro, 1978; and Simon, 1979.)

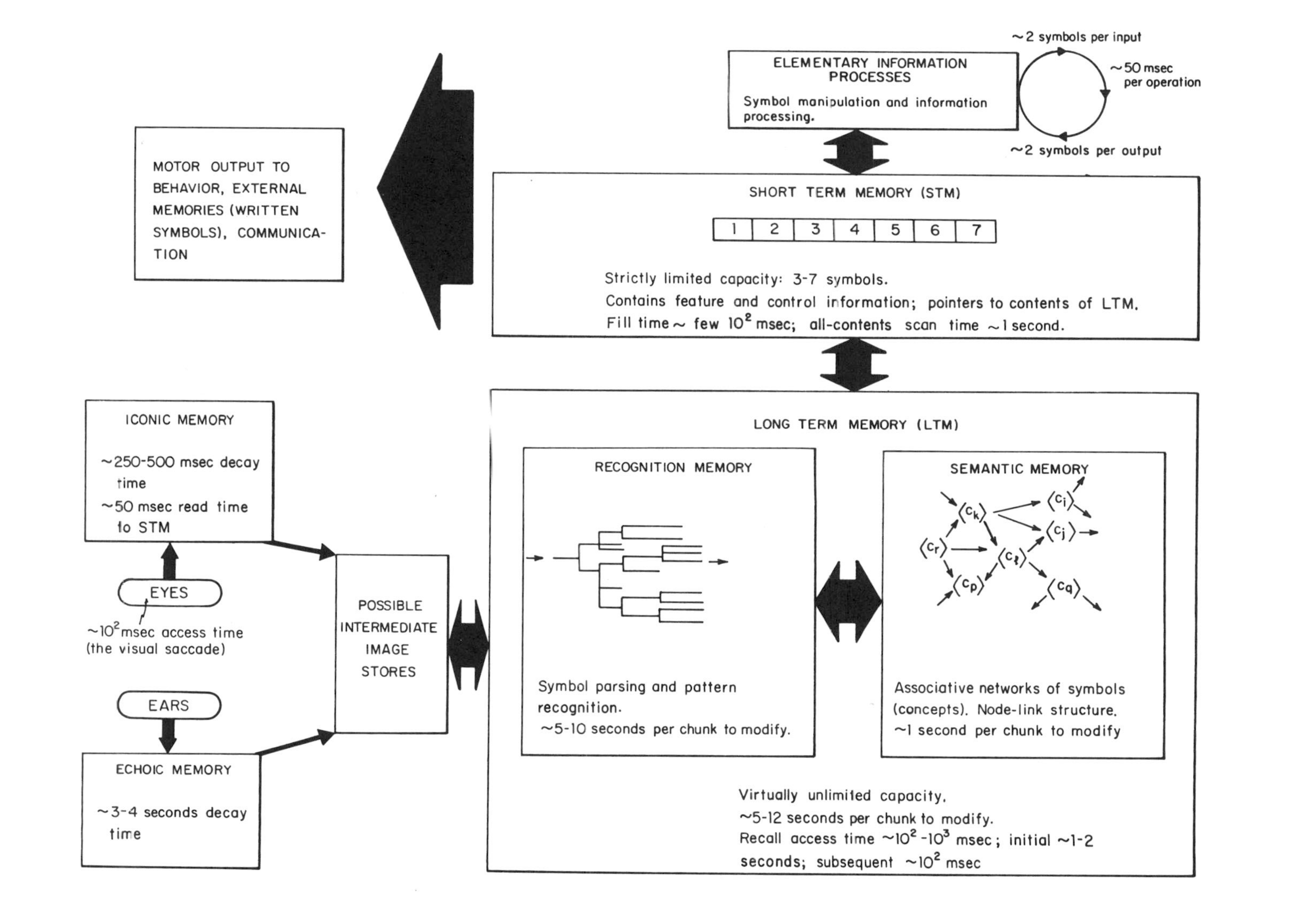
~2 symbols per input
~50 msec per operation
~2 symbols per output
ELEMENTARY INFORMATION PROCESSES
Symbol manipulation and information processing.
SHORT TERM MEMORY (STM)
1 2 3 4 5 6 7
Strictly limited capacity: 3-7 symbols.
Contains feature and control information; pointers to contents of LTM.
Fill time ~ few 10^2 msec; all-contents scan time ~1 second.
MOTOR OUTPUT TO BEHAVIOR, EXTERNAL MEMORIES (WRITTEN SYMBOLS), COMMUNICATION
LONG TERM MEMORY (LTM)
RECOGNITION MEMORY
Symbol parsing and pattern recognition.
~5-10 seconds per chunk to modify.
SEMANTIC MEMORY
Associative networks of symbols (concepts). Node-link structure.
~1 second per chunk to modify
Virtually unlimited capacity.
~5-12 seconds per chunk to modify.
Recall access time ~10^2-10^3 msec; initial ~1-2 seconds; subsequent ~10^2 msec
ICONIC MEMORY
~250-500 msec decay time
~50 msec read time to STM
EYES
~10^2 msec access time (the visual saccade)
EARS
ECHOIC MEMORY
~3-4 seconds decay time
POSSIBLE INTERMEDIATE IMAGE STORES

underlying brain system, capacities and execution times are tightly constrained. In the early stages of the processing, some of the features are selected and others discarded; those retained are placed into a "preperceptual storage space." Up to this point the processes are genetically fixed with sufficient rigidity to be impervious to learning. They include at least some of the primary epigenetic rules noted in Chapter 2. Recognition begins when the features in the preperceptual space are compared with perceptual units in long term memory. The placement of the stimulus on the perceptual space is achieved by its further integration into the long term memory (LTM), a process sometimes designated as secondary recognition (Massaro, 1975). This step is influenced not just by memory but also by expectation and mental set. These properties depend in part on reentrant signaling from the hippocampal-septal axis and other portions of the limbic system. The secondary recognition process translates this information into an additional, abstract code that matches the signals against symbols and imparts to them certain qualities beyond simple recognition. Thus the synthesized perception of a friend's voice results in the identification of the words and the source, while the abstract encoding procedure defines the voice in terms of such properties as low, harsh, unnatural, and so forth.

In a series of ingenious experiments Oden and Massaro (1978) demonstrated that the auditory information process employs what is often called fuzzy logic. Rather than make sharp distinctions among phonemes on the basis of all-or-none features, the brain evaluates phonetic qualities of the sounds, particularly the place of articulation and voice onset time, by sorting them into overlapping sets spaced along continuous scales. Since the qualities vary continuously, they are represented as predicates that are more or less true rather than absolutely true or false. Phoneme classification is made up of referrals to phoneme prototypes in long term memory. As explained in Chapter 2, these prototypes are learned, but the learning follows primary rules that automatically space out the phonemes and limit their number. The forebrain then classifies the sound as belonging to the phoneme prototype providing the closest match. The data on phoneme recognition in fact fit Luce's model (1959), which states that the probability of identifying a sound as a particular phoneme is equal to the goodness of the match of the sound to that phoneme relative to the sum of the goodness-of-match values for all phonemes being considered in the perceptual space. Another apparent particularity in phoneme identification is the noninteraction of the acoustic features during the feature evaluation stage of processing.

These elementary procedures may be fundamental to the human mind. Cognitive psychologists have recently assembled an array of evidence suggesting that the use of prototypes and fuzzy sets characterizes the processes of recognition and discrimination in activities additional to speech perception, including vision, gestures, and concept organization (Brown, 1978; Rosch and Lloyd, 1978; Wickelgren, 1979a).

Further insight into the procedures used to match stimuli with symbols in long term memory can be gained by considering the evaluation of paired as opposed to individual stimuli. In experiments conducted by Getty and coworkers (1979), subjects were asked to evaluate visual representations of eight complex sounds. They judged the pictures pairwise according to subjective degree of similarity and also identified them individually in various subsets. The similarity judgments were placed in a three-dimensional perceptual space by means of a multidimensional scaling procedure. The data closely fitted a decision model based on weighted interstimulus distances, in which confusability between stimuli falls off as an exponential function of the distance between them. When given subsets of the pictures, the subjects evidently placed varying weights on the dimensions in such a way as to maximize the percentage of correct identification. When a separate set of subjects were given the identification task alone, they employed the same set of dimensions.

These results from experiments on visual discrimination support the validity of the use of multidimensional scaling analysis in mapping perceptual space, since the space revealed by the procedure could be used to account for behavior in an independent task. The fit to the confusability model indicates that in distinguishing visual patterns the brain uses a fuzzy set procedure similar to that employed in discriminating phonemes.

Information recall also has several particular qualities that can be tentatively characterized as human specific, pending the accumulation of further data on nonhuman cognition (see for example Lauer and Lindauer, 1971; Griffin, 1976). Unrehearsed short term memory lasts approximately thirty seconds. A delay of that length wipes out most of the ability to recall sets of unrelated words or signs that have been memorized hastily. The capacity of short term memory is approximately three to seven symbols of control and feature information. To use the original formulation of Miller (1956a,b), this is the equivalent of the "magical number" of seven elements, or, more precisely, seven plus or minus two. People can rapidly memorize approximately seven unrelated syllables, integers, or words. They can also discriminate with two to three bits of precision the tone and loudness of a sound, the salinity of a salt solution, and the position of a pointer in a linear interval. Such resolution on a continuous scale translates into approximately seven equiprobable symbols in a discrete scale. Miller wondered if it is only a "pernicious, Pythagorean coincidence" that there are seven wonders of the ancient world, seven seas, seven deadly sins, seven daughters of Atlas in the Pleiades, seven ages of man, seven levels of hell, seven days of the week, and so forth. We suspect not, although evidence from non-Western cultures is conspicuously lacking.

In contrast, long term memory is all but permanent, potentially infinite, and indestructible. It is achieved through repetition and strong reinforcement, especially that which engenders emotion. Storage and recall are greatly enhanced by the process of chunking, a packaging of information symbols that then require only the recollection of a single element to be retrieved. If short term memory permits only about three to seven symbols to be activated and kept in the conscious mind for a short period during the give-and-take of ongoing behavior, large amounts of additional information can still be brought into play by the retrieval of associated symbols. The symbols are the neural representations of information and upon recognition or activation evoke their stored designators (Miller, 1956a,b; Newell and Simon, 1972). Chunking is a cortical function that serves semantic memory, configuration formation, and processing efficiency.

These aspects of human cognition function within two epigenetic constraints. One is the minuscule symbol storage capacity of short term memory. Insofar as all deliberated symbol manipulation and computation take place on structures temporarily active in short term memory, the three- to seven-symbol limit creates a severe bottleneck between the essentially inexhaustible long term memory and the brain processes subserving the conscious mind. This basic restriction greatly limits processing speed and efficiency and reduces the number of processing strategies that are practical at any given time.

A second major constraint, which can be called the $10J$ Rule, is created by a marked asymmetry inherent in the operation of long term memory. No less than $5J$ to $10J$ seconds of neuronal processing time are needed in order to store in the long term memory the symbols necessary to recognize and interpret a chunk composed of J familiar chunks or subpatterns (Newell and Simon, 1972; Simon, 1979). Such writing times are very long on the behavioral time scale and represent a real cost to the organism in terms of time and opportunities lost. In this way they limit the assimilation of new information. The times required to retrieve chunks *from* long term memory are shorter by one to two orders of magnitude. These two constraints, the capacity of short term memory and the 10*J* Rule of long term memory, are unexplained from the viewpoint of evolutionary theory, and we shall return to them in Chapters 6 and 7.

On the basis of an analysis of the hippocampalamnesiac syndrome, Wickelgren (1979b) concluded that the hippocampus plays a critical role in the physiology of cognition by differentially priming free neurons that symbolize and link the elements of the chunk. Current information on learning and particularly the consolidation process of long term memory appear to us to be consistent with Edelman's hierarchical model of brain action, in which selected "recognizer" cells, such as groups of complex neurons in the striate cortex, are activated by stimuli to which they are preprogrammed to respond (Edelman and Mountcastle, 1978). Then "recognizers of recognizers," which are neurons in the temporal, frontal, or prefrontal cortex, respond to signals from recognizer groups. The candidate neuronal groups among the recognizer of recognizers form a degenerate subset of all such groups. That is, more than one group can recognize a particular assembly of recognizer cells. Recognizer-of-recognizer groups can interact in a hierarchical manner to create increasingly abstract representations. Thus learning is to some extent programmed, and its rate thereby increased, but the total content of symbolic information eventually stored in the brain is highly variable.

On the basis of his model Edelman (in Edelman and Mountcastle, 1978) envisages consciousness as "a form of associative recollection with updating, based on present reentrant input, that continually confirms or alters a 'world model' or 'self theory' by means of parallel motor or sensory outputs. The entire process depends upon the properties of group selection and reentrant signaling in a nervous system that is already specified by embryological, developmental, and evolutionary events." Griffin (1976) has suggested that human-like elements of consciousness exist in the more intelligent animals, and he has outlined a series of tests by which this possibility can be explored.

Our improved understanding of the constraints on the learning process and the physical devices of storage and recall have made it increasingly clear why the human mind operates primarily by means of symbols and hierarchical classifications of stimuli (Bartlett, 1932; Miller, 1956a,b; Simon, 1979). It is reasonable to speculate that any living system in evolutionary transit from the protocultural to the eucultural state would encounter similar constraints, diminish them in the same way, and thereby assemble a conscious mind operating largely with symbols and chunks.

General Properties of Secondary Epigenetic Rules

We are now in a better position to consider the secondary epigenetic rules themselves. Each rule contributes to two components in the bias curves. The first is penetrance: the propensities to use some culturgen—any culturgen—of a given category, regardless of whether the choice is made from among many or few. The second component is selectivity among the available culturgens. In the idealized choice curve of Figure 3–5 a high penetrance is measurable as a low frequency of individuals in the null category, while a high degree of selectivity is reflected in the concentration of those individuals who have made a choice on one or a relatively few of the culturgens (see also Appendix 3–1). Penetrance and selectivity in culturgen transmission can evolve independently of each other. Furthermore, both can be separately modulated by the actions of no less than three major cellular systems: the primary sensory receptors and coding interneurons, which determine the ease with which particular culturgens are perceived; the associative centers of the cerebral cortex and the attention and recall mediating centers in the hippocampus, which determine learning capacity; and the centers of the limbic system and midbrain's substantia nigra, which affect reinforcement (Oades, 1979).

Data from developmental psychology indicate that epigenesis has the least penetrance at infancy. Later in life, the action of context-dependent epigenetic rules allows high penetrance while maintaining high selectivity. For purposes of model building, we believe it is valid in the case of at least the simpler culturgen categories to reduce the relevant part of the epigenetic rules to a single filter expressed as a profile of transition probabilities. Most of these profiles, or bias curves, change as a function of the surrounding culture and major features in the physical environment. Those of greatest analytic tractability are sufficiently robust to retain an easily recognizable form across cultures.

Wolff (1970) and Hess (1973) have argued that the time course of genetically directed learning in infants and children is too long for the learning to qualify as true imprinting. (Imprinting is defined by ethologists as an event that occurs within a very short period during development.) In the case of offspring attachment to mothers, the susceptible period is two hours in sheep, thirty-six hours in ducks and chickens, one week in dogs, and three days in rats. Few human behaviors have learning regimens limited to periods of less than three months. Human infants and children engage in undoubted directed learning, but it is concentrated in longer, less well-defined periods. Thus the primary socialization of children, espe-

PENETRANCE

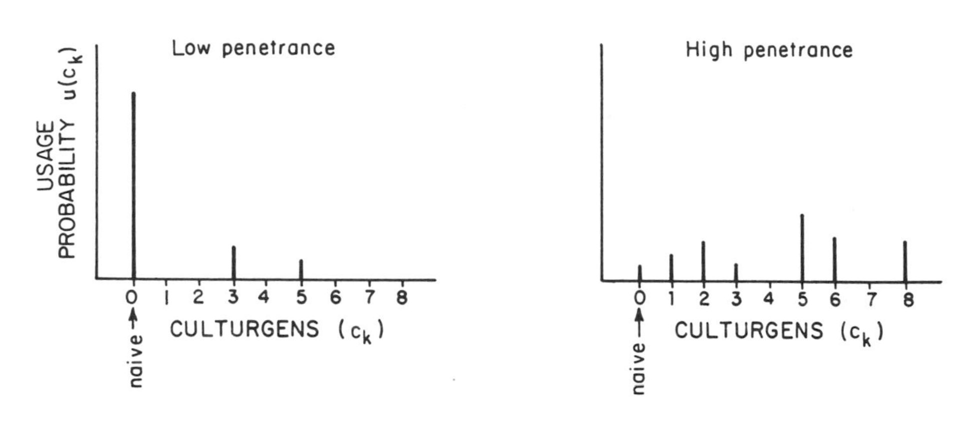

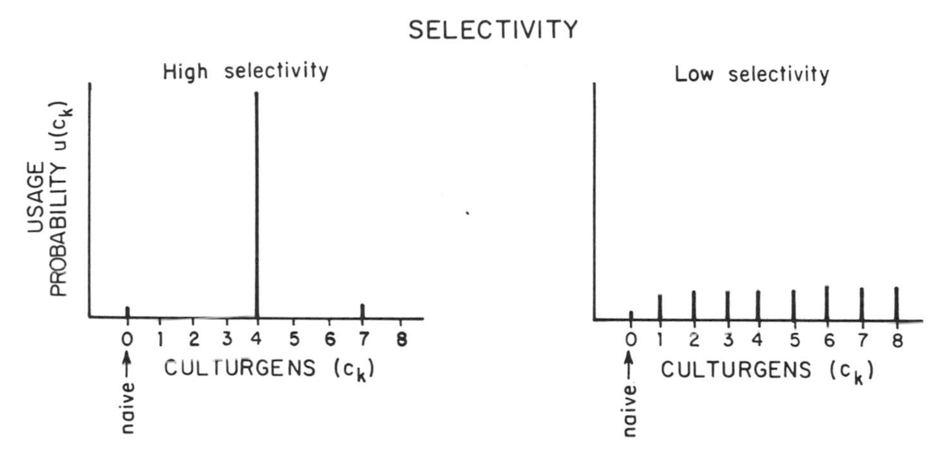

Figure 3–5 Four imaginary usage bias curves chosen to illustrate the two principal properties of epigenetic rules: penetrance, the extent to which any culturgen c_k of a given category is used by members of a society (those who learn no culturgen remain "naive") and selectivity, the degree of preference for some culturgens over others among those available. The distributions are consistently displayed by individuals of particular genotypes within specified environments.

cially attachment to particular adults, is thought to occur between six weeks and six months (Gray, 1958). Bonding in the reverse direction, the attachment of mothers to infants, does possess the properties of imprinting, since it occurs within a sensitive period of hours or days and produces aftereffects that last for years.

Early human social development is usefully regarded as a process of focusing—a passage from general to more specific classes of stimuli during periods that individually last, according to the stimulus category, from days to years. The epigenetic rules are expressed as a stepwise change in the degree of directedness during the focusing process. The first stage is often little more than an automatically greater attraction for a particular set of stimuli as opposed to others. As Fantz and his colleagues (1975) have expressed the matter in their excellent account of early visual development, perception precedes action. In other words, preference for certain cues exposes infants and children to the most relevant processible information. Furthermore, infants fixate on images, even small portions of images, longer than do older children (Kagan, 1970; Salapatek, 1973). Attention is thus an important component of the early epigenetic rules. And because it is also experimentally the most tractable, it provides a first, valuable glimpse of the basic epigenetic process. In Table 3–1 we have summarized some of the results obtained by Fantz and his associates. The sequence from top to bottom gives the visual choices in the order of their appearance. Figures 3–6 and 3–7 present some of the configurations used during the experiments, along with the time curves of

Table 3–1 Sequential epigenetic rules in the development of infant visual preference. The favored choice in each competing pair is given first. (Based on data from Fantz et al., 1975.)

Visual preference	Postnatal age at first development (weeks)
Simple transitive differences	
Larger elements	1
More numerous elements	1
Curved vs. straight lines	1–8[a]
Distinction among patterns	
Bull's-eye design vs. parallel stripes	8
Touching elements vs. those that are separated	8–10
Nonlinear arrays vs. linear-latticed arrays	8–12
Three-dimensional vs. two-dimensional objects	8–12
Recognition of novelty	
Novel visual patterns (general) vs. learned patterns	16–30[a]
Novel faces vs. familiar faces	20–24
Novel orientation of face vs. conventional orientation	20

[a] Timing depends on patterns compared.

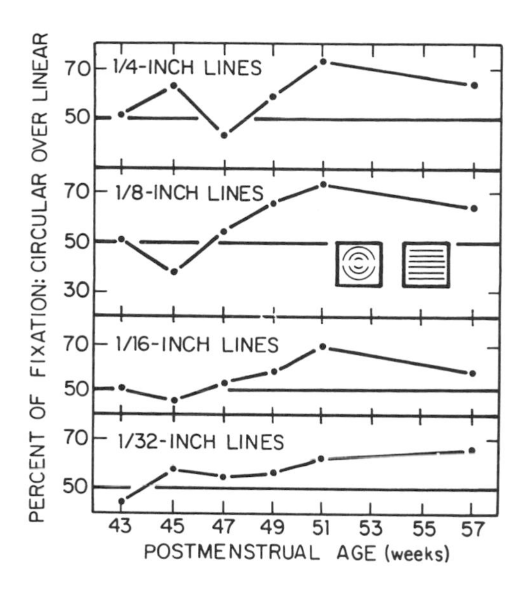

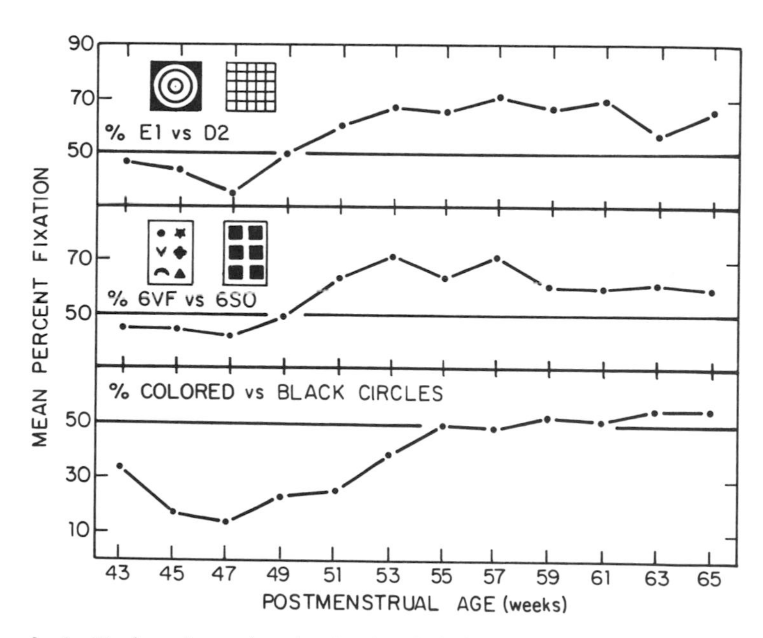

Figure 3–6 Early epigenetic rules in visual choice. The curves represent the relative attention shown by infants to various competing and simultaneously presented designs. The age cited is the time since the last menstruation of the mother. This interval is approximately the time since conception, and its use has the advantage of eliminating differences in term. (Modified from Fantz et al., 1975.)

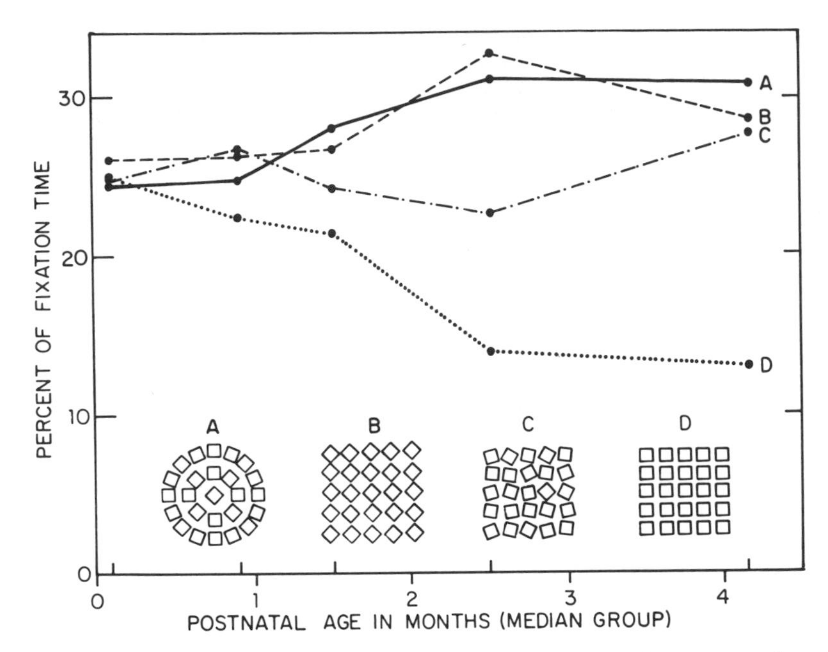

Figure 3–7 Choice by infants among members of a set of more complex designs. (Modified from Fantz et al., 1975.)

visual preference. These data indicate the existence of at least three key steps in visual choice by infants. At the beginning, choices are made on the basis of simple transitive distinctions, such as a liking for larger elements or for more numerous elements. Next, the infant chooses among related complex patterns, with preference being given to intermediate degrees of complexity and symmetry. Finally, in a schedule overlapping that of the previous two strategies, the infant develops a slight preference for novel images.

These and similar studies indicate that in addition to the transition from attention to learning action there is a focusing sequence with reference to the stimuli discriminated. This sequence begins with an automatic restriction to stimuli filtered by the sensory receptors and coding interneurons and a preference for certain stimuli as opposed to others. It passes to the biased learning of a class of objects. It ends with a preference for (or aversion to) particular objects. Thus in the case of visual development the infant directs its attention preferentially to elementary visual designs found in a very broad range of objects, including faces and similar configurations. It simultaneously narrows its preference to faces with normally arranged features as opposed to similar patterns. Finally, it learns and comes to prefer its mother's face. Within the first two of these three levels the learning is biased genetically toward certain classes of visual patterns. This conception of programming in human learning rules is at least consistent with current models of hierarchical organization in the nervous system and with decision-making apparatus in animal behavior (Dawkins, 1976b; Bentley and Konishi, 1978).

It is our impression (one that cannot be rigorously tested by the available data) that the human species has followed a rule of parsimony in the evolution of the epigenetic rules. In the hierarchy of procedures just cited, leading to a final experiential focus on culturgens, evolution of the epigenetic rules stops when the rules reach the least degree of selectivity that will suffice. Thus infants possess an innate preference for the principal features of the human face over other, similar patterns, but there is as yet no evidence that they prefer a female face, or one of any other particular size, shape, or color (Jirari, 1970). Genes prescribing an early cessation of lactase production occur at high frequencies in Oriental populations, and milk is generally avoided in Oriental cuisines. But the aversion appears to be based on gastrointestinal distress rather than an innate lactose aversion programmed to appear in late childhood (Rozin, 1976). Sexual fixation is a powerful form of prepared learning activated during adolescence and early maturity. It ordinarily leads to heterosexual pair bonding, but idiosyncratic experiences during the susceptible period of development can divert individuals in part or in whole to homosexuality or to deviant practices such as pedophilia, fetishism, urolagnia, coprolagnia, and necrophilia, which are difficult to alter later (Goleman and Bush, 1977; VanDeventer and Laws, 1978).

If the principle of parsimony is valid, it parallels the rule of metabolic conservation deduced in studies of other modes of evolution. For example, in the course of evolutionary time certain amino acids become "essential" and other molecules become vitamins if they are supplied so abundantly in the normal diet that the need for their independent synthesis is removed. In analogous fashion most species of obligatory cave animals tend to lose their eyes and bodily pigment. The conventional explanation given by geneticists is that when the products of particular routes of biochemical synthesis are no longer necessary for survival and reproduction, the materials and energy they consume constitute a deficit. Any mutations eliminating the biosynthetic routes will then have a selective advantage; the phenomenon has been discussed by E. O. Wilson (1975:160–161). Similarly, special coding devices in the development of social behavior can be expected to decline whenever the requirement for selectivity is reduced or eliminated. Nevertheless, for reasons explained in Chapter 1, coding devices are extremely unlikely to be eliminated altogether.

A low degree of selectivity in the epigenetic rules can be expected to create several important effects in social interaction. We have already cited the rarity of imprinting and critical periods in infant development. Another anticipated result is a greater dependence of the development of normal behavior on high levels of socialization and hence experience in day-to-day encounters, rather than on more nearly automatic, inborn responses. The classic studies of Harlow and others on rhesus monkeys revealed that when infants of this species are deprived of maternal and peer stimulation during early life, they are later incompetent in sexual performance and parenting (Harlow et al., 1966; Hinde and Spencer-Booth, 1969). In what may be a parallel relation, Steele and Pollock (1968) found that all parents involved in severe child abuse in sixty families had themselves been deprived of parental physical affection during their childhood. They also generally suffered from depression and had difficulty establishing ego identities appropriate to their ages.

A final anticipated effect of low selectivity is the existence of supernormal stimuli, which elicit stronger responses than the signals normally produced by individuals during communication. For example, herring gulls prefer dummy wooden eggs provided by experimenters over their own eggs as long as the substitutes are larger in size—in fact, the larger the better. This appears to be a simple innate rule of decision making that in the vast majority of cases permits the correct separation of eggs from stones and other inanimate objects. In a similar manner, human beings sometimes prefer stimuli that lie beyond the norm of those experienced during socialization. In initial encounters on the part of adults, for example, they respond more strongly to supernormal visual images that evoke sexual or parental interest, such as those showing exceptionally large breasts in women or abnormally wide eyes and small noses in children (Eibl-Eibesfeldt, 1975). The phenomenon appears to ensure greater attention to the more ordinary features that do in fact characterize the vast majority of women and small children.

Another trend in the evolution of epigenesis might be called the transparency principle: the more the effect on genetic fitness of a category of behavior depends upon environmental circumstances, the more clearly the conscious mind perceives that relation and the more flexible its response. In the extreme case, the behavior is modified to fit each particular contingency, following conscious reflection on the circumstances. It is possible that once the context is specified, the form of the response is completely predictable. We can speak of such behavior as being both flexible (varying according to context) and selective (invariant within a given context), or the behavior can be both flexible and relatively nonselective—in other words, it varies both between and within different contexts.

The important point is that when the impact on genetic fitness of such behavior depends on the context, the mind is more likely to perceive the relation and to make decisions according to this perception. Thus economic behavior, including the procedures of energy harvesting and reciprocal transactions, have effects that impinge directly on survival according to the peculiarities of the surrounding environment and social organization. These relations are intuitively understood with more or less clarity and are subject to conscious deliberation. Economic behavior is comparatively flexible, and culture varies greatly in its expression (Haggett, 1972; Boehm, 1978; Clarke, 1978). At the opposite extreme, the impact on genetic fitness remains approximately constant in most or all possible circumstances. In this case the mind is typically unaware of the relation between the behavior and genetic fitness, and the behavior does not vary according to context. Accordingly, the adaptive significance of deep grammar (consistent, rapid sentence formation), incest avoidance (reduction of inbreeding depression), and the consumption of sugar (exceptionally high caloric content) are understood only by the small number of societies that have studied them scientifically during the past two hundred years (Katz et al., 1974; Rozin, 1976; Katz, 1980). The behaviors are incorporated automatically into most or all human societies under the guidance of strong, selective epigenetic rules. If this correlation between transparency and context dependence proves to be general, it is consistent with the principle of parsimony suggested earlier, namely that epigenetic rules evolve until they achieve the least sufficient degree of selectivity.

A series of case studies in cognitive development and adult decision making will illustrate the diversity of the secondary epigenetic rules and the principles just suggested. While this mode of analysis is in the earliest stage of development, its results clearly are destined to be exceptionally important to the future of the social sciences.

Facial Recognition

The human face is an early object of fixation and serves as a source of comfort to infants (Argyle and Cook, 1976). For both older children and adults it contains the principal set of features used in individual recognition, nonverbal communication, and a substantial part of artistic expression. It is therefore no surprise to find relatively selective epigenetic rules in the utilization of facial features. Experiments by Jirari (1970) revealed that even newborn infants fixate more on facial designs than on simpler patterns. They also prefer normal features over scrambled features of the same level of complexity, with the eyes being especially important parts of the attractive schema (see Figure 3–8). Similar results have been obtained in experiments on four-month-old infants by Haaf and Bell (1967) and by McCall and Kagan (1967). It is difficult to ascribe the results to any form of learning; Jirari repeated the first series of comparisons shown in Figure 3–8 with forty infants whose average age was *10 minutes* (range 2–17 minutes), and observed the same rank order of preference. Some data indicate that relative attraction to faces is strengthened still further during the ensuing six months. Fantz (1963) found a 15-percent increase in average fixation time during this interval and a concomitant decrease in fixation on concentric circles. By four months, infants no longer prefer normal to scrambled faces (Kagan, 1970), but this is the time in which novel geometric patterns and novel faces come to be preferred over familiar stimuli (see our Table 3–1).

A rapid focusing of learning capability follows the maturation process in facial recognition. By five months, infants discriminate and recall differences among classes of faces such as those of men versus women and those of women versus children. By seven months, infants are able to distinguish individual persons and to utilize a variety of facial angles in achieving recognition (Fagan, 1979). This increase in resolving power parallels the general growth in ability to recognize abstract geometric designs of similar complexity, but it remains an independent faculty. In fact, the capacity to distinguish faces is dependent on specialized regions on the undersurface of the temporal and occipital lobes. Lesions in this area cause prosopagnosia, a remarkably selective disability. The patient can identify objects by sight and persons by their voices, but he cannot recognize persons by looking at their faces (Geschwind, 1979).

Visual Pattern Complexity

Although human beings respond to supernormal stimuli in some categories of culturgens, the preference does not increase monotonically with a continued intensification of the stimuli. In all categories known to us to have been tested, there exist intermediate values or ranges of values preferred over others more extreme. This principle has been well estab-

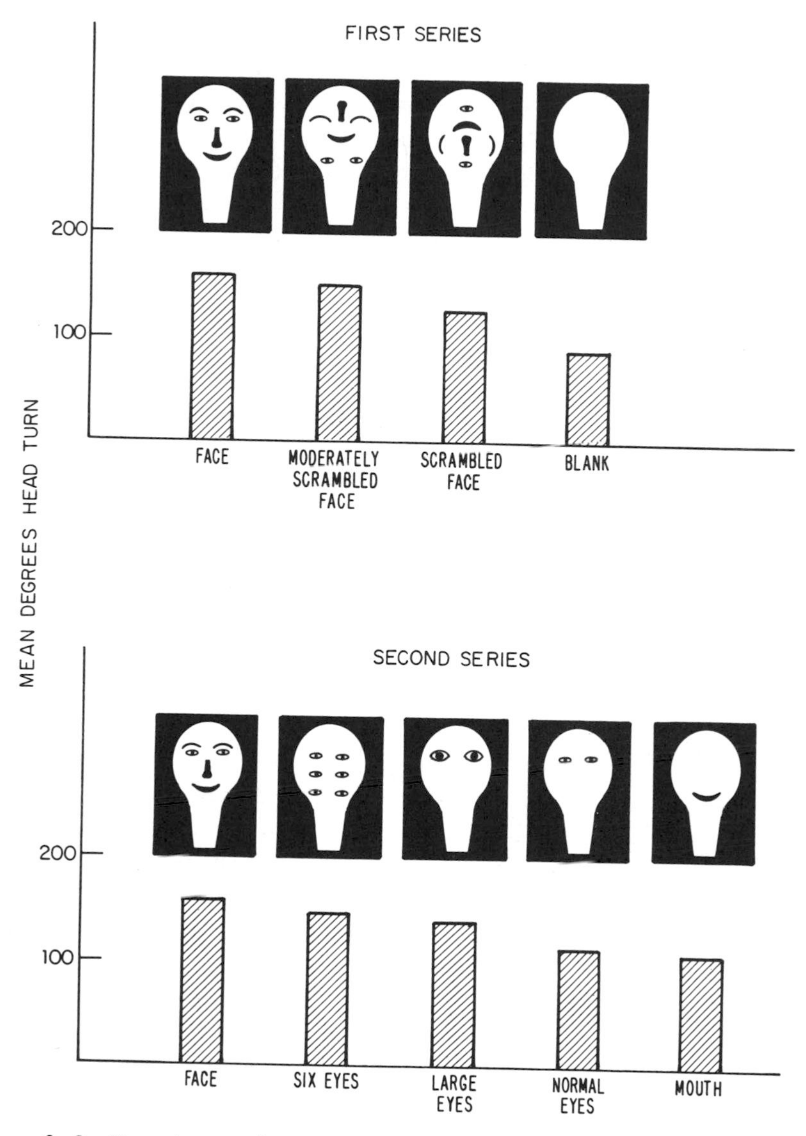

Figure 3–8 Experiments demonstrating the occurrence of secondary epigenetic rules early in facial recognition. Diagrams containing a normal array of facial features were presented to newborn infants in competition with diagrams containing scrambled or reduced features. Fixation was measured as the degree of head turning by infants presented with moving images; thirty-six newborns were used in each of the two series. The infants preferred a normal array of facial features, and the eyes were an important component of the preferred pattern. (Based on Jirari, 1970, in Freedman, 1974.)

lished, for example, in the study of motivation. Schroder and colleagues (1967) have documented the maximization of emotional reward and information processing at intermediate levels of complexity. They have constructed a general model that attempts to predict these peaks as the maximum sum of information content, reward, and the negative effects of computational difficulty and emotional stress.

The same principle has been adduced independently during studies of cognition. People tend to break down information into a relatively small number of categories. As mentioned earlier, Miller (1956a) suggested that the number is the "magical seven" of mythology and folklore, plus or minus two. Reviewing the literature to that time, he showed that the ability to discriminate pure tones, tastes, sizes, colors, and temperatures, without special training and aids and with trivial error, is in fact usually limited to this range. More recently, Pendse (1978) demonstrated that the optimal number of categories in communication systems generally is a function of the signal-to-noise ratio. The larger the number of times a signal is transmitted in a noisy channel, the smaller is the optimal number of categories into which any sample of the signal can be broken down. If this mathematically deduced relation has been obeyed during organic evolution, the design of the brain itself can be expected to specify rather precisely the degree of complexity employed in ordinary, intuitive classifications. This hypothesis can be tested in part through the following prediction generated by the mathematical model: the left brain hemisphere, which must verbalize and hence retransmit information more extensively than the right hemisphere, should perform better in handling systems with small numbers of categories. Superiority should shift to the right hemisphere as the number of categories is increased. This prediction has been confirmed by Pendse's experimental studies.

The emotional components of pattern discrimination and preference become paramount in the experimental studies of esthetics. In studies on the choice of polygons (squares versus octagons versus six-pointed stars, and so forth), Eysenck (1968) found that preferred figures have the following qualities: less familiar, symmetric, nonright angles, and numerous nonparallel sides. Rashevsky (1960) went so far as to devise a neurological model to account for such data. He obtained a close fit when the assumption was made that pleasure centers are maximally stimulated by the summed excitation of many redundant elements in repeating but nonidentical sets.

The polygons used in early esthetics experiments were limited in complexity, with the result that the data reveal only the correlates of esthetic response at the lower end of the scale of complexity. Subsequent studies by Smets (1973), using the indirect but more precisely measurable response of alpha brainwave blockage as a signal of arousal, indicate the existence of intermediate maxima when the competing designs are made more complicated (see Figure 3–9).

It is of interest that the preference for designs of intermediate complexity can be traced back to earliest infancy. Hershenson and colleagues (1965) presented newborns with an array of randomly constructed figures classified according to whether they contain five, ten, or twenty turns. The infants gazed most consistently at the set with ten turns (see Figure 3–10). Later, more precise epigenetic rules of pattern choice, entailing complexity as a key variable, were disclosed by the experiments on visual development by Fantz and his associates (see Table 3–1).

Knowledge of the epigenetic rules governing the development of complexity choice, problem solving, and esthetics is still rudimentary. Berlyne (1971) and Bortz (1978) concluded that on the basis of present knowledge no simple link can be made between arousal curves and subjective esthetic judgment. It is unlikely that preference curves in esthetic judgment are of the unimodal form typifying the arousal curves. Yet it seems probable that strong indirect links do exist and have exercised significant effects on cultural evolution. Furthermore, the mean optimum levels of complexity and the higher moments of frequency distribution of the levels within populations can be tentatively interpreted as species-specific traits of *Homo sapiens*. They will be understood more clearly when comparable choice experiments are conducted with other intelligent animal species, especially the Old World monkeys and apes.

Nonverbal Communication

Movements used in nonlinguistic communication offer promising examples for the study of epigenetic rules. Some of the signals are relatively invariant, so that a great deal of convergence in their form and meaning occurs across cultures. Yet nearly all are also subject to modifications peculiar to the individual cultures. For example, Ekman (1973) found relative uniformity in the use of facial expressions to denote fear, loathing, anger, surprise, and happiness. He photographed Americans acting out these emotions and New Guinea highland tribesmen as they told stories in which similar feelings were emphasized. When individuals were then shown portraits from the other culture, they interpreted the meanings of the facial expressions with an accuracy greater than 80 percent.

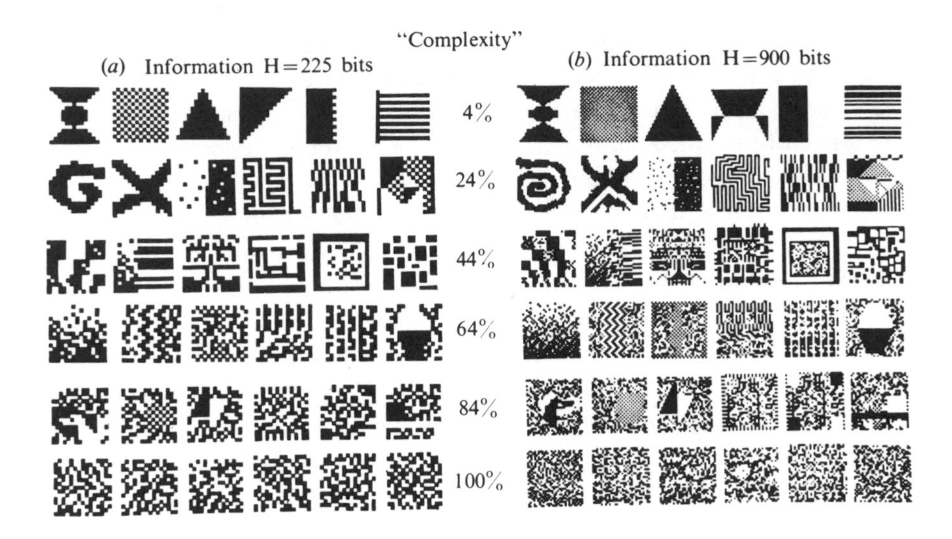

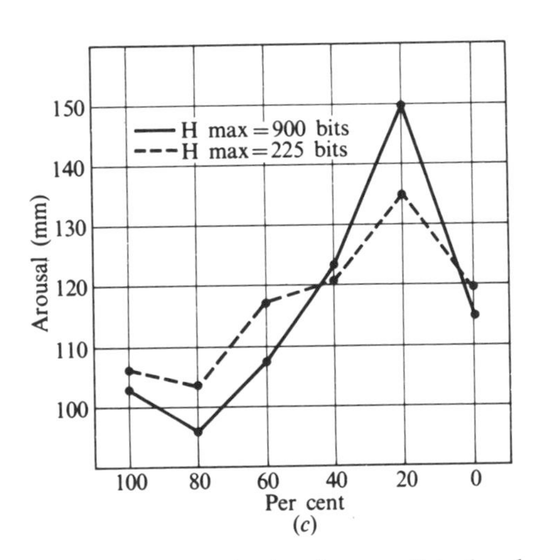

Figure 3–9 Maximum arousal reached at intermediate levels of complexity in visual patterns. Complexity in this instance is measured by the percent of redundancy in two series that differ in the number of elements (one series generates 225 bits, the other 900 bits of uncertainty). The arousal measure is the time the alpha wave of the electroencephalogram was blocked (desynchronized) following presentation of the figure; the number given is the corresponding distance along the polygraph record. Each point in the lower diagram gives the average response of sixty-seven persons. (After Young, 1978; from Smets, 1973.)

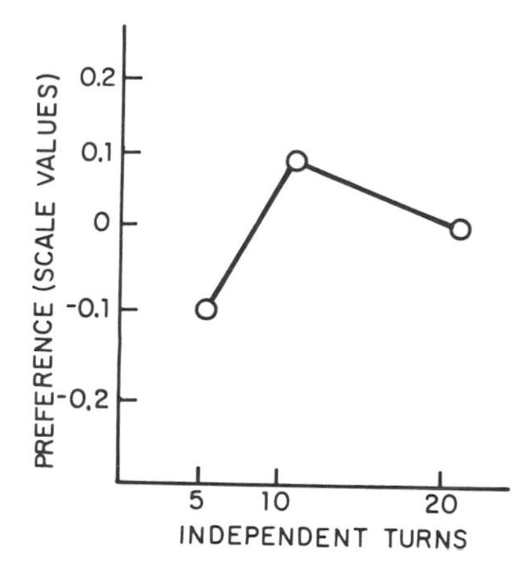

Figure 3–10 An epigenetic rule in the choice of complexity. The preferences given measure the relative length of eye fixation by newborn infants shown sets of randomly generated figures. Complexity in the figures is measured by the number of turns along their edges. (Modified from Hershenson et al., 1965.)

In his continuing field research in human ethology, Eibl-Eibesfeldt (1975, 1979) has documented varying degrees of convergence in other forms of nonlinguistic communication, including phallic displays, eyebrow flashing, threat staring, pouting, gaze aversion, and kissing. The evidence is strong enough to leave little doubt that much of this behavior is specific to man and distinguishes him narrowly but clearly from other Old World primates.

Studies of smiling conducted independently by psychologists and anthropologists have revealed a substantial amount of canalization in its development and uses. Smiles are first displayed by infants between two and four months of age. They at once evoke an abundance of affection on the part of attending adults. The infants of the !Kung San of the Kalahari are nurtured under very different conditions from American infants. They are delivered alone by their mothers without the aid of anesthetic, kept in almost constant physical contact with the mothers or other adults, nursed several times an hour, and trained rigorously to sit, stand, and walk. Yet their smile is identical in form to that of American infants, appears at the same time, and serves the same social function (Konner, 1972, 1977). Smiling also appears on schedule in deaf-blind children and even in thalidomide-deformed children who are not only deaf and blind but crippled so badly they cannot touch their own faces. Under these extreme circumstances it is virtually impossible to implicate learning in the early development of the behavior (Eibl-Eibesfeldt, 1979).

Throughout life smiling is used primarily to signal friendliness and approval, and secondarily (and erratically) to indicate a sense of pleasure (Kraut and Johnston, 1979). Each culture molds the precise meaning of the behavior into a series of nuances determined by its exact form and the context in which it is displayed. Of course, among sophisticated adults smiling can be turned to the uses of irony and light mockery. But even in such cases its meanings still span only a tiny fraction of those contained in all facial expressions.

Eibl-Eibesfeldt has gone on to trace both the ontogenies of nonverbal signals in children and the modification of the signals to acquire new meanings in the course of cultural evolution. One example, the elementary eyebrow raise and its culturgenic derivatives, is given in Figure 3–11. Another case is the ritualization of body movements to signal a "no." The most widespread movement is head shaking, a standard signal in cultures as separately evolved as the Papuan highlanders, the Yanomamö of Venezuela, the Himba of southern Africa, and the Kalahari San. Many Mediterranean and Mideastern peoples signal a "no" by jerking the head back while closing the eyes, sometimes turning the head sidewise and lifting one or both hands in a gesture of refusal. The Ayoréo Indians of Paraguay wrinkle their noses as if they were reacting to an unpleasant odor, close their eyes, and often push their lips forward in a pout. The Eipo of New Guinea indicate a factual "no" with a headshake, and a refusal in a social encounter with a pout. Eibl-Eibesfeldt points out that virtually all of these signals can be interpreted as ritualizations of the more direct, motor rejection of unpleasant physical stimuli—a "shaking off" of objects on the head or closure of the eyes, nostrils, and mouth. The pout has been additionally ritualized to respond to an insult and to cut off contact.

The principal conclusion we have drawn from these studies is that much of nonverbal communication has been built upon the ritualization of elementary behavior patterns, with the ritualized versions of the behavior being guided by new epigenetic rules. In some categories such as head shaking, the elementary patterns are little more than undifferentiated motor movements serving other, noncommunicative functions as well. But the patterns of smiling, additional basic facial expressions, laughing, and crying appear to have been limited to signal functions from their evolutionary origin, and they were subjected to more rigid canalization in the ritualization process afterward. Eyebrow raising is intermediate in its degree of specialization. To use the imagery of gene-culture theory, the epigenetic rules vary among the categories of nonverbal signals in narrowness and specificity, but in all cases these qualities are strong enough to

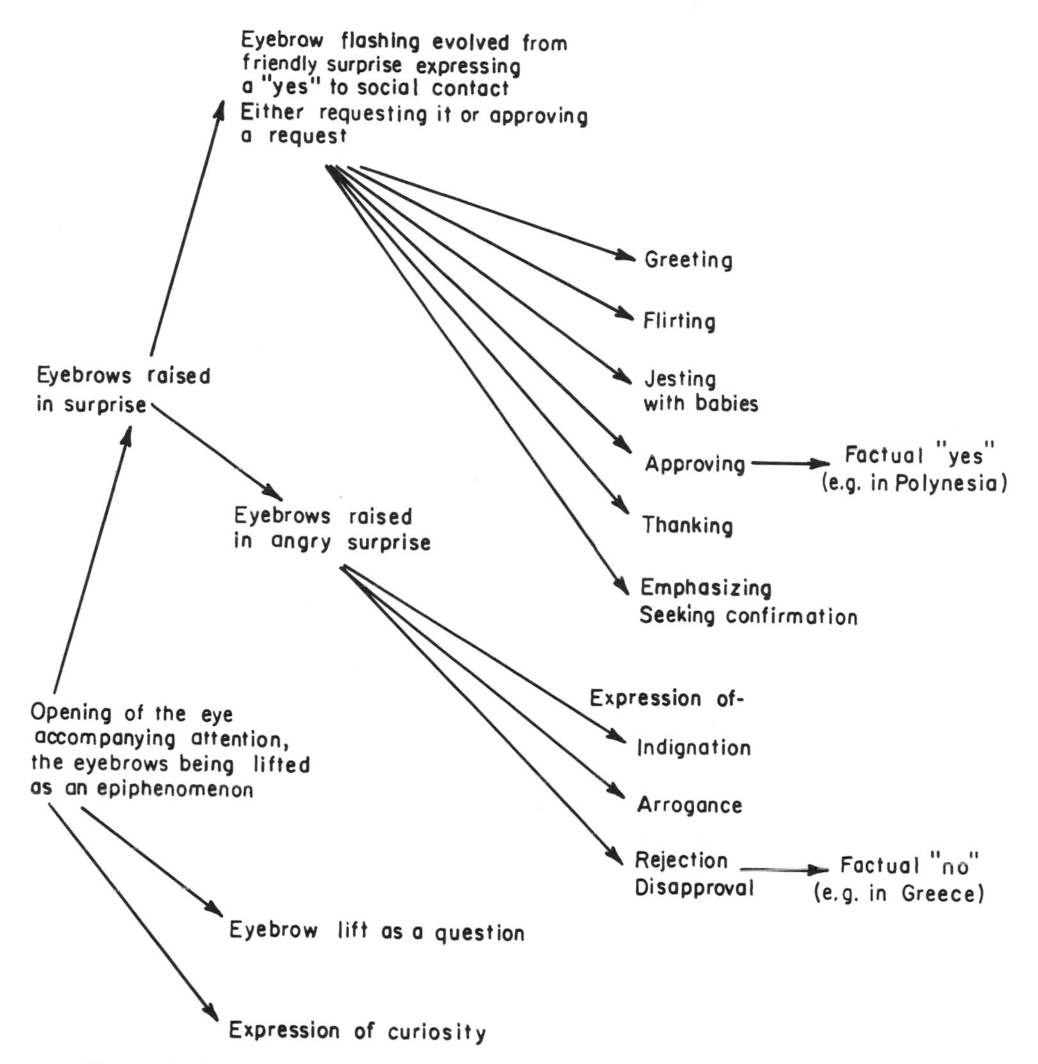

Figure 3–11 Eyebrow lifting and its culturgenic derivatives. (Modified from Eibl-Eibesfeldt, 1975.)

restrict substantially the array of nonverbal signals generated during cultural evolution.

Mother-Infant Bonding

Examples of epigenetic rules operating in adults can be found in the attachment of mothers to their newborn infants (Adrienne Harris, 1979).

The strength of attachment has been examined during twelve studies in four different cultures (American, Swedish, Guatemalan, and Brazilian), including both lower and middle socioeconomic classes. Eleven of the studies produced essentially the same result. When mothers were given frequent contact with their babies during the first hours and days, they provided closer attention to the infants during subsequent months than did control mothers allowed only intermittent contact in accordance with standard hospital practice (see the review by Kennell and Klaus, 1979). The infants also enjoyed slightly but significantly better health and freedom from later abuse and neglect.

Two of the studies can be cited to illustrate this effect in more detail. Klaus and coworkers (1972) observed 28 primiparous mothers and their normal, full-term infants in Ohio. The 14 mothers in the control group followed the traditional hospital schedule of contact with their infants: a glimpse of them immediately following birth, a brief contact at six to twelve hours, followed by visits for twenty to thirty minutes every four hours thereafter for bottle feedings. The experimental group of mothers followed the same routine, but in addition they were handed their nude infants for one hour during the first three hours following birth, and received five extra hours of contact each afternoon for the three days after delivery. Mothers in the experimental group were subsequently observed to stand closer to their infants during medical examinations, bestow more soothing actions, and engage in a higher frequency of eye contact and fondling during feeding. They were also more reluctant to leave their infants with other adults.

O'Connor and colleagues (1977) conducted a parallel study of 301 mothers in Tennessee. One group was given its babies for twenty minutes every four hours for feeding, in accordance with standard hospital procedure. The other group was given six hours of additional contact for two days. During an observation period twelve to twenty-one months later, serious parental mistreatment was suffered by only 1 among the 158 infants exposed for a longer time to their mothers, while 9 of the 143 infants involved in routine hospital contact were hospitalized for parenting disorders, including abuse, neglect, abandonment, and nonorganic failure to thrive.

Kennell and Klaus ask, "How can changes made during just a few hours around delivery profoundly alter the later behavior of a woman who has already lived 160,000 to 180,000 hours?" They suggest that during the early sensitive period a cascade of reciprocal interactions begins between the mother and her baby, which interlocks them and ensures the further development of attachment. A close contact between the mother and her infant during the first hours following birth appears to be crucial for the formation of subsequent strong bonding.

Possible additional evidence of a sensitive period of genetically prepared learning is provided by the remarkable ability of women to recognize the cries of their babies shortly after birth. Morsbach and Bunting (1979) found that mothers can readily distinguish tape-recorded sounds of their own babies from the recorded cries of four other, randomly selected infants. Even though the infants were only three to eight days old and had been contacted briefly in a hospital setting, 22 of 27 women made correct identifications. Of equal interest is the apparent robustness of this response. No factor analyzed was found to have any effect on the performance, including the age of the mother, age and sex of the infant, presence or absence of siblings, medication during labor, method of delivery, and method of feeding. Because of the small size of the sample, this last, important result needs to be confirmed.

Infants possess an equally remarkable ability to recognize their mother's voice. In tests conducted by DeCasper and Fifer (1980), ten neonates were allowed to choose between a nonnutritive nipple that triggered a tape recording of their mother's speech and one that triggered a recording made by another woman. As early as twenty-four hours the infants displayed a preference for the tape of their mother's voice.

Still another nonuniform epigenetic rule affects the way mothers carry their infants and leads to a directly perceived bias curve. In an American study Salk (1973) discovered that 25 of 32 left-handed mothers, or 78 percent, held their newborn infants on their left side. Approximately the same fraction of right-handed mothers, 212 out of 255, or 83 percent, also held their infants on the left side. (Left-handedness occurs in 5–10 percent of adults and is apparently to a large extent genetically determined.) In 400 works of art randomly selected from four cultures and depicting mother and infant, the infant was held on the left side near the heart 80 percent of the time. The bias has been confirmed in a separate study of early Christian art and in impressionist and postimpressionist paintings. On the other hand, no bias existed in representations of men holding children or in women nursing infants (Finger, 1975). Lockard and associates (1979) found the same predisposition among 79 women observed carrying infants in the vicinity of Seattle, Washington: 77 percent carried their children on the left side. And of 85 women in the Wolof tribe watched at Dakar by the same authors, 59 percent carried infants on the left side. In contrast, men observed in the Seattle area were about equally likely to carry infants on either side.

Salk suggests that the adaptive function of left-side carrying is to bring the infant closer to the sound of the mother's heartbeat. Recordings of hearts beating at a normal rate (72 cycles/minute) soothed infants, whereas the same recordings run at a faster rate (125 cycles/minute) had a disturbing effect. The Lockard group noted that the Dakar infants shifted their heads closer to the position of the heart.[1] Other studies have revealed that newborn infants are quieter and cry less when rocked at approximately the rate of a normal heartbeat than when rocked at slower or faster rates. Furthermore, they never habituate to the stimulus, so that it remains effective indefinitely (Ambrose, 1969). Because advanced fetuses are capable of hearing at least sharp, loud noises, Morris (1971) and Salk (1973) have suggested that the infant becomes conditioned to the sound of the mother's heart prior to birth. The heartbeat then serves as a postpartum bonding stimulus, which the mother unconsciously enhances by holding the infant on her left side.

Particularities in infant holding are paralleled by those in book carrying. Eighty-two percent of male college students observed by Spottswood and Burghardt (1976) carried their books in their right hand with their arm straight along the body, while 79 percent of women students carried their books pressed to their side with the forearm or clasped to their breast.

The same sex difference has been noted in other widely scattered communities in the United States. It has also been traced back into preadolescence, before anatomical differences can play a role in determining the physically most comfortable positions (Jenni and Jenni, 1976). Thus the usage bias curve of object carrying appears to have a genetic component, although the sex-linked specificity needs to be traced back into still earlier stages of development. At the same time, nuances certainly arise from pure cultural evolution. For example, almost all of the students of both sexes observed in Costa Rica tucked one or two books in the waistbands of their pants or skirts, a behavior rarely if ever seen in the United States.

Fears and Phobias

Most animal learning is moderately to strongly directed. Individuals are genetically *prepared* or *contraprepared* to learn to respond to particular conditioned stimuli in those behavioral categories of greatest importance to their survival and reproduction, while in other behavioral categories they are typically unprepared or neutral (Seligman, 1972a,b). Perhaps the

1. There is some evidence that orangutan infants also prefer to be carried over the left side of the mother's chest (Horr, 1977).

closest parallels within the corpus of epigenetic rules of human beings are certain classes of fears and phobias. These offer the additional analytic advantage of being translatable directly into bias curves.

A fear of snakes, for example, is widespread in human populations, even where snakes are seldom if ever seen in the wild. The fear develops with very little negative conditioning in a large proportion of children (one-third of British children acquire it), appearing as a mild aversion between the ages of three and four and intensifying into an active dislike by the age of four. In most cases it then declines to the age of fourteen, but in a few individuals it hardens into a permanent phobia (Marks, 1969). The ophidiophobic propensity is a general primate trait that appears to be correlated with the presence of poisonous snakes in the natural environment. When Chalmers Mitchell (cited by Morris and Morris, 1965) carried snakes through the London zoo, the monkeys ran away shrieking; but the lemurs, which come from Madagascar, one of the few parts of the world in which poisonous snakes do not occur, ran forward to the front of their cages.

In an experiment of related interest, Yerkes and Yerkes (1936) and Haslerud (1938) searched for visual stimuli that are naturally frightening to captive young and adult chimpanzees. The ones they found to be most effective included movements that are especially intense, abrupt, and rapidly changing, qualities presented in nature by snakes, but not much else. Thus the chimpanzee (and the child as well?) does not inherit an internal image of a snake with which it associates danger. Rather, it possesses a more general schema bracketing snakes and few other objects.

Anxiety in the presence of strangers is another general human trait, and one much better documented. Sliding easily into fear and hostility, it contributes to the tendency of people to live in small groups of intimates and has important effects on cultural evolution. The response is intensified when the strangers are intruders who stare (Argyle and Cook, 1976). Eyes and eye-like patterns have been found to have a generally higher arousal effect than other facial features (Coss, 1972). Aversion to strangers has been noted in very young children in all of the many cultures studied by Eibl-Eibesfeldt (1979). The baby turns away, buries its face in its mother's shoulder, and often commences crying. The response first appears in infants when they are six to eight months old and peaks sometime during the subsequent year (Morgan and Ricciuti, 1973). It does not depend on previous aversive experience with strangers, nor is it linked in any obvious way to anxiety over separation from the mother. The latter, distinctive response first emerges when the infant is about fifteen weeks old

(Hess, 1973). Eibl-Eibesfeldt has suggested that the following relatively parsimonious rule guides the development of the behavior: young children respond automatically with anxiety and fear to the features of another human being, but the fear-releasing quality of the key stimuli are canceled when the person is seen often enough to become familiar.

Other apparently innate fears emerging at specific ages during childhood are directed at heights, the dark, and unfamiliar kinds of animals (see Table 3–2).

The preparedness of human learning is most clearly manifested in the case of the phobias, which are fears defined by a combination of several traits. They are first of all extreme in response, often entailing autonomic reaction. They typically emerge full-blown after only a single negative reinforcement. They are exceptionally difficult to extinguish, persisting

Table 3–2 The principal classes of phobias in adults and their relative frequencies, as observed at Maudsley Hospital and Institute of Psychiatry, England. (Based on Marks, 1969.)

Phobia class	Description	Percent of phobia patients	Principal age at onset (years)
Agoraphobias	Extreme fear of open places, closed spaces, crowds, traveling, or combinations of these	60	15–35
Social phobias	Extreme fear of eating, drinking, blushing, speaking, writing, or vomiting in the presence of other people	8	5–35
Animal phobias	Extreme fear of one or the other of a limited set of kinds of animals (including rats, snakes, spiders, dogs, cats, and horses)	3	1–5
Miscellaneous specific phobias	Extreme fear of heights, wind, darkness, thunderstorms, running water, and a few other (mostly natural) phenomena	14	5–40

even when the subject is repeatedly presented with the conditioned stimulus in the absence of the unconditioned stimulus and the harmlessness of the fear object is carefully explained. Finally, phobias are highly specific; a few objects or qualities induce them easily, while the great majority of other stimuli rarely or never cause such an extreme reaction.

The specificity of the phobias is well illustrated in the data of Table 3–2. It is a remarkable fact that the phenomena that evoke these reactions consistently (closed spaces, heights, thunderstorms, running water, snakes, and spiders) include some of the greatest dangers present in mankind's ancient environment, while guns, knives, automobiles, electric sockets, and other far more dangerous perils of technologically advanced societies are rarely effective. It is reasonable to conclude that phobias are the extreme cases of irrational fear reactions that gave an extra margin needed to ensure survival during the genetic evolution of human epigenetic rules. Better to crawl away from a cliff, nauseated with fear, than to casually walk its edge. Finally, some phobias are sex biased. Sixty percent of the adult patients with social phobias and 75 percent with agoraphobias studied by Marks (1969) at Maudsley Hospital were women.

Incest Avoidance

Incest taboos are a cultural universal; all of the hundreds of societies that have been studied ethnographically permit or even encourage marriages between first cousins but forbid it between siblings and half-siblings. A very few societies have institutionalized brother-sister incest for some of its members. These include the Incas, Hawaiians, some Thais, ancient Egyptians, Monomotapa (Zimbabwe), Ankole (Uganda), Bunyoro (Uganda), Buganda (Uganda), Nyanza (Zaire), Zande (Sudan), Shilluk (Sudan), and Dahomeans. In each case the practice is (or was) surrounded by ritual and limited to royalty or other groups of high status. Van den Berghe and Mesher (1980) note that in all of the known incestuous arrangements polygyny is (or was) practiced by the incestuous males in addition, leading to outbreeding and an overall increase in personal genetic fitness. The ruling families are (or were) patrilineal. Therefore the maximum fitness payoff to a high-ranking male would be to mate with his own sister, producing children who share with him 75 percent of their genes by common descent, as well as with women who are genetically unrelated but more likely to produce normal children. Because of the general trend in hypergamy, royal women are less likely to marry downward in rank and hence are more susceptible to matches with their brothers.

The well-documented result of brother-sister mating is a higher frequency of genetic deformation in the offspring (Seemanová, 1971; Stern, 1973). Ember (1975) concluded from a cross-cultural survey that avoidance of these deleterious effects is the only explanation that fits the detailed patterns of same-generation incest avoidance. It accounts not only for the inhibition of mating between brothers and sisters but also for the detailed patterns observed in the variable tolerance of cross-cousin marriages. The competing hypotheses that were discarded or reduced to secondary explanatory power are the Freudian psychoanalytic model, the perception that incest disrupts family bonds, and the perception that exogamy serves as a bonding device among families.

Ember believes that the consequences of inbreeding have been serious enough to be observed directly and hence to serve as the basis of deliberately contrived taboos. In terms of the transparency rule we tentatively formulated, the effects of incest on genetic fitness permit reliance on flexible epigenetic rules and rational calculation. However, analyses by Shepher (1971) and others on the development of sexual preferences of children in Israeli kibbutzim indicate that this is not the case. There exists instead a relatively specific epigenetic rule, whereby an automatic sexual inhibition between persons who lived intimately together ("used the same potty") emerged as one or all grew to the age of six. Among the 2,769 marriages reviewed by Shepher, none was between members of the same kibbutz peer group who had lived together since birth. There was not even a single known case of heterosexual activity, despite the fact that the kibbutz adults were not opposed to it. Closely parallel data have been provided by a study of Taiwanese families who adopted very young girls for the purpose of later marriage to the hosts' sons. In the great majority of cases the couples refused to accede to the marriage, because of probable sexual inhibition based on early domestic contact (Wolf, 1966, 1968, 1970; Wolf and Huang, 1980). The evidence suggests the existence of a genetically based bias curve in which the preference for outbreeding as opposed to incest is very strong. In American families brother-sister intercourse does occur, but it is still relatively rare, transient, and ordinarily a source of shame and recrimination (Weinberg, 1976). This behavior will be analyzed in greater detail in Chapter 4, where it will be used as an example in the translation from epigenetic rules to cultural patterns.

Epigenetic Rules of Valuation and Decision Making

In the process of reaching a decision, the conscious mind does not use the ideas of the genetic cost and benefits of each potential response. The evidence emerging from cognitive psychology and cognitive anthropology indicates that the mind relies instead on relatively simple heuristics or rules of deliberation that can be applied quickly and effectively to a wide diversity of contingencies (Wason and Johnson-Laird, 1972; Hallpike, 1979; Hutchins, 1980; Nisbett and Ross, 1980). If natural selection has been effective, then epigenetic rules guide both the active choice and the use of these heuristics, acting through directed cognition as well as affects from the limbic system. The large number of decisions so taken lead to patterns of individual behavior and social structure that enhance genetic fitness if summed through the entire life cycle. Yet the result may fall far short of a theoretically conceivable maximum in which the ultimate genetic consequences of every individual response are weighed by a perfect calculating device.

One of the major goals of contemporary psychology is a cognitive algebra, in which valuation and decision can be described as a set of precise, even mathematical, rules. Fechner's Law (1860) was an early partial success: the psychological sensation associated with the loudness of a sound was found to be proportional to the logarithm of the actual physical intensity. Thus we measure sound intensity in decibels instead of units arrayed on a strictly linear scale. More recently, psychologists have been attempting to parameterize and measure complex cognitive qualities of greater importance in social behavior, such as personal likableness, subjective probability, and amount of approval deserved (Fishbein and Ajzen, 1975; Anderson, 1979). One well-documented heuristic used at least by North Americans is that disparate components of such qualities are summed additively and without prominent interaction effects. Thus when components such as degrees of level-headedness, sophistication, boldness, and good-naturedness are combined pairwise, they contribute to the overall impression of attractiveness in an additive fashion, even though different weights are assigned to each.

Another heuristic demonstrated during cognition experiments on North American subjects is the use of multiplication logic in judging gambles. In making decisions on whether to take a risk, people employ subjective expected value. They judge the payoff and multiply it by their subjective estimate of the probability of winning. Thus the value of a lottery ticket is considered to rise more rapidly with an increase of probability of payoff if the reward is a gold watch instead of a pair of sandals. The same multiplicative rule is used when people evaluate the truth of a compound statement, such as the following: How true is it that a sparrow is a bird, or a penguin is a bird, or both? The subjective truth of the compound form is approximately the product of the subjective truths of the two separate propositions (Oden, 1977).

Still another principle of cognitive algebra is the Matching Law (Brown and Herrnstein, 1975; Rachlin, 1976): the proportion of responses equals the proportion of rewards. Subjects asked to monitor a panel of meters in order to record deflections of the indicator needle came to devote time to individual meters in the same proportion that the meters actually registered deflections. When other subjects were asked to guess the next symbol in a random sequence of *X*'s and *O*'s, in which 70 percent were *X*'s, they did not select the best strategy of always choosing *X*, even when the preponderance of this outcome became apparent to them. Instead, they followed the Matching Law and thus obtained a score of 58 percent instead of the 70 percent that would have resulted from the simpler, optimal strategy. The Matching Law, which has been found to apply with equal strength in rats and pigeons, may reflect a still undisclosed fundamental heuristic in the cognitive systems of all higher animals. However, other decision procedures are employed. When rhesus monkeys are trained in a red-green discrimination at ratios of 70:30 and 30:70, they maximize; that is, they begin to choose the more frequently presented color and totally ignore the alternative. When given a 50:50 ratio, they tend to choose the color opposite to the one most recently rewarded (Bitterman, 1975).

Other studies have disclosed that human beings are especially poor intuitive statisticians when dealing with major events of life and death. People tend to confuse low-probability/low-consequence events with low-probability/high-consequence events. They seldom adopt an integrative approach in the assessment of risk and tend to underestimate the effects of catastrophes. As a result they consistently misjudge the future effects of warfare and technological risk, as well as floods, windstorms, droughts, and volcanic eruptions, even when such events are periodically experienced and remembered over many generations (Reijnders, 1978; Orr, 1979).

People also employ a simple heuristic which consists in matching problems against a representative prototype in long term memory. In so doing they tend to override their own knowledge of probabilities based on personal experience (Tversky and Kahneman, 1971, 1974). Thus when an observer is asked to guess the occupation of another person who is shy, helpful, and obsessed with detail, he is more likely to choose librarian over other occupations, regardless of his previous personal experience. Most people—including even some trained statisticians—intuitively expect small random samples to reflect faithfully the large population from which they are drawn, even though this is demonstrably untrue in a large percentage of cases. An additional bias prone to error is the tendency to make judgments on the basis of relevant instances according to the ease with which the instances come to mind (Peterson and Beach, 1967; Tversky and Kahneman, 1973). Such elementary heuristic strategies work most of the time, because they are correlated reasonably well with the realities of contingencies in the real world. For example, the reliance on the most familiar, hence most frequently occurring, representative cases as expressed through conventional stereotypes in the culture is a reliable procedure much of the time, and this is especially true in the most stable, traditional societies. Nevertheless, it often falls far short and creates major difficulties in the most complex, rapidly evolving societies (Nisbett and Ross, 1980).

Other cognitive shortcuts have been discovered. During decision making an activity resembling the chunking procedure occurs. Faced with a challenge, the mind explores a "problem space" and selects possible programs that can be used for solving the problem (Newell and Simon, 1972; Simon, 1979; Brainerd, 1979). In many instances the possible routes to the solution are legion, but the mind reduces the options to a binary choice and decides whether to proceed. When formulations fail to match each other, in other words when the obverse of one element (not-*x*) follows from the other (*y*), the result is cognitive dissonance. The disparity creates an emotion-provoking "noxious" effect that the mind attempts to eliminate, either by adding new perceptions and problem-solving procedures or by reducing the relative importance of those that generated the dissonance (Zajonc, 1968).

In spite of increasingly sophisticated experimental analysis, the manner in which the conscious mind makes moment-to-moment decisions is far from clearly understood (Ajzen and Fishbein, 1977; Bentler and Speckart, 1979). Human "will" might be nothing more than the resolution of competition among schemata. The organism could be guided by feedback loops consisting in a sequence of messages that lead from the sense organs to the brain schemata back to the sense organs and on around again until a sufficiently close fit to the schemata is achieved. Consciousness could be a republic of such schemata, programmed to compete for control of the final decision centers, their individual strength growing or declining according to the relative urgency of physiological needs of the body signaled to the decision centers through the brain stem and midbrain.

The peculiarities and constraints of valuation and decision making observed by cognitive psychologists are at least loosely consistent with the characterization of bounded rationality developed by Simon (1957a,b; 1979) and other economists and political scientists. In essence, this view holds that human groups do not work toward solutions based on omniscient rationality and profit maximization. They make decisions aimed at "satisficing," that is, attaining at least a certain minimal return perhaps accompanied by auxiliary rewards of security and social interaction. Alternative responses are examined on a relatively simple two-valued scale to determine whether they are satisfactory or unsatisfactory. The rules of valuation are considered to change only slowly with time. Results from still other studies in economics have led to a growing emphasis on the emotionally influenced properties of effort, esprit, and cooperativeness in decision making and performance. The mathematical models of microeconomics are being revised with the use of more realistic assumptions about the distinctive operations of the human mind (Winter, 1971; Becker, 1976; Leibenstein, 1976; Hirshleifer, 1977, 1978; Navon and Gopher, 1979).

A view of decision making comparable to satisficing has been independently reached by economic anthropologists. People in primitive economies are overwhelmingly averse to taking risks during resource harvesting (Johnston and Selby, 1978). They adopt strategies that can be characterized as *maximin,* which means that the tactics they entail guarantee a certain minimal, life-sustaining yield of food in every season regardless of how bad conditions become during downward fluctuations of the environment. But the strategies also surrender the possibility of exceptionally large harvests during the good years, and they lower the average yield taken over all years. For example, the Jamaican fishermen studied by Davenport (1960) would be able to maximize their average catch if they were to fish exclusively in the open water beyond their home lagoon. But they would occasionally go broke if they persisted in this strategy, because when the unpredictable current runs fast, their pots are lost and their time and energy are wasted. Instead, they mix their forays in accordance with the knowledge that the outer current runs hard an average of one day out of four. This conservative procedure yields 12 percent less poundage than the maximax strategy of exclusive open-water fishing, but the fishermen never go broke.

For the people of most economically simple societies, to go broke is to perish. It is therefore not surprising to find, given the poor abilities of human beings as intuitive statisticians, not only risk aversion and neophobia among such people but also considerable sophistication in devising maximin strategy. In his study of the Ghanaian village of Jantilla, Gould (1963) employed linear programming to estimate the theoretical yields of various mixtures of yams, maize, hill rice, and millet across the observed combinations of wet and dry years. The maximin combination of acreage deduced, 77 percent maize and 23 percent hill rice, is close to that actually planted by the Jantilla villagers. Comparable results have been obtained in studies of the Round Lake Ojibwa hunter-fishermen of Ottawa by Jochim (1976) and of the Netsilik Eskimos by Keene (1979).

The maximin strategy is sometimes loosely referred to as the "law of minimal risk." It can be related to the more speculative "law of least effort" supported by some anthropologists, which states that people put only enough work into the productive process to maintain culturally determined levels of satisfactory consumption. In most hunter-gatherer and primitively agricultural societies these levels lie close to the maximin. As a result, many such societies stay well below the carrying capacity of the environments defined in terms of potential energetic yield. However, as Cohen (1977) has cogently argued, other societies permit their population size to creep upward, forcing them either to expand their ranges or to exploit new typically less desirable food sources. Western observers are often able to recommend simple, available procedures by which productivity can be increased, but these changes are likely to be resisted if they are perceived as affecting the culture in any significant way. And when adopted, technical improvements frequently lead to rapid social disintegration. Steel axes, to take one example, are three to six times more efficient than stone axes in time expenditure and net energetic yield, but their use by Stone Age peoples typically led to a destruction of the environment together with profound negative effects on socioeconomic organization, including the disruption of intertribal trade relations, breakdown of status systems, and increased dependence on colonial administration (Salisbury, 1962).

Consequently, the propensity toward economic conservatism provides a buffer that allows innovations to be absorbed gradually and less destructively. In the case of the Amazon Indians, population densities may be below carrying capacity, in the sense of not being limited in a density-dependent manner by protein deficiency (Chagnon and Hames, 1979). Although the shotgun has permitted groups such as the Jívaro, Siona-Secoya, and Ye'kwana to hunt with far greater efficiency than was possible with the bow and arrow, it has not increased their catch to anywhere near the potential of the tropical forests in which the Indians live. Their needs remain governed to a large extent by older cultural conventions and have not yet yielded to the insatiable demands of the international fur and feather market. As a result, the advent of the shotgun has resulted most notably in an increase in leisure time (Hames, 1979).

How do preliterate people, and indeed people in general, arrive at such solutions in the absence of calorimeters and optimality theory? At present there is no detailed answer. In at least some cases, time is on their side. Over generations they can employ trial-and-error combinations that gradually converge to the most practical solution consistent with a given degree of risk aversion (Gould, 1963; Haggett, 1965). Overall their principles of thought show a remarkable capacity to "beat natural selection to the draw" by anticipating the consequences of alternative actions for survival and reproduction, a process that Boehm (1978) has called rational preselection and has illustrated in detail. The fact that so much of the human behavior documented by ethnographers falls on or near fitness-maximizing optima provides evidence for the existence of epigenetic rules that shape conscious decisions and thus channel the development of the mind throughout adult life.

One can only speculate on the details of the mechanisms by which the epigenetic rules of valuation and decision making lead to such manifestations as economic conservatism and cultural neophobia. Anthropologists lack rigorous measures of the intensification of these propensities as people within primitive economies come to live closer to the subsistence line. It is to be expected that as the current studies of cognitive and developmental psychology progress, the core epigenetic rules will be understood in a way that links them directly to general rules of economic behavior.

Other Epigenetic Rules

It can be conservatively predicted that when additional developmental studies examine multiple choice in other classes of culturgens, many more epigenetic rules will come to light. One of the most promising domains of behavior is tool use. Connolly and Elliott (1972) have distinguished seven possible handgrips by which a paintbrush can be held, nine movements of the brush, and six basic strokes. Very young children display a characteristic and relatively narrow frequency distribution in the initial choice of each of the components, and the bias curve is constricted still more with practice and imitation, coming in the end to center on the "adult" forms. Comparable trends have been described by Connolly (1973) in the development of cylinder fitting. Underlying the more complex sequences of motor action are certain basic epigenetic rules of proprioception in muscular action. For example, conscious perception on the part of the individual of effort by his individual muscles increases as an exponentially decelerating function of the actual force applied (Banister, 1979).

Another category of behavior in which epigenetic rules can be of importance is the propensity to learn certain kinds of social networks as opposed to others. In a suggestive early account of this subject, De Soto (1960) reported that individuals can best learn the details of relationships (such as "Jim likes Ray, who likes Stan, who likes . . . ") if the structures are asymmetric, transitive, and form completely connected arrays. The development of these irregularities in learning capacity through childhood have not been investigated.

The teaching rules remain little known. One apparent example was cited earlier in the case of infant holding by mothers. Teaching could also be important in fixing sex-role differences in children. Very young children display sex differences in temperament and patterns of attachment to adults and peers that foreshadow the later differentiation of roles. These early forms of divergence are sufficiently widespread, distinctive, and well correlated with hormonal activity to indicate that they are biological in origin (Money and Ehrhardt, 1972; Blurton Jones and Konner, 1973; Maccoby and Jacklin, 1974; Symons, 1979).

In a cross-cultural study Barry and others (1957) found a remarkable degree of consistency in the way adults promote further role differentiation. In all of the societies sampled, the adults trained boys more than girls in "self-reliance" and girls more than boys in "nurturance." In a large majority of cases they also directed boys with greater consistency toward "achievement" and girls more toward "obedience" and "responsibility." A subsequent study by Bearison (1979) indicates that in American families the influence of other-sex parents is the more effective in role formation. An underlying epigenetic rule might be implied in his generalization that "mothers tended to regulate daughters' behavior by appealing to the psychological attributes (needs, intents, feelings, etc.) of the self and others, while fathers tended to appeal to the positional aspects of social conduct with their daughters. The opposite relation held between parents and their sons." It will prove singularly difficult to design experiments capable of quantifying the epigenetic rules that operate in sex-role training and in other teaching behaviors, but the results should prove worth the effort.

The Reification Learning Rules

In Chapter 1 we identified reification as a diagnostic activity of human consciousness. Higher mental process consists to a large degree in sorting vast quantities of aphasically timed and nearly chaotic stimuli into categories, labeling the categories with metaphors and symbols, and freighting them with emotional qualities that emanate from the limbic system. Most human communication involves the transmission of these symbols as words, which are strung together in combinations to convey a virtually unlimited diversity of meaning. When the information is inserted into certain contexts and accompanied by facial expressions, changes in voice tone, and other paralinguistic embellishments, it gains in precision and emotional strength.

The anthropological literature contains many examples of the growth of culture through the telescoping abstraction of complex phenomena in everyday affairs. Often this process of ritualization is immediately clear to observers. To take one case, the Bemba of Zambia and neighboring countries live in a harsh area of uncertain seasonal rainfall and depend almost exclusively on finger millet, which is made into a porridge called *nbwali.* The word *nbwali* occurs repeatedly in metaphorical form in proverbs, puns, jokes, and folktales. It stands for life and health in the Bemba ceremonies of tribal political action, female initiation, marriage ceremonies, and kinship relations (Richards, 1939).

The Dasun of Borneo invest the interior arrangements of their houses with intense meaning. Each room or area and each piece of furniture is associated with calendric rituals and magical and social beliefs. The house is further reified into a "body" possessing arms, a head, a belly, legs, and other parts. The house is believed to "stand" properly in one direction, to be upside down if built on a hill slope, and to be variously young and strong, old and worn out, and fat or skinny. In Dasun folktales and riddles the house parts are routinely anthropomorphized (Williams, 1972a).

In numerous cultures handedness has been employed as a metaphor to create dichotomous classifications. Approximately 10 percent of human beings have been naturally left-handed in most or all populations since prehistory (Hardyk and Petrinovich, 1977), a minority status that has usually been translated into inferiority. This trait evidently has a partial genetic basis (Carter-Saltzman, 1980). It is typical to identify right with men, left with women; right with good, left with bad (hence, "sinister"); right with good omens, left with evil omens; right with physical strength, left with weakness; and so on. These distinctions permeate ceremonies and religious beliefs even to the most sacred levels (Needham, 1973).

So distinctive and powerful is the process of reification that it is reasonable to postulate the existence of special epigenetic rules that guide the clustering and objectification of both perceived stimuli and new knowledge acquired through mental activity. These reification rules are in fact metarules. They govern the processing of information that has already been filtered and channeled in a primary manner by the ordinary epigenetic rules. Perhaps they also direct the processes by which some of these rules assimilate information. One of their most important consequences is to provide a more easily perceived and interpreted measure of the collective behavior of other members of the society. With the aid of reification, the individual can respond quickly to the complex patterns of mass behavior occurring around him.

Despite many excellent but largely inspirational studies of the process of symbolization—for example by Cassirer (1944, 1946), Langer (1967, 1972), and Lévi-Strauss (1969a,b)—the properties of the reification process are not well understood. In addition to the tendency to group objects and propensities into artificial categories, there is a nonrational proneness to use two-part classifications in treating socially important arrays, such as in-group versus out-group, child versus adult, kin versus nonkin, sacred versus profane, and so forth, and to invest the boundaries between the two domains with taboo and ritual.

Lévi-Strauss and other writers, whose hypotheses concerning brain action are sometimes referred to collectively as structuralism, suggest that precise, often complicated rules govern the dichotomization. The mind conceives of the universe to a large degree in terms of binary oppositions, such as (man : woman), (endogamy : exogamy), and (earth : heaven). These pairings create contradictions that must then be resolved, often by myth. Thus the concept of life necessitates the concept of death, which is resolved by the creation of the myth of death as the gateway to eternal life. Binary oppositions are linked still further into complex combinations by which cultures are constructed as integrated wholes. The structuralist approach is innovative and stimulating, but it has been weakened by inconsistencies and even disagreements within the ranks of the structuralists concerning the fundamental procedure of the analysis itself, as noted in the essentially sympathetic reviews by Kaplan and Manners (1972) and Kronenfeld and Decker (1979). We suspect that the problem is not the fundamental conception but the lack of an adequate foundation in cognitive psychology. In their recent account of metaphor, Ortony and his coworkers (1978) in fact note that the psychological study of the phenomenon has not even begun to utilize standardized tests and objective measures.

Summary

Following primary perception, the brain extracts certain features and places them into perceptual space by an integration process within long term memory. Then it evaluates them and decides upon an action. Using experimental studies and associated models from the literature on auditory and visual cognition, we show how each of these procedures in mental activity follows distinctive rules that profoundly affect final behavior patterns. Among the phenomena are fuzzy logic, the arithmetic and multiplicative procedures of cognitive algebra, chunking, the Matching Law, cognitive dissonance, satisficing, and general risk underestimation.

The secondary epigenetic rules affecting specific patterns of behavior are constructed in part from these more basic cognitive processes. In identifying the rules that can be inferred from psychological studies, we have drawn several generalizations of potential importance in gene-culture coevolution. First, usage bias curves have the least penetrance and the greatest rigidity during infancy. The time course of genetically directed learning in young children is too long to qualify as animal-like imprinting. (A single case resembling imprinting, the attachment of mothers to infants, does nevertheless occur in adults.) During early social development the young undergo a process of focusing, a passage from general to more specific classes of stimuli. Thus infants begin with a visual preference for larger and more numerous elements and curved lines, then refine the preference to bull's-eye designs, nonlinear arrays, and so forth, and finally (at 16–30 weeks) come to prefer novel configurations. A principle of parsimony appears to have been followed during genetic evolution: the epigenetic rules stop at the highest level of generalization that will suffice and seldom prescribe an animal-like recognition of sign stimuli. The human genotype, in other words, does not adopt two rules where one will suffice. Still another generalization is the transparency principle: the more the effect on genetic fitness of a category of behavior depends upon environmental circumstances, the more clearly the conscious mind perceives that relation and the more flexible its response. Thus economic behavior is typically flexible, while deep grammar, incest avoidance, and sugar consumption by children are more tightly programmed.

Particular secondary epigenetic rules and their associated bias curves are identified in the cases of facial recognition, visual complexity preference, nonverbal communication, mother-infant bonding, fears and phobias, incest avoidance, and other behaviors. The special category of reification learning rules, entailing the selection of symbols and metaphors, is also examined.

Some usage bias curves, especially those derived from the primary epigenetic rules, are relatively rigid. Others vary greatly, to the extent that some can be reversed (see Figure 3–3). But even in the case of those most modifiable, the context dependence cannot be taken as prima facie evidence of cultural determinism, or more precisely of pure cultural transmission. Instead the relation between the curves and context can equally well be under the control of genetically determined core rules of cognition and decision. Only appropriate developmental studies can decide the presence or absence of such rules.

Appendix 3–1

The Measurement of Selectivity in Epigenetic Rules

Developmental selectivity can be quantified in any one of numerous ways. For the two-culturgen case the most direct measures are the difference in the probabilities of adoption of the two culturgens, $\phi = u_2 - u_1$, as illustrated in Figure 3–3, and the ratio u_j/u_i, where $u_j \leq u_i$.

For categories with more than two culturgens the intuitively most satisfying measure may be the entropy of the usage bias curve. The frequencies of culturgen choice u_j are first normalized by defining quantities p_j as follows ($j = 0$ labels the naive state; see Figure 3–5):

$$p_j \triangleq u_j \Big/ \sum_{j=0}^{C} u_j,$$

where C is the number of culturgens and $\Sigma_j \, p_j = 1$. We now define the learning curve entropy as

$$-\frac{1}{\ln (C + 1)} \sum_{j=0}^{C} p_j \ln p_j$$

and selectivity S as

$$S = 1 + \frac{1}{\ln (C + 1)} \sum_{j=0}^{C} p_j \ln p_j .$$

In the case of *no selectivity,* where all culturgens are equally preferred and all p_j are the same, $S = 0$. In the case of *maximum selectivity,* where one p_j is 1 and the rest are 0, $S = 1$.

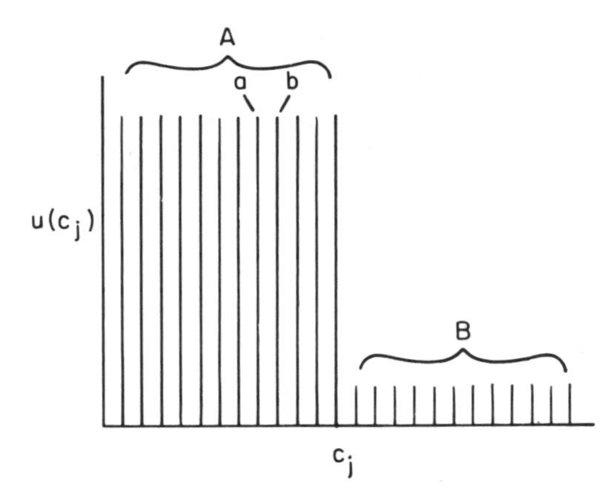

Figure 3–12 Culturgens divided into pairs of clusters (*A* and *B*) for convenience of analysis, instead of being treated as full arrays.

Modeling can be facilitated by clustering culturgens that are similar in quality and adoption probability into groups, as illustrated in Figure 3–12. On some occasions the question of interest might be cultural evolution of two very similar culturgens, for example *a* and *b* in the figure, where $u(a) = u(b)$. On other occasions, especially those entailing long term genetic models, it will be more useful to treat the entire potential culturgen array but to recognize two clusters, such as *A* and *B*, where $u(A) > u(B)$. When culturgens are mapped onto knowledge structures in long term memory, a procedure that will be discussed in Chapter 6, it is possible to define clusters in terms of common features that are most prominently employed by the mind in recognition and valuation. Thus the modeling of larger numbers of culturgens can be made more biologically realistic.

CHAPTER FOUR

Gene-Culture Translation

The central tenet of human sociobiology is that social behaviors are shaped by natural selection. In spite of perturbations due to time lag and random effects, those behaviors conferring the highest replacement rate in successive generations are expected to prevail throughout local populations and hence ultimately to influence the statistical distribution of cultures on a worldwide basis.[1] Sociobiological theory to the present time has consisted largely in application to this proposition of the principles of population genetics and ecology. Biologists and social scientists have attempted to deduce strategies of kin relationships, life-cycle dynamics, predator avoidance, and energy harvesting, then test them with the observed phenomena of social life. In its most general form conventional theory can be applied to all three modes of behavioral transmission—pure genetic, pure cultural, and gene-culture. It allows but does not presuppose the existence of genetic bias in the development of individual categories of social behavior.

As illustrated in the writings of Chagnon and Irons (1979), Freedman (1979), and van den Berghe (1979), such analysis has gone far toward explaining the peculiar forms of aggression, polygamy, hypergamy, incest avoidance, and other categories of social behavior in human beings. But it

1. In the case of discrete, nonoverlapping generations the replacement rate R_0 is simply

$$R_0 = \sum_{x=0}^{\infty} l_x m_x$$

where l_x is the probability of survival to age x and m_x is the expected number of female offspring per female at age x. Thus either higher relative survivorship, higher relative fecundity, or both in combination can win in natural selection. The fitness can be made inclusive by adding the effects of the patterns on the replacement rates of relatives, devalued by the coefficients of relationship of each relative in turn.

cannot link genetic and cultural evolution directly. In order to carry human sociobiology to this next logical stage, we must insert the intervening epigenetic rules, most of which are distinctively human. If successful this procedure will disclose the true properties of *gene-culture translation,* defined as the effect of the genetically determined epigenetic rules of individual cognitive and behavioral development on social patterns (Lumsden and Wilson, 1980a,b).

We suggested earlier that natural selection shapes the epigenetic rules, the products of which are expressed in abstract form as bias curves. The goal now is to create a stoichiometry of these individual, unitary processes capable of predicting the "compounds" and "mixtures" of mass behavior that compose social systems.

The approximate form of usage bias curves is known in a few categories of cognition and behavior, including color vision, infant carrying, incest avoidance, and phobias. Others will certainly be revealed when more attention is given to the development of choices among competing stimuli. It is also apparent that bias curves can be derived from one or a very few epigenetic rules, but are more likely to be compounded from many. Similarly, patterns of culture can conceivably be derived from one epigenetic rule, but are more likely to be the result of many. Thus color vocabularies could be under the dominant influence of the primary color coding of the visual neurons (see Chapter 2), but the patterns of territorial defense (Dyson-Hudson and Smith, 1978) are likely to be constructed from such components as infant aversion to strangers (Eibl-Eibesfeldt, 1979), competitive aggression among groups of children at play (Maccoby and Jacklin, 1974:257), and others. We expect the complete patterns of causation, from genes to behavior, to be commonly multivalent and reticulate (Figure 4–1). In the initial analysis it is nevertheless most efficient to examine two-culturgen epigenetic rules and a single linkage to the cultural patterns. The simplification is further justified by the tendency of the mind to chunk information into binary alternatives, in the case of both elementary economic decisions and myth formation (Simon, 1957a,b, 1979; Lévi-Strauss, 1969a,b; Tversky and Kahneman, 1974), as indicated in our discussion in Chapter 3.

Socialization and the Propagation of Culturgens

We shall begin the analysis of gene-culture translation with a consideration of the peculiarities of human socialization.[2] In order to develop a

2. Although a distinction can usefully be made between *socialization* as the transmission

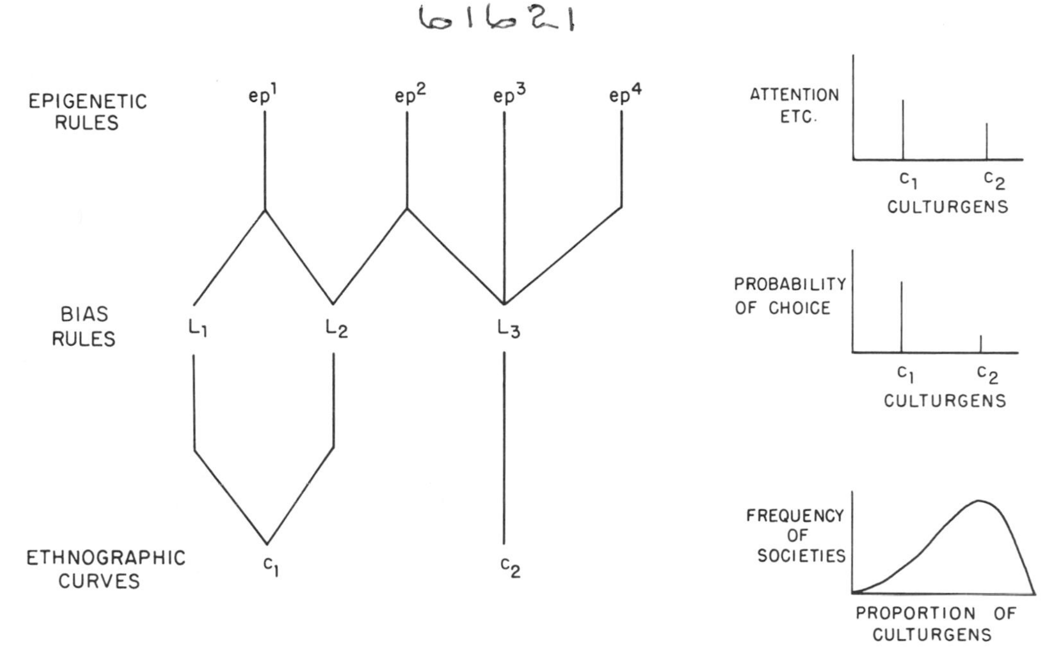

Figure 4–1 Causation from genetically determined epigenetic rules to the final ethnographic curves, a pattern considered to be often reticulate. That is, both multiple causation and multiple effects occur between adjacent levels, as for example between the basic epigenetic rules and the more derivative rules of learning, valuation, and decision. However, simple linkages entailing only single causes and single effects are possible (such as $ep^1–L_1$ and $L_3–c_2$) and provide the starting point of translation theory. The ethnographic curves are the frequency distributions of cultures possessing various proportions of alternative culturgens.

sound theory it is necessary to know how culturgens are transmitted, and the manner in which the psychological development of individuals is affected by the choices already made by other members of the society.

Three consistent features of socialization have been recognized in ethnographic data (Williams, 1972a,b): (1) the patterns of kinship provide for a very large number of parent surrogates, or alternates, with the result that culturgens are diffused widely through smaller societies in each generation; (2) the competing demands on children by different kin groups are met during enculturation by means of concepts such as totems and life spirits; and (3) grandparents play a major role in teaching myths, folktales, and various other justifications of the culture. Of 128 cultures reviewed by Williams, 99 (or 77 percent) possess all three of these traits.

of general, species-wide traits and *enculturation* as the transmission of traits specific to individual cultures (Mead, 1963), we use the two expressions interchangeably to mean the transmission of culture in the broadest sense (see Chapter 1).

The remaining 29 either have one or more of the features missing or else have been insufficiently studied to permit a judgment. Only 6 of the societies (American, English, French, German, Israeli, Russian), all of which belong to the Western industrialized grade of culture, are known to lack all three of the traits. In other words, human enculturation generally results in a broad propagation of information through each generation, and special mechanisms exist that enhance the transmission of the culture as an integrated whole. These qualities are especially prevalent in economically primitive societies, of the kind in which most human genetic evolution is believed to have occurred.

The spread of culturgens has been studied intensively by social scientists with the aid of diffusion, epidemiologic, and information-theoretic models. In adapting their results to gene-culture coevolutionary theory, it is necessary to distinguish propagation occurring between groups from that occurring between members of the same group. Intergroup propagation entails interesting phenomena of its own (including the geography of advancing wave fronts and enhancement by military dominance), but for purposes of the first, elementary models of gene-culture theory it can be treated as a source of innovations.[3] Intragroup propagation, on the other hand, determines the fate of novel culturgens and influences the dynamics of enculturation in a given society, which for the moment can be considered as a discrete entity.

A society in the midst of cultural evolution is itself a learning system, in which individuals communicate and observe the consequences of their own actions and the behavior of others. The introduction of new behaviors, modes of thought, and artifacts is a continuing process. Some are invented, either by design or accident, while others are imported from neighboring societies. Each is tried out and either discarded or incorporated into the culture of the society. Individuals decide on their own which culturgens to adopt, but in most instances they are influenced heavily by the experiences of others. Imitation and vicarious learning, deliberately by the procedures of formal education and religious indoctrination, constitute one of the primary mechanisms of culturgen propagation (Rosenthal and Zimmerman, 1978). People observe the effects of the usage in terms of economic and affective costs and benefits, then make their choices in accordance with their heuristics of valuation and decision (see

3. The reader interested in geographic diffusion may wish to consult the reviews by Haggett (1972), Ammerman and Cavalli-Sforza (1973), Clarke (1978), and Renfrew and Cooke (1979).

Chapter 3). It is clear from the literature of social psychology (Berelson and Steiner, 1964; Freedman et al., 1978) and consumer research (Moschis and Moore, 1979) that in most cases people are also influenced by the numbers of their peers who adopt one culturgen over another. While they use the observed consensus as a rough measure of utility, they are also susceptible to direct pressures from their peers to conform to general usage. Under many circumstances both of these forces cause adoption probabilities to rise disproportionately with an increase in the observed level of usage by other members of the society.

Social contagion is enhanced by the relatively high rate at which individuals observe the behavior of their peers. Hunter-gatherer bands, in which most genetic evolution of the epigenetic rules took place, usually contained between fifteen and seventy-five individuals, all of whom probably knew one another intimately (Wobst, 1974; Hage, 1976; Buys and Larson, 1979). In living bands such as the G/wi and !Kung San, important decisions concerning foraging are made only following intense communication aimed at reaching a consensus. During the discussions large amounts of information about the environment are passed back and forth. The exchange is facilitated by the toleration of substantial ambiguity in attitude prior to the consensus (Biesele, 1978). Even in large industrial countries the network of personal communication is remarkably extensive. In the United States the average number of intermediates needed to link any two persons through personal acquaintance is only about five (Travers and Milgram, 1969).

One result of rapid cultural transmission within groups is that individuals actively employ certain culturgens while storing the alternatives in a passive state in long term memory. Persons are aware of alternative tools, forms of dress, word usage, attitudes toward brothers-in-law, ways of opening mollusks, and so forth, but they select only one or a relatively few on the basis of their personal histories and day-by-day evaluation of the benefits, under the biasing influence of the epigenetic rules.

Although for the purpose of analyzing propagation it is often sufficient to assume a ready accessibility of each culturgen to all group members, the total flow of culturgens in complex societies is seldom that envisioned in the extreme limit of the perfect information market or "semantic horde." The relations that partition social groups, including division of the sexes and labor, kin lineages, marriage and residence rules, alliances, age classes, dominance hierarchies, and special-interest groups, sink channels through which culturgens move preferentially (Haggett, 1972; Leinhardt, 1977; Hamblin et al., 1979). Privileged information is in fact

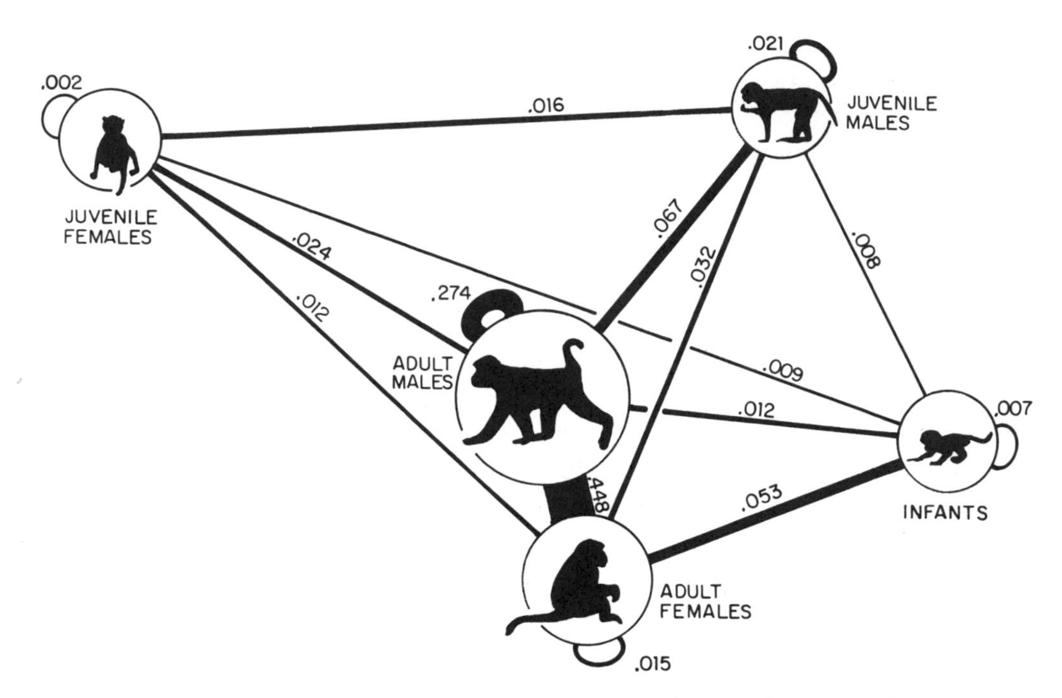

Figure 4–2 The pathways of communication in a free-ranging troop of rhesus macaques. Information, including that leading to protoculture and socialization, is strongly channeled; the probabilities of social interaction are represented here as roughly proportional to the thickness of the lines and the proximity of the figures. (Modified from Altmann, 1968.)

virtually a defining feature of human and perhaps higher primate sociality. An example from the Old World monkeys is given in Figure 4–2. The full potential network of communication among individuals is usually attained only after long periods, even in groups as small as those characterizing hunter-gatherer bands (Figure 4–3). The possible consequences of such structure have been considered only to a limited extent by social anthropologists (see for example Zachary, 1977).

Figure 4–3, opposite The pathways of direct culturgen flow in a group of seventy-three high-school boys. The dots in *A* connect individuals who named others as "friends." The sparse structure and clustering of friendships into subgroups is apparent. In *B* the contacts have been projected in time to the limit, in order to produce a "matrix of ultimate social contact." The procedure shows that in this case individuals have access to most of the group via chains of contact, although conspicuous isolates among the group members still occur. (Modified from Coleman, 1964:450–453.)

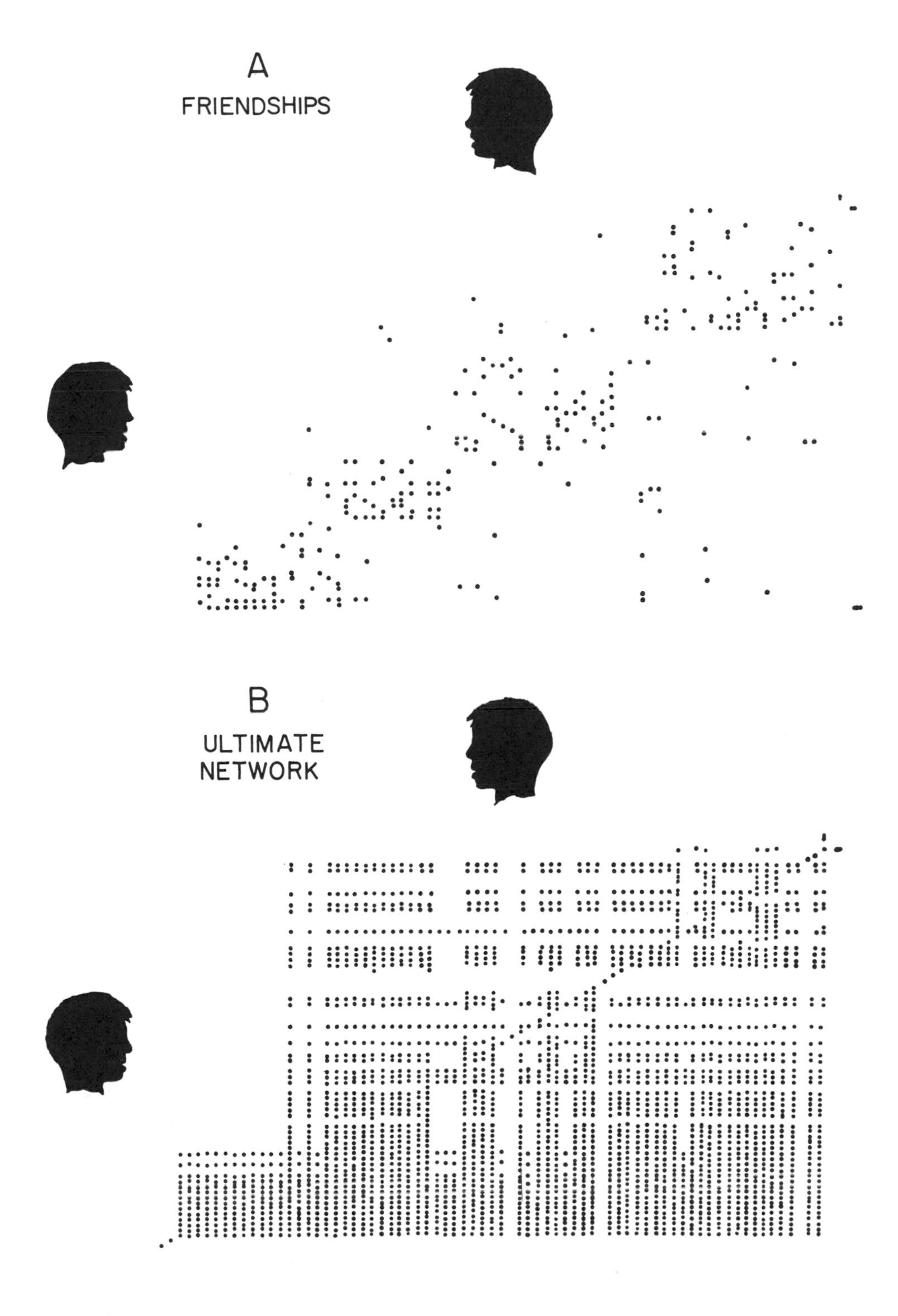
A
FRIENDSHIPS
B
ULTIMATE
NETWORK

Research on within-group propagation indicates that even in very structured societies the differential rate of imitations (dI/dt) is often some positive proportion p of past imitations I, such as

$$\frac{dI}{dt} = pI, \qquad 0 < p \leq 1. \tag{4-1}$$

Building on this basic law of change, Hamblin (in Hamblin et al., 1979) constructed a propagation model in which new culturgens of equivalent function are occasionally introduced. Culturgens are evaluated, and people reduce the use of particular versions or desert them altogether in favor of less costly alternatives. When the desertion rate is assumed to be proportional to the cumulative number of desertions, in effect a negative form of imitation, the result is a Gompertz equation with the form

$$U(t) = ca^{b^t} \tag{4-2}$$

where $U(t)$ is the number of users of a given culturgen at time t, and a, b, and c are coefficients derived from the proportionality constants of the adoption and desertion rates. Hamblin and his coworkers fitted data from seventeen instances of use to the equation, representing such diverse activities as automobile registration, helicopter passenger miles, and movie attendance. The correlation coefficient in all cases was 0.99 or better, whereas the fit to the pure logistic curve, which is the S-shaped cumulative curve used in most prior diffusion models, was 0.98 or less.

An interesting prediction of the model is that the usage of any given culturgen will eventually become zero, provided there is a continuing innovation process. The principal merit of the model is that it is based on a perception of how people actually acquire information and make decisions. Whether the Gompertz-Hamblin or logistic equations are accepted, the Hamblin studies indicate that imitation and reinforcement are key properties of individual choice behavior, and relatively simple assumptions concerning their relation to mass phenomena can be made fairly precise. This interpretation is supported by Griliches' 1957 study of hybrid corn usage in the United States, which contains one of the most complete accounts of cultural propagation published to date. Griliches obtained 163 empirically determined logistic equations from the spread of hybrid corn through 132 crop-reporting districts in 31 states. He was able to account for almost all of the variance in the starting date of usage by two exogenous variables: the percentage of total farm acreage planted in corn prior to the introduction of hybrids, and the date at which adjacent districts reached moderate levels of planting. About 70 percent of the variance in the rate of increase in usage and final level of usage resulted from what can be interpreted broadly as reinforcement: the percentage of farm acreage planted in corn, and the profit margin of corn prior to the introduction of hybrids. Additional variance was accounted for by the difference in yield between hybrid and open-pollinated corn. Comparable S-shaped adoption curves have been obtained in Hägerstrom's classic early studies of the adoption of agricultural innovations in Sweden, Jones's analysis of the conversion to tuberculin-tested milk production in England, and Bowden's analysis of the shift to irrigation by cattle farmers in Colorado (reviewed by Haggett, 1972).

In summary, the emerging consensus is that information concerning culturgens spreads in a predictable way within groups of various sizes and even in the presence of privileged channels. The adoption of preferred culturgens proceeds through a time curve consistent with the assumption covering elementary rules governing imitation and valuation.

How can these conclusions be incorporated into gene-culture coevolutionary theory? The key determinants of imitation and valuation are themselves abstracted and greatly reduced expressions of the epigenetic rules of cognitive development. The proportionality function of imitation can be translated into rules of culturgen transmission and peer pressure, which express changes in the probability of culturgen adoption by individual members of the society as a function of the percentage of use by other members of the society and of valuation measures placed on the culturgen.

Although seemingly a commonplace, the capacity of individuals to perceive the patterns of usage by whole groups and to use the result in decision making is actually quite remarkable. We suggest that two mechanisms collaborate to produce this sensitivity to group and cultural patterns. The first is a strategy in the decision process that seeks to reduce uncertainty by the observation and imitation of others. The second is the process of reification, the strong mental propensity to transform complex patterns and processes, including those arising in group organization, into real objects. The objects are often given animal or human form as well. Properties at the level of whole cultures, such as institutions and group norms, are "shrunk down" by means of reification, anthropomorphized, and added to the roster of group members. They become concrete, human-like entities to which each individual must adapt. The mind has a tendency to go further in producing mentifacts that have even less immediate material basis. It spins the reveries and dreams from which fiction, mythology, and art are born.

The existence of reification has implications for social theory that are far-reaching and far from obvious. The translation of individual behavior into societal patterns generates levels of organization that feed back upon one another. In pristine hierarchies or multilevel structures of classical conception, the various levels of organization are relatively independent and sealed off from one another (Lumsden, 1977). In contrast, cultures are *heterarchies*—mixed-level systems where the individual perceives and responds to macrocultural features, including institutions, social norms, and usage patterns (Figure 4–4).

In the remainder of this chapter we shall show that a knowledge of the two classes of psychobiological response within the heterarchy, the epigenetic rules of cognitive development and the patterns of culturgen propagation, permit a first approximation of the translation from individual behavior to cultural pattern. Other processes must of course be kept in mind. Innovation, for example, expands the array of culturgens and speeds the rate of replacement. Old culturgens can either be lost entirely or transferred from an active to a passive state, perhaps to be reactivated later as the environment changes or revitalization movements restore old social orders. The possession of one active culturgen can inhibit the adoption of an alternative, while major culturgens are less likely to be supplanted than minor ones (Berelson and Steiner, 1964:541). Such culturgenic neophobia, ranging in degree from mild to severe, is a pervasive, everyday phenomenon (see Figure 4–5). Some biases change with the

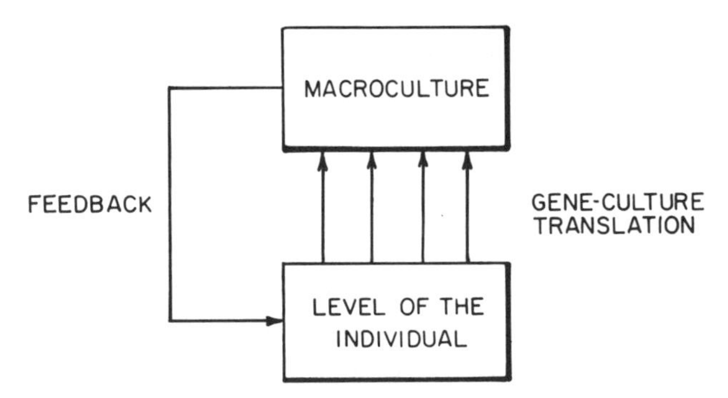

Figure 4–4 Cultures as heterarchies, or mixed-level hierarchies. Gene-culture amplification generates institutions, social norms, and other macrocultural patterns, which are codified through reification learning and fed back into the processes of individual decision making.

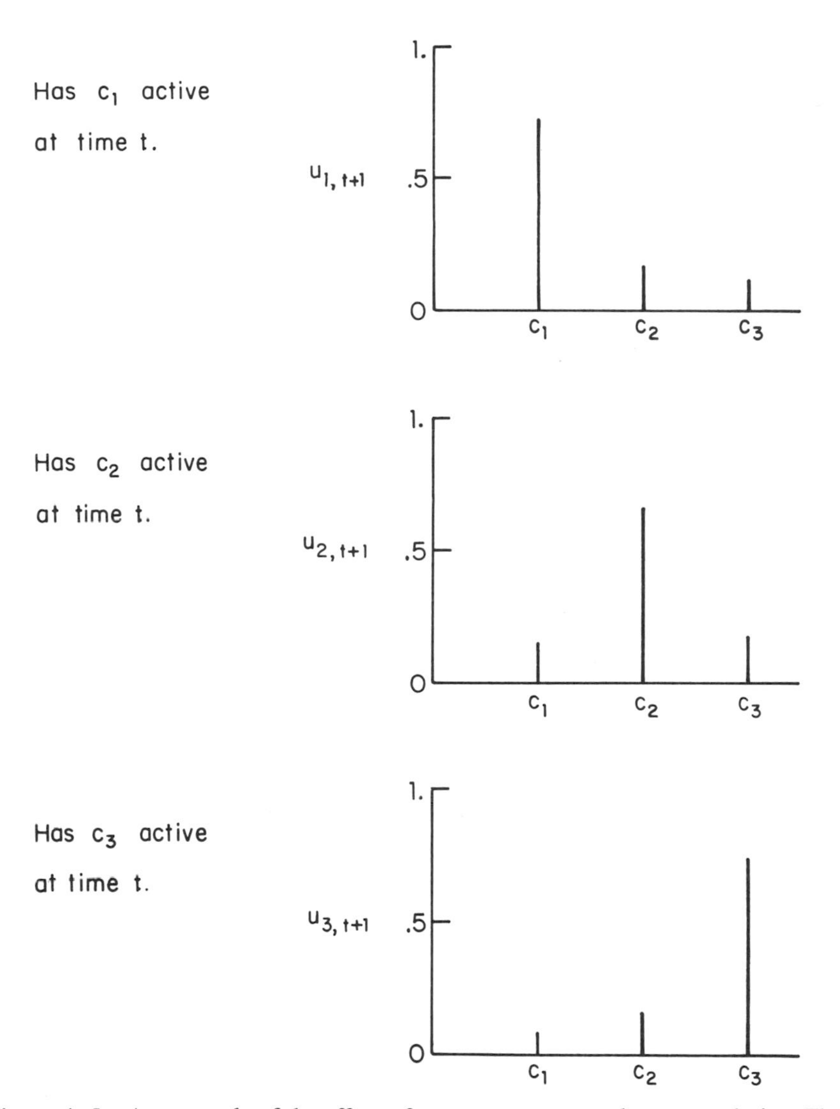

Figure 4–5 An example of the effect of current usage on subsequent choice. The histograms give the probabilities that a woman using one brand of pancake flour (c_1, c_2, or c_3) will choose the same or a different brand on her next shopping trip. (Based on data from Coleman, 1973:12.)

environment, but as we suggested earlier, even these shifts are likely to follow discernible core rules. Finally, some culturgens lock together to create a more stable holistic pattern, resistant but not impervious to change.

All of these effects can be subjected to postulational-deductive modeling and incorporated into gene-culture theory, and they will be considered again in Chapters 6 and 7. For the moment, however, we shall put them aside in order to evaluate the most basic aspects of the translation process.

Individual Decision Making

The premises derived from cultural anthropology and gene-culture theory are sufficient to establish a link between individual response and cultural pattern. In the next two sections we characterize the properties of decision making, expressed in the form of the rates (designated as v_{ij}*) at which individuals change back and forth between two culturgens (*c_1 *and* c_2*). In subsequent sections this elementary cognitive description is incorporated into an analysis of the cultural patterns that result from the interactions of many individuals engaged in similar enculturation and decision making. From this vantage point we develop our principal conclusion about gene-culture translation: that relatively small changes in the epigenetic rules can force profound changes in the overlying cultural patterns.*

Our use of the simplest, two-culturgen case is a strategic decision that follows a common procedure in first theoretical formulations. It would be a straightforward exercise to employ an arbitrary number of culturgens and later, in more complete models of coevolution, to incorporate arbitrary numbers of cognitive processes and controlling genes. However, the resulting equations would be too complex to impart much immediate understanding. We have chosen to begin with an elementary formulation that captures in a crude manner the essential properties of genes, mind, and culture while providing maximum clarity in the postulates. Indeed, so new are the problems involved in this area of investigation that even the simplest models of gene-culture coevolution remain unexplored. Once the first steps have been taken, the postulates and the models can be made indefinitely more complex, in order to move gene-culture coevolutionary theory closer to the "thick description" that characterizes most of the traditional social sciences.

The formal argument concerning the properties of the rates of culturgen change are as follows. Consider the simplest social system, class free, with N egalitarian members at or close to a demographic steady state. We treat the classic anthropological case of *binary choice,* in which two culturgens, c_1 and c_2, compete for use in the group. They can be envisaged as concrete entities, such as competing horticultural practices, verbal color classifications, or blowgun designs. After enculturation group members are aware of c_1 and c_2, and they hold both in long term memory. Within the group n_1 individuals currently use c_1 and n_2 individuals use c_2.

Each group member possesses both a set of learning procedures and a cognitive decision system characteristic of the species and perhaps of a particular genotype (Figure 4–6). Each individual evaluates his personal usage repeatedly, and he is influenced by the innate epigenetic rules on these occasions either to maintain current usage or to switch to the competing culturgen. Because of a combination of circumstances, which include reliance on fuzzy heuristics (see Chapter 3), the uncertainties inherent in the environment, the limited information available to a group member at each decision point, and the possible occurrence of mixed strategies (see Chapter 5), the epigenetic rules set *probabilities of transition* rather than fixed usage patterns. Let u_{ij} be the likelihood that an individual at a decision point will adopt culturgen c_j, given that he has just been using culturgen c_i. Then the epigenetic decision rules can be expressed in terms of the u_{ij} (Figure 4–7), and a decision point is described by the transition matrix

$$\begin{bmatrix} u_{11} & u_{12} \\ u_{21} & u_{22} \end{bmatrix}. \tag{4-3}$$

The mean lifetime between sequential decision points, at which the epigenetic rules take effect, is τ_1 for a c_1-user and τ_2 for a c_2-user. The decision mechanism is thus characterized by mean rate constants $r_1 = 1/\tau_1$ and $r_2 = 1/\tau_2$ respectively. The matrix of mean rates is

$$\begin{bmatrix} r_1 & 0 \\ 0 & r_2 \end{bmatrix}. \tag{4-4}$$

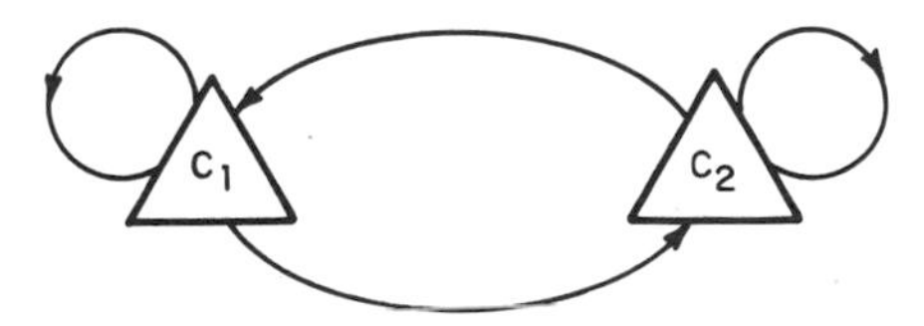

Figure 4–6 The minimal decision structure of an individual group member. The two alternate culturgen states are represented by c_1 and c_2, and the arrows symbolize the epigenetic rules that control transitions between the states.

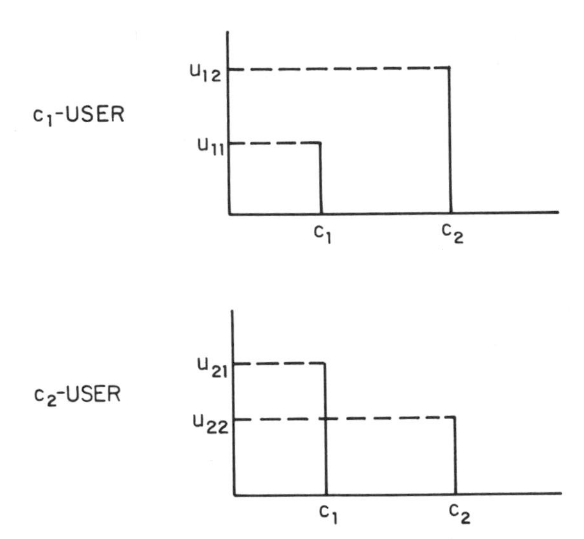

Figure 4–7 Epigenetic rules for the decision process. Here u_{ij} is the likelihood that an individual at a decision point will adopt culturgen c_j, given that he has just been using culturgen c_i.

We note that the r_k are accessible to observation by informant analysis and can in some cases be prearranged by the investigator (see for example Dodd, 1955; Rachlin, 1976).

The parameters τ_1 and τ_2 are the means of the probability distributions for the holding times in states c_1 and c_2. We shall model these distributions with exponential densities (Figure 4–8), meaning that the underlying cognitive system is a Markov learning and decision process. This is a useful approximation in many cases (Bush and Mosteller, 1955; Kemeny and Snell, 1962; Coleman, 1964, 1973; Atkinson et al., 1965; Bartholomew, 1973; Greeno, 1974). It can also be shown that deep connections exist between the laws of such Markov systems and the theories of information processing in which the cognitive and ethnosemantic mechanisms of learning and decision are naturally expressed. This important relation is discussed further in Appendix 4–1.

The transition rates of the decision process are then v_{ij}, where

$$\begin{bmatrix} v_{11} & v_{12} \\ v_{21} & v_{22} \end{bmatrix} = \begin{bmatrix} r_1(u_{11} - 1) & r_1 u_{12} \\ r_2 u_{21} & r_2(u_{22} - 1) \end{bmatrix} \qquad (4\text{-}5)$$

(Howard, 1971b:769 ff.). We have now acquired not only the matrix of transition rates for individual group members but also a recipe for calcula-

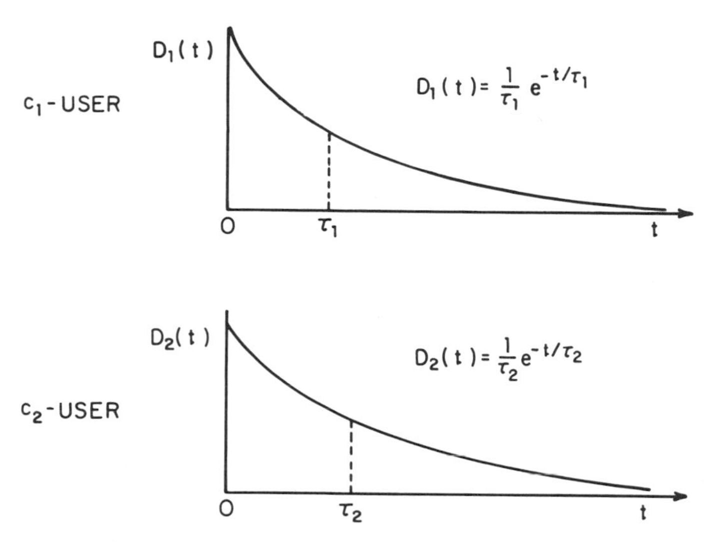

Figure 4–8 Holding time distributions for the c_1- and c_2-user states. The probability densities are exponential with means τ_1 and τ_2 respectively.

tion of the v_{ij} in terms of the somewhat more intuitive parameters r_i and u_{ij}. Furthermore, in some circumstances the r_i and u_{ij} will be more accessible to direct measurement than the v_{ij} (for instance Rachlin, 1976: 544–603).

Social contagion and the reification learning rules guarantee that the group members do not act in isolation. Instead they form a composite learning and decision system (Figure 4–9). The goal of models of gene-culture translation is to predict the organizational and functional properties of such systems, given the rules of individual epigenesis. In the present instance we have a natural macrocultural pattern of the simplest form, namely the two *institutions* formed respectively by the users of culturgen c_1 and by the users of culturgen c_2. Heterarchic feedback appears as a dependence of the properties r_k, u_{ij}, and v_{ij} upon these two institutions. Of the various institutional properties that could induce the dependence, we shall focus on the sizes of the two user groups. Although other dependencies are conceivable, including rates of growth of the two groups, group size or "social pressure" per se has been most intensively studied. Its importance to human decision making, particularly in such small groups as committees and hunter-gatherer bands, has been thoroughly documented (examples are Berelson and Steiner, 1964; Biesele, 1978; Lee, 1979). Thus in general we have $r_k = r_k(n_1, n_2)$, $u_{ij} = u_{ij}(n_1, n_2)$ and $v_{ij} = v_{ij}(n_1, n_2)$.

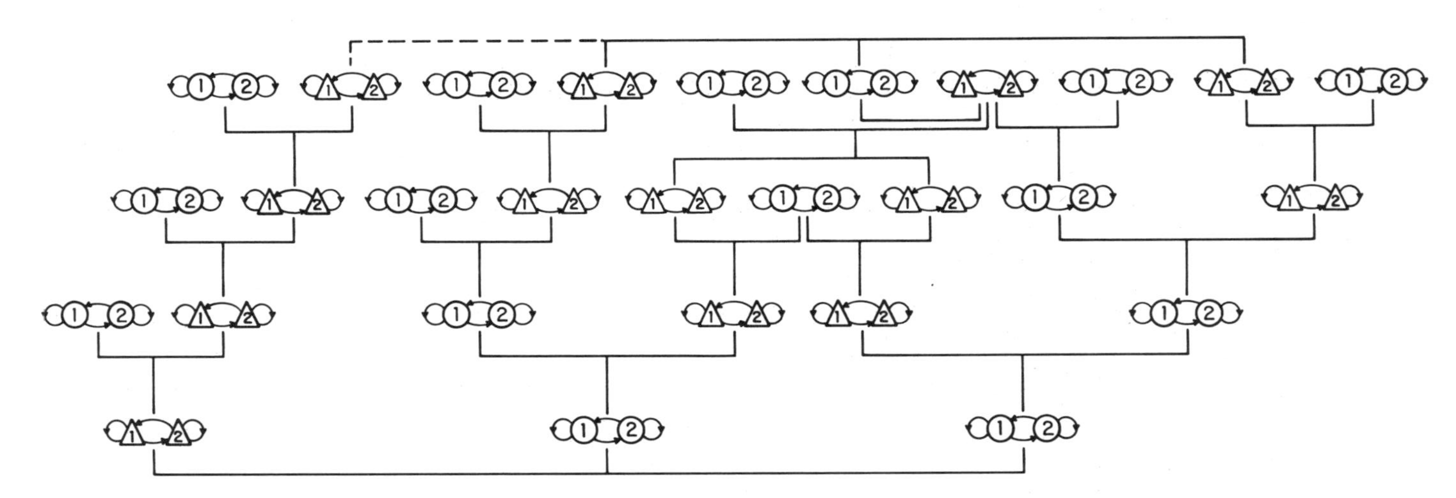

Figure 4–9 A kin group as it would appear acting on a two-culturgen semantic field. Information and social influence flow along kin lines and other components of the matrix of social relations. The structure depicted is an actual Yanomamö lineage, from the data of Chagnon (1977). Males are denoted by triangles and females by circles.

A more detailed analysis will include the full matrix of social contacts between individuals, an important step when information flow, social pressure, and contagion effects are restricted by social stratification and other kinds of partitioning. This refinement can be achieved by straightforward extension of the methods discussed here; a concise guide to the mathematics has been provided by Lumsden and Trainor (1976).

The Cognitive and Ethnographic Pictures

To the field ethnographer and human ethologist the cognitive states of subjects are often inaccessible and even of secondary concern. Yet both the ethnographic-ethological and cognitive descriptions deal with aspects of real behavior. The switch between competing culturgens is an observable aspect of both ways of looking at human behavior. In a complete theory the cognitive picture and the ethnographic picture must therefore be related. Let us symbolize the former by "Cog" and the latter by "Eth." Then in completing the description of the individual level of the social system, we seek a map ψ which relates Cog to Eth:

$$\psi : \mathrm{Cog} \longmapsto \mathrm{Eth}. \tag{4-6}$$

The generalized inverse map would take the ethnographic picture back into the cognitive picture, so that a full system of relations would be

$$\mathrm{Cog} \underset{\psi^{-1}}{\overset{\psi}{\leftrightarrows}} \mathrm{Eth}. \tag{4-7}$$

For the remainder of this section superscript C will denote quantities in the cognitive picture, while superscript E will label quantities as measured in the ethnographic picture.

An observer who restricts his attention to external behavior when collecting data on a system of the kind illustrated in Figure 4–6 will tabulate only overt switches of culturgens (Figure 4–10). The reentrant looping of the decision process will be invisible to him. Thus in the ethnographic picture the only transitions with nonzero probabilities are u_{12}^E and u_{21}^E. Furthermore, both u_{12}^E and u_{21}^E have unit value: when a behavioral transition does occur, it has only one possible destination, namely the other culturgen. This description is very different from the cognitive picture, in which reentrance in the same culturgen state can delay departure from one culturgen to its competitor. To the ethnographer these delays show

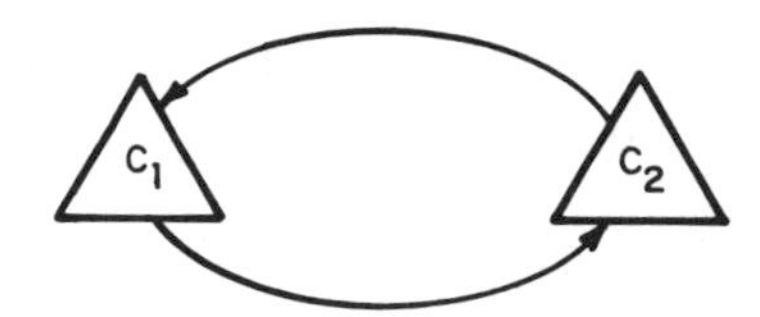

Figure 4–10 Shifts from one culturgen to another, as perceived by an observer of external behavior only. The internal decisions to remain with a given culturgen cannot be observed.

up as increased values of the holding times τ_1^E and τ_2^E in each state. Watching the decision maker in its natural state, the ethnographer sees an a priori transition probability matrix

$$\begin{bmatrix} u_{11}^E & u_{12}^E \\ u_{21}^E & u_{22}^E \end{bmatrix} = \begin{bmatrix} 0 & 1 \\ 1 & 0 \end{bmatrix} \tag{4-8}$$

and a rate matrix

$$\begin{bmatrix} r_1^E & 0 \\ 0 & r_2^E \end{bmatrix} = \begin{bmatrix} 1/\tau_1^E & 0 \\ 0 & 1/\tau_2^E \end{bmatrix}. \tag{4-9}$$

Together these give a matrix for the transition rates

$$\begin{bmatrix} v_{11}^E & v_{12}^E \\ v_{21}^E & v_{22}^E \end{bmatrix} = \begin{bmatrix} -r_1^E & r_1^E \\ r_2^E & -r_2^E \end{bmatrix}. \tag{4-10}$$

Both the ethnographer and the cognitive anthropologist are observing the same individual, so that the system is consistent if and only if $v_{ij}^C = v_{ij}^E$ when the clocks of the two observers are synchronized. For the cognitive anthropologist, we determined that

$$\begin{bmatrix} v_{11}^C & v_{12}^C \\ v_{21}^C & v_{22}^C \end{bmatrix} = \begin{bmatrix} r_1^C(u_{11}^C - 1) & r_1^C u_{12}^C \\ r_2^C u_{21}^C & r_2^C(u_{22}^C - 1) \end{bmatrix}. \tag{4-11}$$

The map between the cognitive and ethnographic pictures in this system is then obtained:

$$r_1^E = r_1^C u_{12}^C, \qquad r_2^E = r_2^C u_{21}^C. \tag{4-12}$$

These findings mean that important data on cognitive processes are open to measurements on the undisturbed system when the model is applied to ethnographic data. Furthermore, if the looping rates r_1^C and r_2^C for the decision process are independent of culturgen state, then $r_1^C = r_2^C$ and

$$r_1^E/r_2^E = u_{12}^C/u_{21}^C. \tag{4-13}$$

From observations on external behavior, the relative structure of the bias curves for the two states can be determined in this type of system.

The Translation Process

We can now state the key problem as follows: how are the properties of the epigenetic rules translated into a social pattern? In the next two sections we shall describe a procedure that models this important step. The heart of the procedure is a balance equation written to specify the rate at which the entire culture changes from one state to another, in particular from a condition in which a certain number of individuals actively use one culturgen (c_1) and the others actively use the other culturgen (c_2), to the condition in which some different proportion of individuals use the two culturgens. For example, we might ask: given a certain epigenetic rule at the individual level, what is the probability at a particular time that in a society of thirty individuals, twenty-six use culturgen c_1 and four use c_2? And how rapidly is this probability increasing or decreasing through time? The cultural pattern of one society is the relative proportion of its members that use each of the two culturgens. The probability distribution, which we call the ethnographic curve, gives the percentages in a sample of many such societies that use each one of the possible cultural patterns. In essence, we are searching for the rate and direction of change of ethnographic curves of various kinds of cultures, and the steady-state ethnographic curves toward which many societies as a whole tend to converge.

Since the time of Coleman's monograph (1964), balance equations of the kind alluded to above have been used to describe the dynamics of human groups. Indeed, they are the subject of a growing literature in mathematical sociology and despite their relatively elementary form appear to capture important mass properties of social groups. (For background and a sampling see Coleman, 1964; Liboff, 1970; Cohen, 1971; Weidlich, 1972; Bartholomew, 1973; West, 1974; Walls, 1976; Haken, 1977; Nicolis and Prigogine, 1977; Bowers, 1978a,b.) In the development that follows we shall draw this cumulative body of formalism into the service of gene-culture theory and carry out the translation from the level of epigenetic rules to the level of cultural pattern. Given the connection between the epigenetic rules and the dynamics of individual behavior that we established in the preceding sections, the step from individual to culture can be taken in a fairly straightforward way by means of established mathematical techniques.

We are then in a position to address the central issue of the gene-culture translation process, namely the impact that epigenetic rules have on pattern and structure at the cultural level. Since the ethnographic curve provides a detailed description of cultural pattern, we can proceed by analyzing its ties to the epigenetic rules. Upon deriving the appropriate new theorems, which we call amplification laws, we are able to show that even small changes in the epigenetic rules can produce large changes in the ethnographic curves. Thus the effects of the epigenetic rules are not lost in the translation process. They are instead greatly amplified. The nature and extent of this amplification is explored further by applying the general theory to specific cases. In particular, we investigate models of sibling incest, of village fissioning among the Yanomamö Indians, and of cycles of fashion change among Western women.

The formal argument proceeds as follows. The society as a whole can be specified at a time t by the vector $\mathbf{n} = (n_1, n_2)$, the number of individuals possessing each of the two culturgens respectively. Since the system is probabilistic, the quantity of interest is $P(\mathbf{n},t)$, the likelihood that at time t the group has n_1 persons with c_1 culturgen and n_2 persons with c_2 culturgen such that

$$\sum_{\mathbf{n}} P(\mathbf{n},t) = 1. \tag{4-14}$$

In other words, the summed probabilities of all conceivable states of culturgen usage at any given moment of time is one. Let us distinguish one particular state $\mathbf{n}$ from all other possible states $\mathbf{n}'$. The probability flux into the state $\mathbf{n}$ is

$$\sum_{\mathbf{n}' \neq \mathbf{n}} P(\mathbf{n}',t) R_{\mathbf{n}'\mathbf{n}} \tag{4-15}$$

where $R^{\mathbf{n}'\mathbf{n}}$ is the probability per unit time of transition from state $\mathbf{n}'$ to state $\mathbf{n}$ for the system as a whole, and $P(\mathbf{n}',t)$ is the probability of being in $\mathbf{n}'$. The probability flux out of the state $\mathbf{n}$ is similarly

$$\sum_{\mathbf{n}' \neq \mathbf{n}} P(\mathbf{n},t) R_{\mathbf{n}\mathbf{n}'} \tag{4-16}$$

and the equation of motion for $P(\mathbf{n},t)$ is therefore the balance equation

$$\frac{d}{dt}P(\mathbf{n},t) = \sum_{\mathbf{n}'\neq\mathbf{n}} P(\mathbf{n}',t)R_{\mathbf{n}'\mathbf{n}} - P(\mathbf{n},t)\sum_{\mathbf{n}'\neq\mathbf{n}} R_{\mathbf{n}\mathbf{n}'}. \tag{4-17}$$

This equation can be linked to events at the individual level by noting that in any differential interval of time dt the probability of two or more simultaneous decisions is of order at most $(dt)^2$ and as a consequence can be ignored. Then if the society moves from state (n_1',n_2') at time t to state (n_1,n_2) by time $t + dt$, only one person changed culturgens. It can only be that

$$(n_1',n_2') = (n_1 + 1,n_2 - 1)$$

or

$$(n_1',n_2') = (n_1 - 1,n_2 + 1). \tag{4-18}$$

Similarly, if the society shifts from state (n_1,n_2) to state (n_1',n_2') by time $t + dt$, it can only be that

$$(n_1',n_2') = (n_1 + 1,n_2 - 1) \tag{4-19a}$$

in which case one person switched from c_2 to c_1,

or

$$(n_1',n_2') = (n_1 - 1,n_2 + 1) \tag{4-19b}$$

where one person switched from c_1 to c_2.

It follows that the whole-group transition rate to or from state $\mathbf{n} = (n_1,n_2)$ is equal to the transition rate $v_{ij}(n_1,n_2)$ for single individuals multiplied by the number of individuals in the culturgen-usage group from which the transition took place:

$$\begin{aligned}
R_{(n_1+1,n_2-1)\to(n_1,n_2)} &= (n_1 + 1)v_{12}(n_1 + 1,n_2 - 1)\\
R_{(n_1-1,n_2+1)\to(n_1,n_2)} &= (n_2 + 1)v_{21}(n_1 - 1,n_2 + 1)\\
R_{(n_1,n_2)\to(n_1+1,n_2-1)} &= n_2v_{21}(n_1,n_2)\\
R_{(n_1,n_2)\to(n_1-1,n_2+1)} &= n_1v_{12}(n_1,n_2).
\end{aligned} \tag{4-20}$$

The cultural dynamics may be visualized as a walk on a one-dimensional lattice, the points of which correspond to societies in which the c_1 and c_2

subgroups have their specific sizes (Figure 4–11). With higher-dimensional lattices, problems involving multiple culturgens can be handled similarly.

Placing these terms in Eq. (4-17) gives the master equation

$$\frac{d}{dt}P(n_1,n_2,t) = (n_1+1)v_{12}(n_1+1,n_2-1)P(n_1+1,n_2-1,t)$$
$$+ (n_2+1)v_{21}(n_1-1,n_2+1)P(n_1-1,n_2+1,t)$$
$$- [n_1 v_{12}(n_1,n_2) + n_2 v_{21}(n_1,n_2)]P(n_1,n_2,t) \qquad \text{(4-21a)}$$

for $0 < n_1 < N$,

$$\frac{d}{dt}P(0,N,t) = v_{12}(1,N-1)P(1,N-1,t)$$
$$- Nv_{21}(0,N)P(0,N,t) \qquad \text{(4-21b)}$$

for $n_1 = 0$, and

$$\frac{d}{dt}P(N,0,t) = v_{21}(N-1,1)P(N-1,1,t)$$
$$- Nv_{12}(N,0)P(N,0,t) \qquad \text{(4-21c)}$$

for $n_1 = N$. The probability density distribution of all combinations of culturgen frequencies at any given time can be called the *ethnographic curve* of the population; the curve provides the expected frequency distribution for many populations considered concurrently or for one or a few popula-

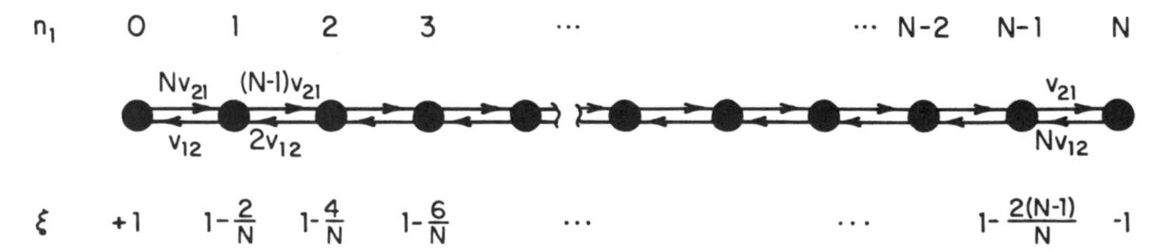

Figure 4–11 The state space for the cultural dynamics of a social group of the kind represented in Figure 4–9. The number of c_1-users is denoted by n_1; ξ is the order parameter $1 - 2n_1/N$.

tions observed repetitively through time. For the state space shown in Figure 4–11, this distribution approaches a steady state given by

$$P(n_1, N - n_1) = P(0,N) \binom{N}{n_1} \prod_{i=1}^{n_1} \frac{v_{21}(i - 1, N - i + 1)}{v_{12}(i, N - i)} \qquad \text{(4-22a)}$$

$$= P(0,N) \binom{N}{n_1} \exp \left[\sum_{i=1}^{n_1} \ln \frac{v_{21}(i - 1)}{v_{12}(i)} \right] \qquad \text{(4-22b)}$$

such that

$$P(0,N) = \left[1 + \sum_{n_1=1}^{N} P(n_1, N - n_1) \right]^{-1}. \qquad \text{(4-22c)}$$

After a transient phase the pattern of use frequency in the group will be fully characterized by the probability density with frequencies $P(n_1, N - n_1)$. This distribution is the steady-state ethnographic curve (Figure 4–12). A key objective is to assess the sensitivity of the ethnographic curve to changes in the underlying epigenetic rules.

Parameterization of the ethnographic curve in terms of n_1 is inconvenient when comparing groups of different size. We therefore seek a new independent variable that will scale all ethnographic curves to the same interval. One convenient scaling variable is

$$\xi \triangleq \nu_2 - \nu_1 = 1 - 2n_1/N \qquad \text{(4-23)}$$

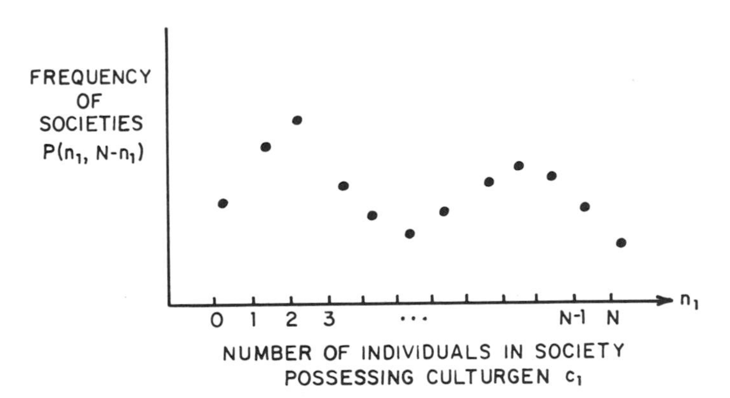

Figure 4–12 The ethnographic curve. In this literal representation the $P(n_1, N - n_1)$ are the frequencies of appearance of culture groups with c_1-institutions of size n_1, that is, the frequencies of groups of N individuals of which exactly n_1 individuals possess culturgen c_1 as opposed to culturgen c_2.

where ν_1 and ν_2 are the frequencies of c_1-users and c_2-users in the group:

$$\nu_1 = n_1/N, \qquad \nu_2 = n_2/N = 1 - n_1/N. \tag{4-24}$$

Regardless of the absolute size of N, ξ ranges from $+1$ to -1 as n_1 ranges from 0 to N (Figure 4–13).

The variable ξ measures an important aspect of cultural patterning. When $n_1 = n_2$, ξ vanishes and when $n_1 \sim n_2$, $|\xi| \ll 1$. Thus ξ is small when the two culturgens have similar frequencies. At such points the group is highly disordered in terms of culturgen usage, with large numbers of both c_1 and c_2 active. As $|\xi|$ approaches unity, one culturgen dominates and the group becomes increasingly ordered. We show later that under certain conditions there are concise laws of motion for ξ, in terms of which the group macroculture is endowed with covering laws.

In the case of groups the size of hunter-gatherer bands, the ethnographic curves are readily calculated from Eq. (4-22) if specific forms for the transition rates $v_{12}(n_1, n_2)$ and $v_{21}(n_1, n_2)$ are given. The selection of models for the $v_{ij}(n_1, n_2)$ is the crucial step, after which standard methods can be applied. Concise, exact solutions of Eq. (4-22) in terms of known functions occur rarely, although we shall obtain such forms in later specific applications to anthropological case histories. The exact equations (4-21) and (4-22) thus are somewhat inefficient tools for producing rapid insight into the dependence of the ethnographic curves on the epigenetic rules. We therefore introduce an approximation to Eq. (4-21), which becomes very accurate for large groups ($N \to \infty$) and is often good to within several percent for low-order moments of the ethnographic distribution for groups as small as $N = 25$—in other words, near the mean of

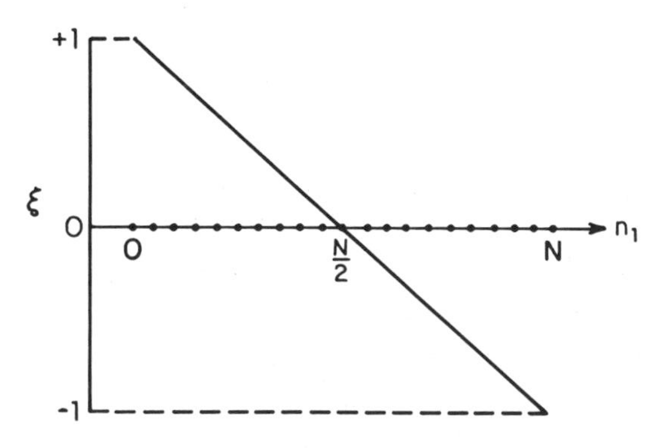

Figure 4–13 The order parameter ξ as a function of n_1, the number of c_1-users.

hunter-gatherer groups. The approximation can be invoked by noting that as $N \to \infty$, the separation $\Delta\xi$ between adjacent values of the scaling variable ξ becomes vanishingly small. Since $\Delta\xi$ is the difference between $\xi(n_1 \pm 1)$ and $\xi(n_1)$, if we recall Eq. (4-23), $\Delta\xi$ must equal N^{-1} and therefore has the limiting behavior

$$\lim_{N\to\infty} \Delta\xi = 0. \tag{4-25}$$

Hence ξ behaves as a continuous variable. Given this property, the system of ordinary differential equations (4-21) can be mapped into a single partial differential equation for a continuous ethnographic curve $P(\xi,t)$. The necessary steps follow standard methods and appear in Appendix 4–2. We find that the equation of motion for the ethnographic curve is of the forward diffusion, or Fokker-Planck, form:

$$\partial_t P(\xi,t) = -\frac{\partial}{\partial\xi}[X(\xi)P(\xi,t)] + \frac{1}{2}\frac{\partial^2}{\partial\xi^2}[Q(\xi)P(\xi,t)] \tag{4-26}$$

where

$$\begin{aligned} X(\xi) &= (1-\xi)v_{12}(\xi) - (1+\xi)v_{21}(\xi), \\ Q(\xi) &= \frac{2}{N}(1-\xi)v_{12}(\xi) + \frac{2}{N}(1+\xi)v_{21}(\xi), \end{aligned} \tag{4-27}$$

and $$-1 < \xi < 1.$$

A generalization to $M > 2$ culturgens is readily achieved (see Appendix 4–3).

After transients decay, the ethnographic curve for our model culture group approaches a steady state which, as shown in Appendix 4–4, takes the form

$$P(\xi) = \frac{C}{Q(\xi)} \exp\left[2\int_{-1}^{\xi} \frac{X(\xi')}{Q(\xi')}\,d\xi'\right], \qquad -1 < \xi < 1 \tag{4-28}$$

where C is a normalization constant that can be determined by the requirement

$$\int_{-1}^{+1} P(\xi)d\xi = 1. \tag{4-29}$$

Equation (4-28) is the continuous-ξ analog of the discrete-ξ ethnographic curve (4-22).

The Structure of Ethnographic Curves

The ethnographic curve equation (4-28) has interesting properties that contribute directly to an intuitive understanding of gene-culture translation. In this and the following sections we shall elucidate a property that appears to us to have great significance for the application of gene-culture theory to the social sciences. It will be shown that the ethnographic curve is sensitive to small changes in the epigenetic rules of its members. Only a small amount of innate bias in favor of a culturgen, or a genetic or environmentally induced alteration in the sensitivity of the individual to the choice of culturgens made by others, can create large shifts in the ethnographic curve. This principle has potentially major implications for both developmental psychology and anthropology.

The mathematical argument begins in the following way. In Eq. (4-28) $P(\xi)$ can be regarded as the product of two factors, $[Q(\xi)]^{-1}C$ and $e^{2V(\xi)}$ where

$$V(\xi) \triangleq \int_{-1}^{\xi} \frac{X(\xi')}{Q(\xi')} d\xi'. \tag{4-30}$$

The integral $V(\xi)$ is evaluated from $\xi' = -1$ to $\xi' = \xi$ and it is thus the area under the curve $X(\xi')/Q(\xi')$, bounded on the left by $\xi' = -1$ and on the right by $\xi' = \xi$ (see Figure 4–14). The $V(\xi)$ is influenced by the *ratios* rather than by the absolute values of the transition rates $v_{ij}(\xi)$, since

$$\frac{X(\xi')}{Q(\xi')} = \frac{N}{2} \frac{[(1 - \xi') - (1 + \xi')\mathcal{R}(\xi')]}{[(1 - \xi') + (1 + \xi')\mathcal{R}(\xi')]} \tag{4-31}$$

where

$$\mathcal{R}(\xi') = v_{21}(\xi')/v_{12}(\xi'). \tag{4-32}$$

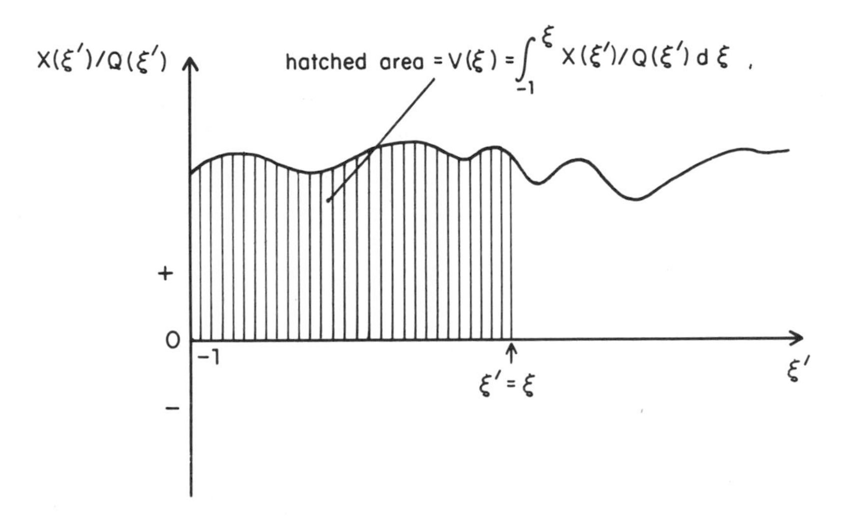

Figure 4–14 The integration of the key functions of decision probability and reification effects, summarized in $X(\xi')$ and $Q(\xi')$. In this case the area integrated is all above the $X(\xi')/Q(\xi') = 0$ axis and hence is positive, but it could also be below the axis and negative, or it could cross back and forth.

This relation is parallel to the conception of fitness in classical population genetics, where relative rather than absolute fitnesses ultimately determine the fate of specific alleles. The $v_{ij}(\xi)$ act as fitness functions for the two culturgens c_1 and c_2.

The function $Q(\xi)$ is a $v_{ij}(\xi)$-weighted sum of the two straight lines $y = 1 + \xi$ and $y = 1 - \xi$. The feedback of cultural patterns and social contagion to individual cognition is expressed by the ξ-dependence of the transition rates. The natural reference model with which to compare more complex systems is that in which each v_{ij} is a constant, independent of ξ. Each group member then acts independently of changing culturgen usage patterns and $Q(\xi)$ is a straight line with either positive, zero, or negative slope in accordance with the relative values of the v_{ij}:

$$Q(\xi) = \frac{2}{N}[(v_{12} + v_{21}) + (v_{21} - v_{12})\xi]. \qquad (4\text{-}33)$$

Such a $Q(\xi)$ is monotonic on $(-1,1)$ and the corresponding $[Q(\xi)]^{-1}C$ cannot induce interior maxima or minima in the ethnographic curve. The existence and location of these must be determined by $X(\xi)$, and indirectly by the $v_{ij}(\xi)$, acting through $e^{2V(\xi)}$.

When the first derivatives $dv_{ij}(\xi)/d\xi$ change sign one or more times on

$(-1, 1)$, $Q(\xi)$ is no longer monotonic in general. However, $e^{2V(\xi)}$ varies exponentially with group size N while $Q(\xi)$ varies as N^{-1}, with the consequence that in many cases of practical interest the factor $e^{2V(\xi)}$ dominates in the behavior of the ethnographic curve over much of $(-1, 1)$. A study of the exponent $V(\xi)$ then provides considerable information about the qualitative structure of the ethnographic curve. Indeed, for well-behaved systems there is a coordinate map $\xi \mapsto z$ such that $Q(z) \triangleq 1$ and the ethnographic curve is a purely exponential functional

$$P(z) = Ce^{V(z)} \tag{4-34}$$

(Goel and Richter-Dyn, 1974:40).

The function $Q(\xi)$ is always positive for $-1 < \xi < 1$. On the other hand, $X(\xi)$ is given on the same interval by the difference rather than the sum of the two positive terms $(1 - \xi)v_{12}(\xi)$ and $(1 + \xi)v_{21}(\xi)$. Thus for appropriate individual transition rates it is possible for $X(\xi)$ to change sign one or more times. As exemplified in Figure 4–15, the effect is to generate one or more peaks in $e^{2V(\xi)}$. This occurrence of multiple modes has key anthropological significance, indicating that two or more distinct patterns of culturgen usage, often widely divergent, have relatively high likelihood for the same culture group.

The number of modes in the function $e^{2V(\xi)}$ is equal to the number of zeros on $(-1, 1)$ at which $X(\xi')/Q(\xi')$ passes from $+$ to $-$ as ξ progresses toward $+1$; and the number of local minima is equal to the number of zeros at which $X(\xi')/Q(\xi')$ passes from $-$ to $+$ during the same change in ξ. All of these zeros are found at the points $\xi_0 \in (-1, 1)$ for which the functions

$$g(\xi) = (1 - \xi)/(1 + \xi) \qquad \text{and} \qquad \mathscr{R}(\xi) = v_{21}(\xi)/v_{12}(\xi) \tag{4-35a}$$

intersect:

$$g(\xi_0) = \mathscr{R}(\xi_0), \qquad \xi_0 \in (-1, 1). \tag{4-35b}$$

The natural reference model with which to initiate a study of the relations of Eq. (4-35) is again that with v_{ij} a constant, independent of ξ. In general, the function $g(\xi)$ diverges to $+\infty$ at $\xi = -1$ and decreases monotonically to zero at $\xi = 1$ (Figure 4–16A). For constant v_{ij} there is exactly one point ξ_0 on $(-1,1)$ at which Eq. (4-35b) is satisfied (Figure 4–16B).

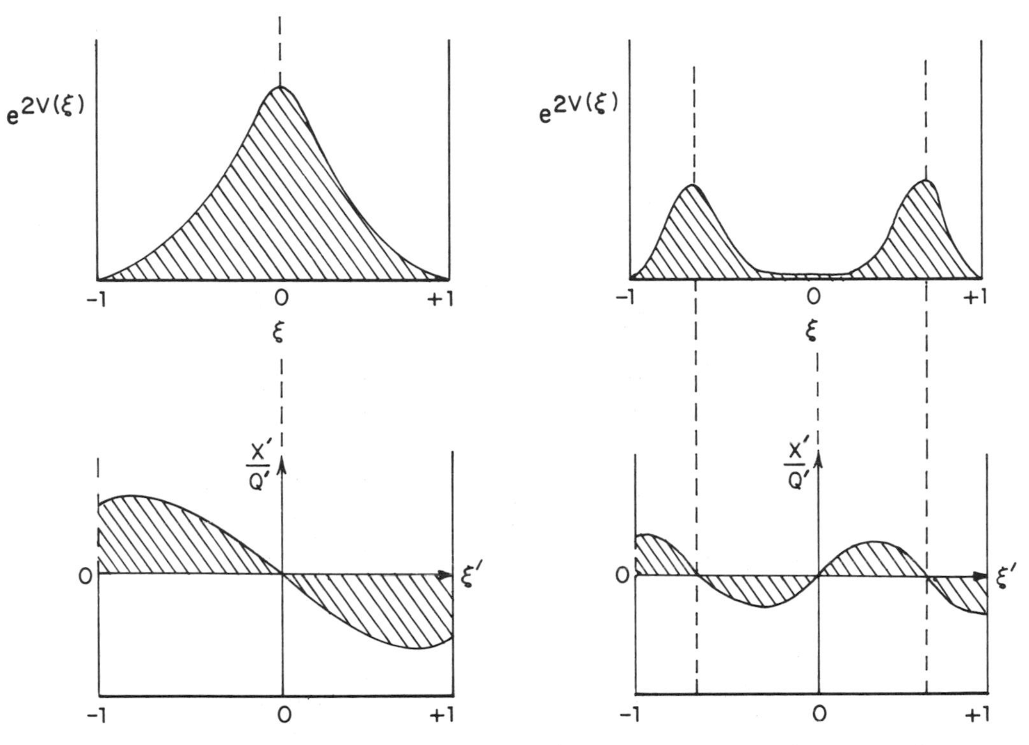

Figure 4–15 The determination of modality in the ethnographic curve. The number and position of the modes in the ethnographic curve factor $e^{2V(\xi)}$ *(upper pair of diagrams)* are determined by the changes in sign of $X(\xi')$ and thus of $X(\xi')/Q(\xi')$ *(lower pair of diagrams)*, since $Q(\xi') > 0$ everywhere on $(-1, 1)$.

For v_{ij} constant, $X(\xi)$ is a straight line of negative slope and passes through $X = 0$ at the point ξ_0. Thus the factor $e^{2V(\xi)}$ associated with constant v_{ij} is necessarily unimodal and its mode is located at $\xi = \xi_0$.

For $v_{ij}(\xi)$ that vary slowly with ξ, the situation is essentially unchanged, since the ratio function $\mathcal{R}(\xi)$ is also a weak function of ξ. Models of this type are appropriate for decision and learning processes that have a relatively slight dependence on social patterns. The resulting factor $e^{2V(\xi)}$ will once more be unimodal (Figure 4–16C). Multiple modes occur only when $\mathcal{R}(\xi)$ varies sufficiently to intersect $g(\xi)$ more than once on $(-1, 1)$. A readily visualized example is provided by the exponential trend-watcher model

$$v_{12}(\xi) = a_2 e^{a_1 \xi}, \quad v_{21}(\xi) = a_3 e^{-a_1 \xi} \tag{4-36}$$

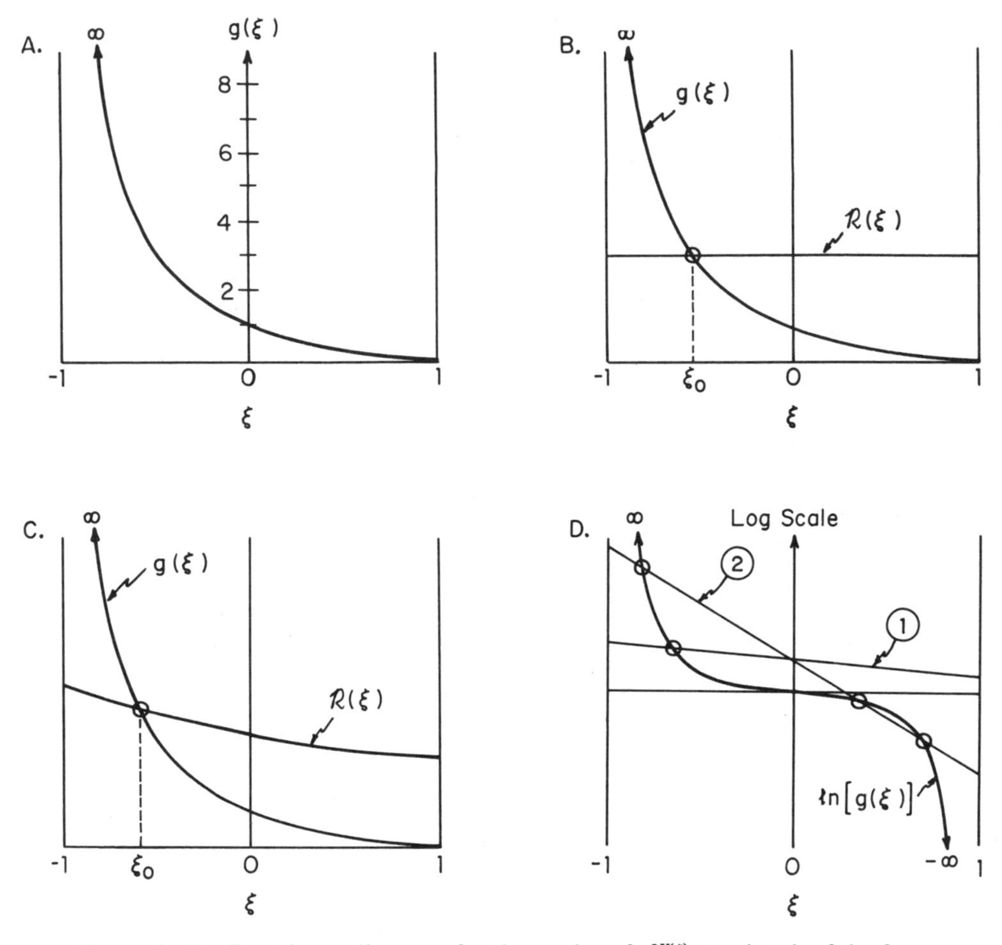

Figure 4–16 Root locus diagrams for the modes of $e^{2V(\xi)}$. *A*, sketch of the function $g(\xi) = (1 - \xi)/(1 + \xi)$. *B*, location of the mode for transition rates v_{ij} = constant. *C*, location of the mode for v_{ij} only weakly dependent on ξ. *D*, mode bifurcation in the exponential trend-watcher model of Eqs. (4-36) and (4-37). For $a_1 \ll 1$, meaning only weak dependence of the epigenetic rules on ξ, there is one root corresponding to one interior mode (*line 1*). For a_1 large (*line 2*), there are three roots and a bifurcation of $e^{2V(\xi)}$ to bimodal structure.

generalized from the one studied by Weidlich (1972), where $a_j \geq 0, j = 1, 2, 3$ and from Eq. (4-35)

$$\mathcal{R}(\xi) = (a_3/a_2)e^{-2a_1\xi}. \tag{4-37}$$

Plotted on a natural log scale, $g(\xi)$ is an antisymmetric function and this

$\mathcal{R}(\xi)$ is a straight line with slope $-2a_1$ and intercept $\ln(a_3/a_2)$ (see Figure 4–16D). For small values of a_1, $\mathcal{R}(\xi)$ and $g(\xi)$ have exactly one point of intersection; but as a_1 increases, the slope of $\ln \mathcal{R}(\xi)$ becomes increasingly negative until finally $\mathcal{R}(\xi)$ cuts $g(\xi)$ in three places. The factor $e^{2V(\xi)}$ becomes bimodal. This is a specific instance of the mode bifurcation process illustrated qualitatively in Figure 4–15.

In the space of ethnographic distributions $P(\xi)$, the sets containing distributions that have a specific number of modes form natural families, all distinct. As the $v_{ij}(\xi)$ change through the action of learning, culture shift, and natural selection, $P(\xi)$ moves both over the membership of a set and between sets. In our models the $v_{ij}(\xi)$ contain adjustable parameters $a_1, \ldots, a_M$ for characterizing these changes—see for example Eq. (4-36)—so that $v_{ij}(\xi) = v_{ij}(\xi|a_1, \ldots, a_M)$, and a natural coordinate system is provided for each set of functions v_{ij} that differ only in the values of $a_1, \ldots, a_M$. Uniquely specifying $\mathbf{a} = (a_1, \ldots, a_M)$ picks out a unique member of the set $\{v_{ij}(\xi|\mathbf{a})|\mathbf{a} \in \mathcal{S} \subset \mathbb{R}^M\}$.

Small changes in the parameters $\mathbf{a}$ or in the analytic form of the $v_{ij}(\xi|\mathbf{a})$ either shift $P(\xi)$ locally within the set of unimodal or multimodal distributions, or take the ethnographic curve from one such set to another. Changes of the latter kind are qualitative and we term them *transition thresholds* for the ethnographic curve. Depending on the structure of the underlying epigenetic rules and the constraints on the parameters $\mathbf{a}$, a culture group can have one or more transition thresholds beyond which its ethnographic distribution takes on qualitatively new properties, an example being two modes rather than one. In most cases of interest here, the subsets of the values a_m associated with the distributions having a specific number of modes are continuous subintervals of $\mathbb{R}$, the transition thresholds forming borders between them. Figure 4–17 provides an example of this phenomenon occurring in a population of exponential trend watchers.

Two general conclusions concerning gene-culture translation can be immediately anticipated on the basis of the general translation equations (4-22) and (4-28) and the graphical models just used to evaluate (4-28). The first is that the epigenetic rules and their associated bias curves influence the positions and heights of the ethnographic curves, because they are effective in moving areas around under the $Q(\xi)$ and $X(\xi')/Q(\xi')$ curves and determine the changes in sign of the $X(\xi')/Q(\xi')$ values. The second generalization is that amplification is likely to be powerful, because of the group size, the couplings between the behavior of group members, and the exponential position of $V(\xi)$. A small change in the epigenetic rules,

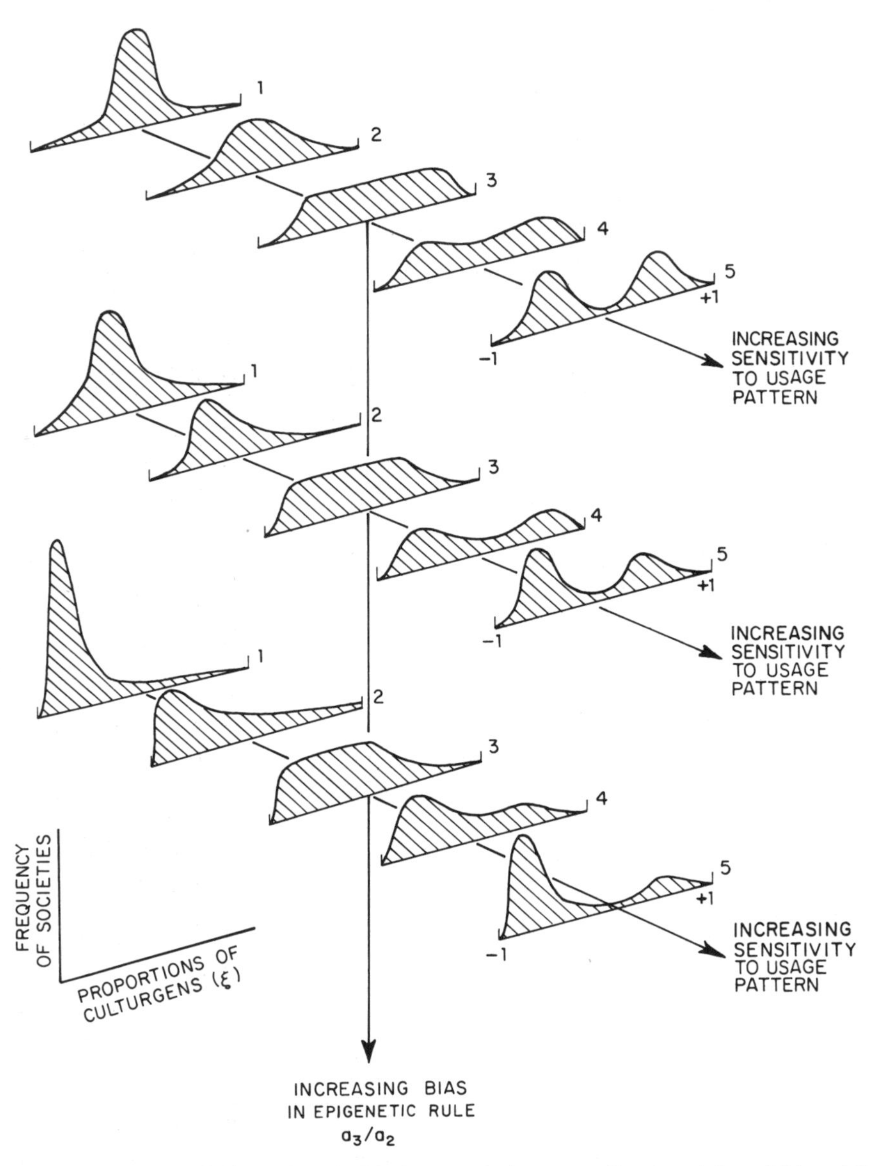

Figure 4–17 Transition thresholds between the sets of unimodal and bimodal ethnographic distributions that occur in the exponential trend-watcher model, Eq. (4-36). For each bias ratio a_3/a_2 there is a transition threshold a_1^*. For $a_1 > a_1^*$ the ethnographic curve is bimodal; for $a_1 < a_1^*$ it is unimodal. The curves are based on the evaluation of Eq. (4-22) for $0 < a_1 < 2$ and $1 \leq a_3/a_2 \leq 5$. (Modified from Lumsden and Wilson, 1980a.)

reflected in the transition probabilities of culturgen choice $v_{ij}(\xi)$, can create a large change in the ethnographic curve. The extent of this amplification will be quantified in the sections that follow.

Human Assimilation Functions

The next step is to evaluate more explicitly the relation between the epigenetic rules, social influence on decision making, and ethnographic distribution. In this section we introduce the assimilation function, which specifies the influence exerted on individual choice of culturgens by the decisions already made by other members of the society. A wide range of possible forms of the function is considered, and the great sensitivity of ethnographic distribution to changes in innate bias and social responsiveness, first deduced in the previous section, will be given rigorous form. We shall derive an amplification law, which poses a simple relation between the sensitivity of the ethnographic distribution and the magnitude of the innate bias in individual choice behavior. We shall also review actual cases of behavioral development that permit estimates of innate bias and the first predictions of the gene-culture process in human beings.

In developing more explicit models, the relation we seek is in effect the gene-culture translation process, T, which maps epigenetic rules and programs of cognitive development into cultural patterns. As Figure 4–18A illustrates, the translation process can be conveniently visualized in geometric form. An objective of gene-culture theory is to estimate the magnitude of the impact that changes in the epigenetic rules, acting through translation, have on the ethnographic distribution (Figure 4–18B).

For convenience the relation $v_{ij}(\xi)$ will be called an *assimilation function;* it gives rise to an assimilation curve, v_{ij} plotted against ξ. Few data exist that can be utilized to characterize real assimilation functions directly. It is therefore useful to develop a summary model of the information-processing stages that produce active assimilation functions, and with it explore a wide range of possible forms for the response of epigenetic rules to patterns of culturgen usage. The model, which is in accord with the principles of human decision and choice psychology (Newell and Simon, 1972; Lindzey et al., 1975), considers that information extracted from the individual analysis of culturgen usage patterns is utilized to revise or update assimilation biases that would otherwise be employed. The values that characterize both the raw biases and the updated assimilation functions are shaped by the epigenetic rules. The

A.

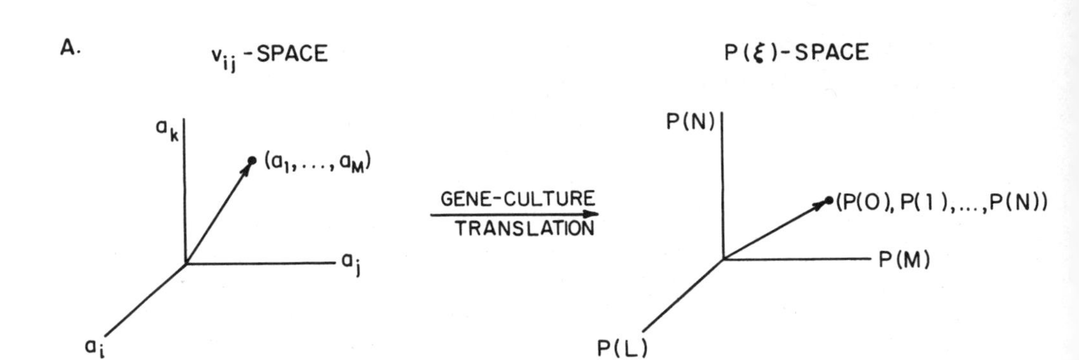

v_{ij}-SPACE
a_k
$(a_1, \ldots, a_M)$
a_j
a_i
GENE-CULTURE
TRANSLATION
P(ξ)-SPACE
P(N)
(P(O), P(1), ..., P(N))
P(M)
P(L)

B.

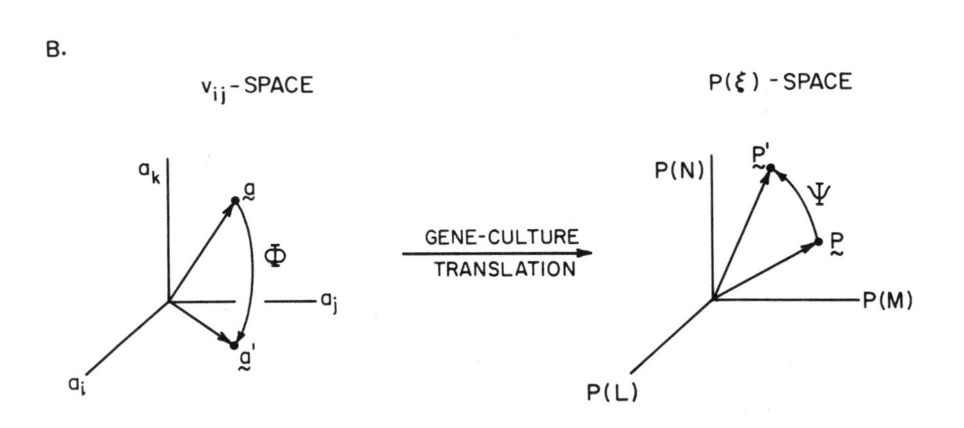

v_{ij}-SPACE
a_k
$\tilde{a}$
Φ
a_j
$\tilde{a}'$
a_i
GENE-CULTURE
TRANSLATION
P(ξ)-SPACE
P(N)
$\tilde{P}'$
Ψ
$\tilde{P}$
P(M)
P(L)

C.

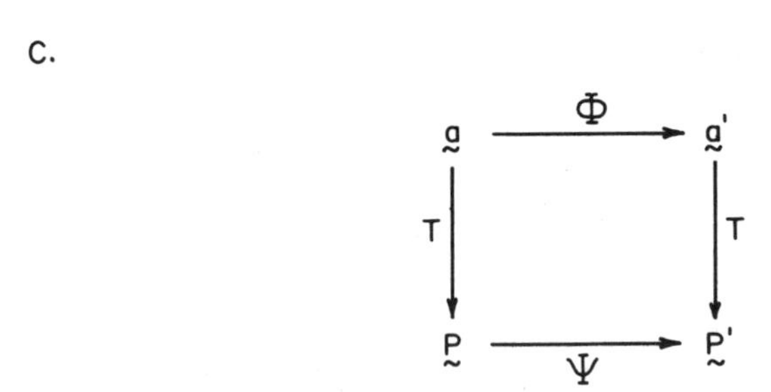

$\tilde{a}$
Φ
$\tilde{a}'$
T
T
$\tilde{P}$
Ψ
$\tilde{P}'$

scheme is outlined in Figure 4–19A and its mathematical model is given in Figure 4–19B. The usage analysis stage assigns updating functions $g_{ij}(\xi|\mathbf{a})$ and $h_i(\xi|\mathbf{a})$, together with specific values of the parameters $\mathbf{a} = (a_1, \ldots, a_M)$, such that the revised and activated versions u_{ij} of the "raw" or "innate" biases u_{ij}^0 are

$$u_{ij}(\xi|\mathbf{a}) = u_{ij}^0 g_{ij}(\xi|\mathbf{a}) \tag{4-38a}$$

giving for $i \neq j$ the transition rates (recall Eq. 4-5)

$$v_{ij}(\xi|\mathbf{a}) = r_i(\xi|\mathbf{a})u_{ij}(\xi|\mathbf{a}) = [r_i^0 h_i(\xi|\mathbf{a})][u_{ij}^0 g_{ij}(\xi|\mathbf{a})]$$

$$\triangleq r_i^0 u_{ij}^0 f_{ij}(\xi|\mathbf{a}) \triangleq v_{ij}^0 f_{ij}(\xi|\mathbf{a}). \tag{4-38b}$$

The categories of assimilation functions (4-38b) that we have considered are shown in Figure 4-20. The most elementary constant function (v_{ij} insensitive to ξ) and the step functions, if combined with high selectivity, will result in relatively uniform, easily predictable species-specific traits. They are most likely to occur during early infancy, when the most robust epigenetic rules direct behavior to certain limited and virtually inevitable choices. In Chapters 2 and 3 we documented possible examples in the vocabulary of color perception, phoneme formation, and several aspects of mother-infant bonding. Such ξ-insensitive rules are also likely to direct behavior under circumstances of low evolutionary transparency at any stage of a life history. Later we shall apply the constant function directly to the ethnography of brother-sister incest. The monotonic, "trend-

Figure 4–18, opposite The gene-culture translation process. The translation equations (4-22) and (4-28) map epigenetic rules v_{ij} into ethnographic distributions $P(\xi)$. Hence there is a natural association between the space of conceivable $v_{ij}(\xi|\mathbf{a})$ rules and the space of ethnographic distributions $P(\xi)$. *A*: The translation process is shown for $v_{ij}(\xi|\mathbf{a})$ differing only in the parameter values $\mathbf{a} = (a_1, \ldots, a_M)$. Every point $\mathbf{a}$ in the multidimensional space $\mathbf{R}^M$, of which a three-dimensional cross-section is shown, labels a unique epigenetic rule. Similarly, the translation equation (4-22) yields ethnographic probabilities $P(0), P(1), \ldots, P(N)$, which form the natural components of a vector $\mathbf{P} = (P(0), P(1), \ldots, P(N))$ in the space $\mathbf{R}^{N+1}$. *B*: When an event, symbolized as Φ, changes the epigenetic rules from $\mathbf{a}$ to $\mathbf{a}'$, the ethnographic distribution changes from $\mathbf{P}$ to $\mathbf{P}'$. Hence the change Φ: $\mathbf{a} \mapsto \mathbf{a}'$ induces a change Ψ: $\mathbf{P} \mapsto \mathbf{P}'$. A key question is how far $\mathbf{P}$ moves when $\mathbf{a}$ is changed. *C*: The maps Φ, Ψ, and T form a commutative system, where T is the gene-culture translation operator realized by Eqs. (4-22) and (4-28).

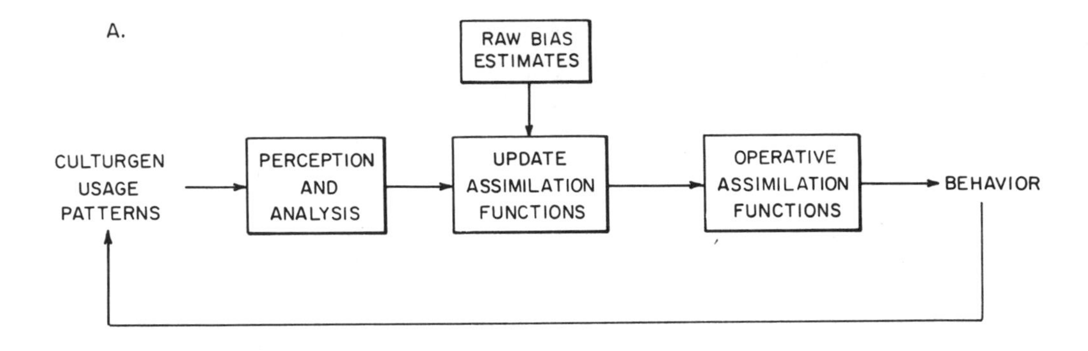

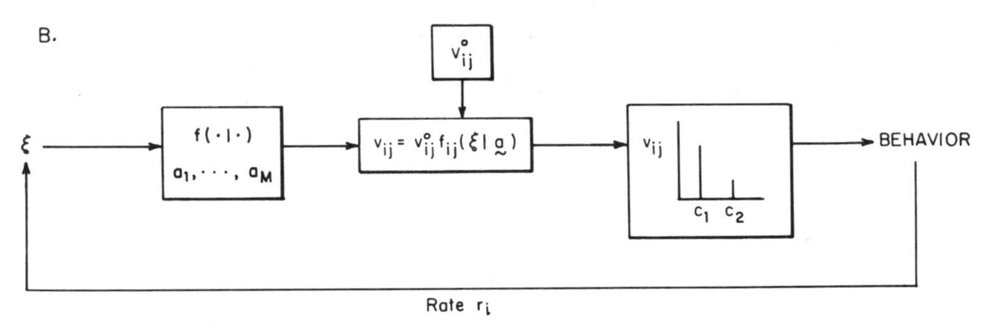

Figure 4–19 Summary of the principal stages of cognitive information processing that lead up to active decision functions.

watching'' cases are also very apt to occur. The existence of curves of this general form is implied by empirical data on intragroup cultural diffusion in Western industrial societies (Haggett, 1972; Hamblin et al., 1979).

At the small-group level, experiments by Asch (1951) showed that if a unanimous majority takes an unusual position contradicted by the evident facts, the percentage of people who prefer to follow rather than be a minority of one rises with group size. In Asch's arrangement conformity leveled off at 30–40 percent when the number of persons in the majority group reached about five. Milgram and associates (1969) obtained similar results in an experiment on the drawing power of crowds. Groups of collaborators of varying size met on a crowded New York street and on signal looked up at a nearby tall building. The numbers of passersby who also looked up were then recorded. The contagion followed a pattern that could fit either of the two trend-watcher curves in Figure 4–20: 4 percent

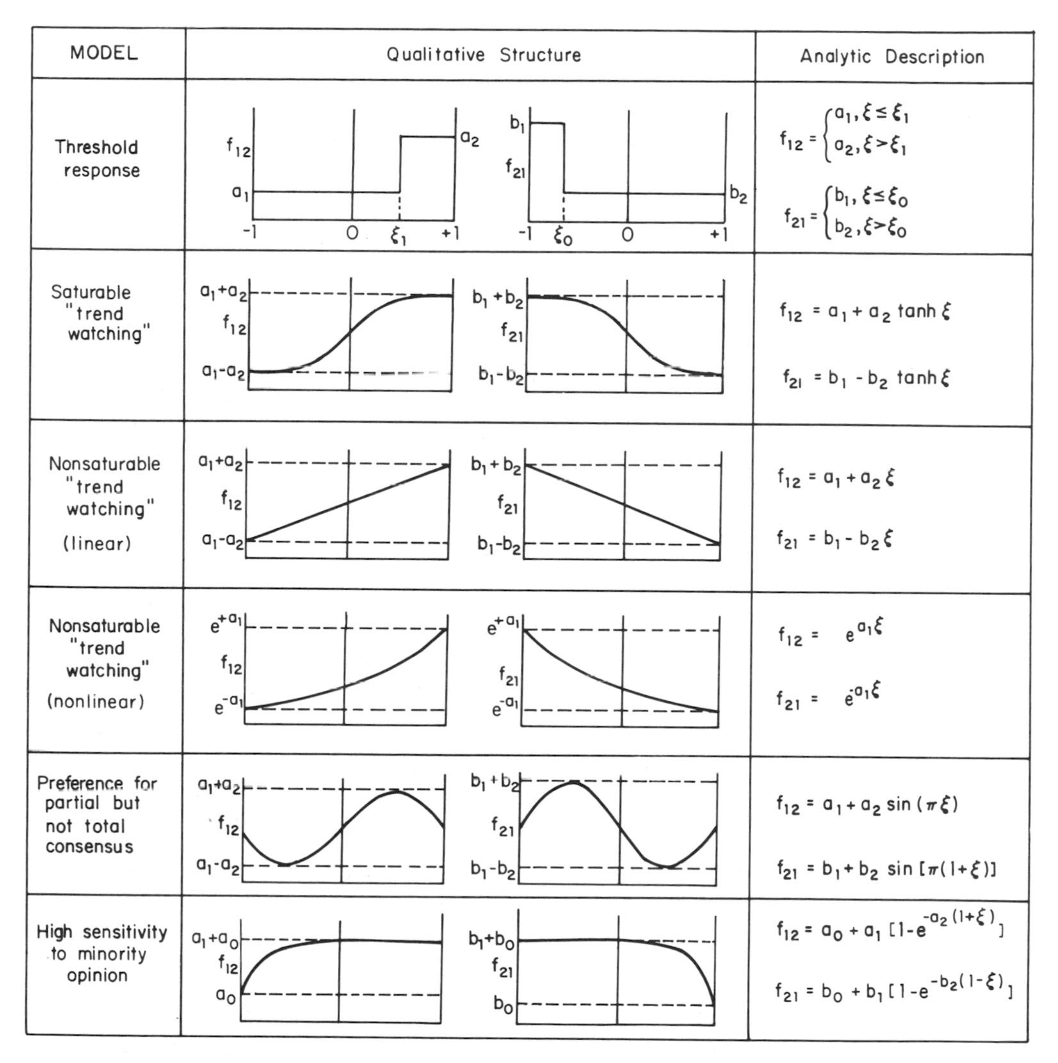

Figure 4–20 The assimilation functions, or dependence of individual transition rates v_{ij} on the usage pattern of other members of the society. (Modified from Lumsden and Wilson, 1980a.)

of the crowd followed when one person in the collaborator group looked up, 16 percent followed five, 22 percent followed ten, and 40 percent followed fifteen.

The gene-culture translation model reveals that ethnographic distributions are sensitive to changes in the assimilation functions; in the geometric picture developed in Figure 4–18, even small displacements Φ in the space of epigenetic rules can produce large displacements Ψ in the space of ethnographic distributions. Documentation was provided in the previous section of the qualitative changes in the structure of $P(\xi)$ that accompany small changes in the sensitivity of the $v_{ij}(\xi)$ to usage patterns and take the ethnographic distributions across transition thresholds. A further important question concerns the general sensitivity of $P(\xi)$ to changes in the genetic constitution of the society members that affect the magnitude of the raw biases u_{ij}^0 and v_{ij}^0. Since evolutionary steps of this type are expected to be small, are they amplified in the process of gene-culture translation, or are they washed out by the culture dependence of the updating process?

The answer is that they are amplified and have dramatic impact on the ethnographic distributions. In order to demonstrate this relation clearly, we take advantage of the structure of the exact translation equations (4-22) and consider the four-way comparison shown in Figure 4–21. The $P^{(0)}[\xi(n_1)]$ and $P^{(1)}[\xi(n_1)]$ are two ethnographic curves; the members of the culture group characterized by $P^{(1)}$ have innate biases v_{ij}^0, while the individuals underlying the distribution $P^{(0)}$ have innate biases v_{ij}^{0*}. Both

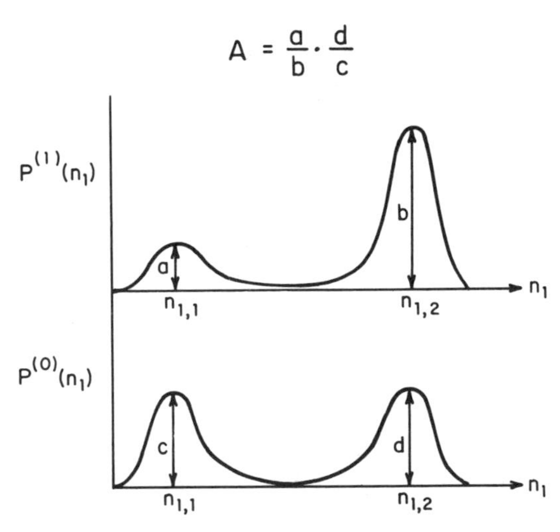

Figure 4–21 Definition of the amplification factor A for innate epigenetic biases. Note that the values of $n_{1,1}$ and $n_{1,2}$ are variable and have been arbitrarily placed under the modes of the ethnographic curves solely for purposes of clarity. Furthermore, the definition of A can be applied to ethnographic curves of arbitrary shape.

populations are steady state, of size N, and have the same repertory of updating functions $f_{ij}(\xi|\mathbf{a})$. In other words, the members of the two culture groups differ only in the values of their innate biases.

Define the amplification factor A as the ratio of the relative magnitudes of $P^{(0)}[\xi(n_1)]$ and $P^{(1)}[\xi(n_1)]$ at two different values of n_1, say $n_{1,1}$ and $n_{1,2}$:[4]

$$A \triangleq \frac{P^{(1)}(n_{1,1})/P^{(1)}(n_{1,2})}{P^{(0)}(n_{1,1})/P^{(0)}(n_{1,2})}. \tag{4-39}$$

Thus A is a relative measure of the difference in the structure of $P^{(0)}$ and $P^{(1)}$. As Figure 4–21 shows, our selection of a two-point comparison allows this measure to express differences in the overall structure of the two distributions. And A appears to be a useful quantifier of changes relative to a preselected reference curve that are induced by changes in the values of the v_{ij}^0. Evaluating Eq. (4-39) through employment of the cognitive model (4-38) and the exact expression (4-22) for the ethnographic distribution yields

$$A = \left[\frac{v_{21}^0}{v_{12}^0}\frac{v_{12}^{0*}}{v_{21}^{0*}}\right]^{n_{1,1}-n_{2,1}} \tag{4-40}$$

A family of natural reference curves is the set associated with $v_{12}^{0*} = v_{21}^{0*}$, namely cultures of innately unbiased organisms. Then v_{12}^{0*}/v_{21}^{0*} equals unity and the amplification equation simplifies to

$$\begin{aligned} A &= (v_{21}^0/v_{12}^0)^{n_{1,1}-n_{1,2}} \\ &= (r_2^0/r_1^0)^{n_{1,1}-n_{1,2}}(u_{21}^0/u_{12}^0)^{n_{1,1}-n_{1,2}}. \end{aligned} \tag{4-41}$$

Let the difference $n_{1,1} - n_{1,2}$ be denoted by δ and the ratio (v_{21}^0/v_{12}^0) of innate biases by $\mathscr{R}_0$—recall Eq. (4-32); then the *amplification law* (4-41) is concisely expressed:

$$A = (\mathscr{R}_0)^\delta. \tag{4-42}$$

When the values of the reentrance rates r_1^0 and r_2^0 are equal, $\mathscr{R}_0$ is given by the ratio of the innate decision likelihoods u_{21}^0 and u_{12}^0. A net measure of

4. Here we use the integer variables appropriate to the discrete distribution (4-22) and recall that $\xi = \xi(n_1) = 1 - 2n_1/N$, where n_1 is the number of c_1-users.

overall relative difference is given by

$$A_{\text{net}} = \sum_{n_{1,1}, n_{1,2}} A. \tag{4-43}$$

Since at least as far back as late Pleistocene times, gene-culture coevolution and translation have occurred in hunter-gatherer bands consisting of fifteen to seventy-five individuals (Wobst, 1974; Hage, 1976; Buys and Larson, 1979). The exponentiation process explicit in (4-42) then permits even slight differences in v_{ij}^0 relative to v_{ij}^{0*} to be greatly amplified and hence to have large effects on the ethnographic curve. Isoclines of A are shown in Figure 4–22 for a group of twenty-five individuals. The effects for larger groups would be exponentially greater.

The amplification law expressed in Eqs. (4-40) to (4-42) is universal in the sense that it covers all organisms whose assimilation functions are approximated by the general model (4-38). The observable consequences for values of A very different from unity will hinge on the structure of the reference distribution $P^{(0)}$. The factor A departs most strongly from unity when the comparison points $n_{1,1}$ and $n_{1,2}$ are widely separated and yield large values of $|n_{1,1} - n_{1,2}|$. This will place either or both $n_{1,1}$ and $n_{1,2}$ in the wings of the ethnographic distribution for groups of hunter-gatherer size. An example of this circumstance can be seen in Figure 4–23. The group is composed of exponential trend watchers with $r_1^0 = r_2^0$. The left panels of the figure illustrate a reference state $P^{(0)}[\xi(n_1)]$ with $v_{12}^{0*} = v_{21}^{0*}$ and the right panels $P^{(1)}[\xi(n_1)]$. Below the transition threshold the ethnographic curves decrease rapidly for $|\xi| > 0$, and the result of a small change in the v_{ij}^0 is qualitatively unnoticeable (upper panels). Beyond the transition threshold, however, $P^{(0)}[\xi(n_1)]$ has substantial mass to either side of $\xi = 0$. The effect of a mere 1-percent change in each v_{ij}^0, for which $A = 1.5$, is then sufficient to give a macroscopic shift in the structure of $P^{(1)}[\xi(n_1)]$ (middle and lower panels).

The amplification factor A obeys a concise and intuitive law (4-42) and is a *relative* measure based on proportions instead of absolute differences. Additional information about the significance of gene-culture amplification is provided by measures that are large when the *absolute* changes in $P(\xi)$ are large. We can begin to quantify the latter type of change directly and include perturbations in the updating functions $f_{ij}(\xi|\mathbf{a})$ by equipping the spaces of epigenetic rules and ethnographic distributions (Figure 4–18) with metric or distance functions. Each of these spaces can be

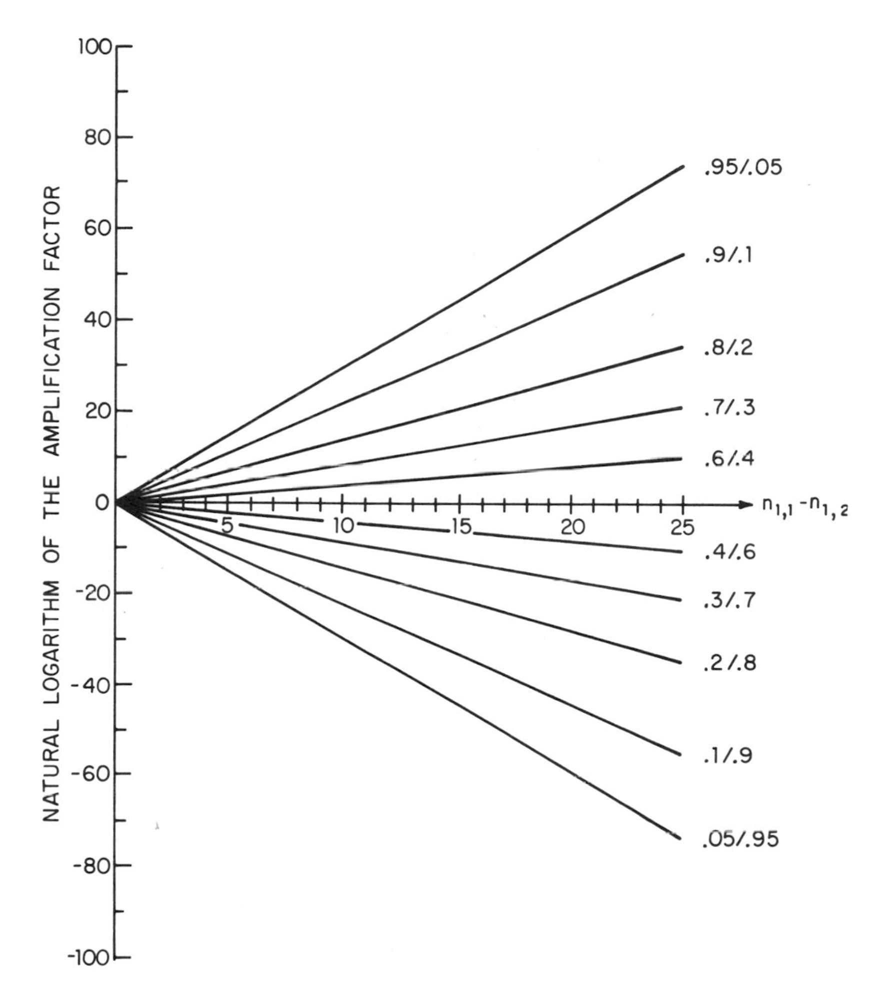

Figure 4–22 The amplification factor A, for a wide range of bias ratios v^0_{21}/v^0_{12}. The isoclines are labeled by the decision ratio u^0_{21}/u^0_{12} assuming $r^0_1 = r^0_2$. Note that the ordinate scale is natural logarithmic. To obtain the A corresponding to $n_{1,1} - n_{1,2}$ in the range $-1, -2, \ldots, -25$, reflect the present diagram around the abscissa axis. The isoclines are calculated for a group of twenty-five individuals.

equipped with the natural geometry of $\mathbf{R}^K$, where $K = M$ in the epigenetic space and $K = N + 1$ in ethnographic space. Let $\mathbf{x}_1$ and $\mathbf{x}_2$ be two points in $\mathbf{R}^K$; then we say that the *distance between* $\mathbf{x}_1$ *and* $\mathbf{x}_2$ is

$$d(\mathbf{x}_1, \mathbf{x}_2) \triangleq \langle \mathbf{x}_1 - \mathbf{x}_2 | \mathbf{x}_1 - \mathbf{x}_2 \rangle^{1/2} \tag{4-44}$$

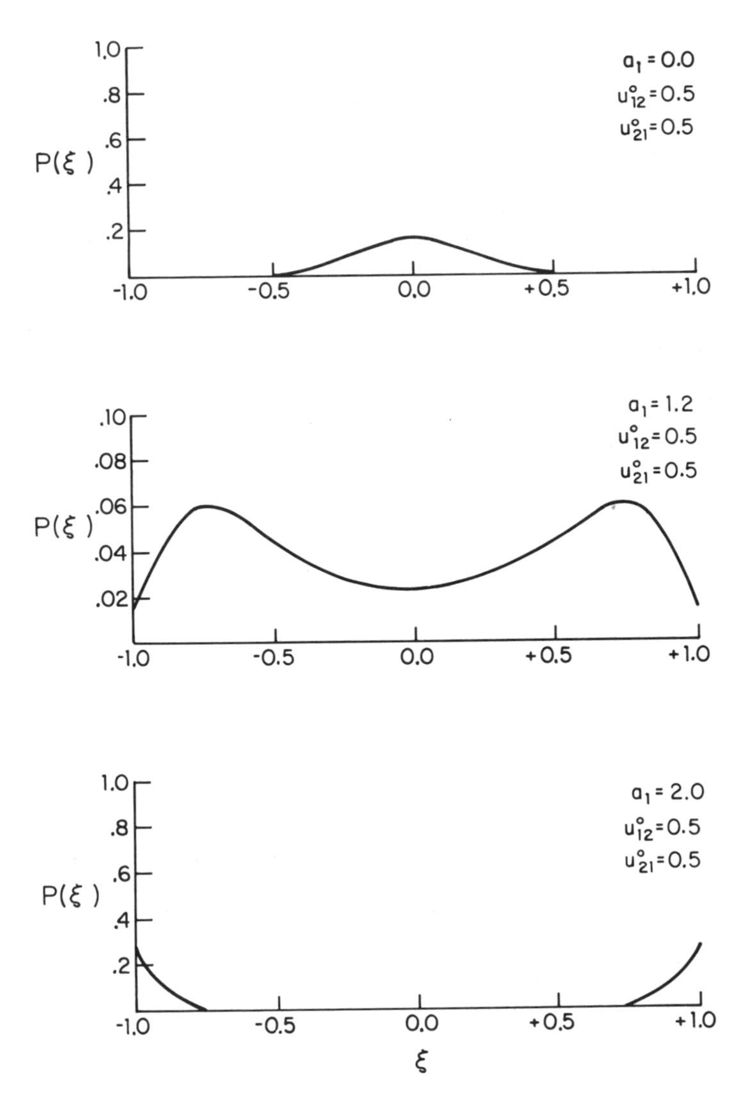

Figure 4–23 The amplifying effect of epigenetic rules on social pattern. This series of ethnographic curves is based on the exponential trend-watcher case, where the probabilities of a change in culturgens are $v_{12} = r_1^0 u_{12}^0 e^{a_1 \xi}$ and $v_{21} = r_2^0 u_{21}^0 e^{-a_1 \xi}$ respectively, with $r_1^0 = r_2^0 = 1$. Above the transition threshold a_1^*, $e^{2V(\xi)}$ bifurcates, and $P(\xi)$ has two readily discernible modes for the threshold $a_1^* \sim 1.0$ when u_{12}^0 and u_{21}^0 are such that $|u_{21}^0/u_{12}^0|$ is on the order of unity. *Left panels:* The $P(\xi)$ series shows the effect of increasing sensitivity to social ambience (culturgen

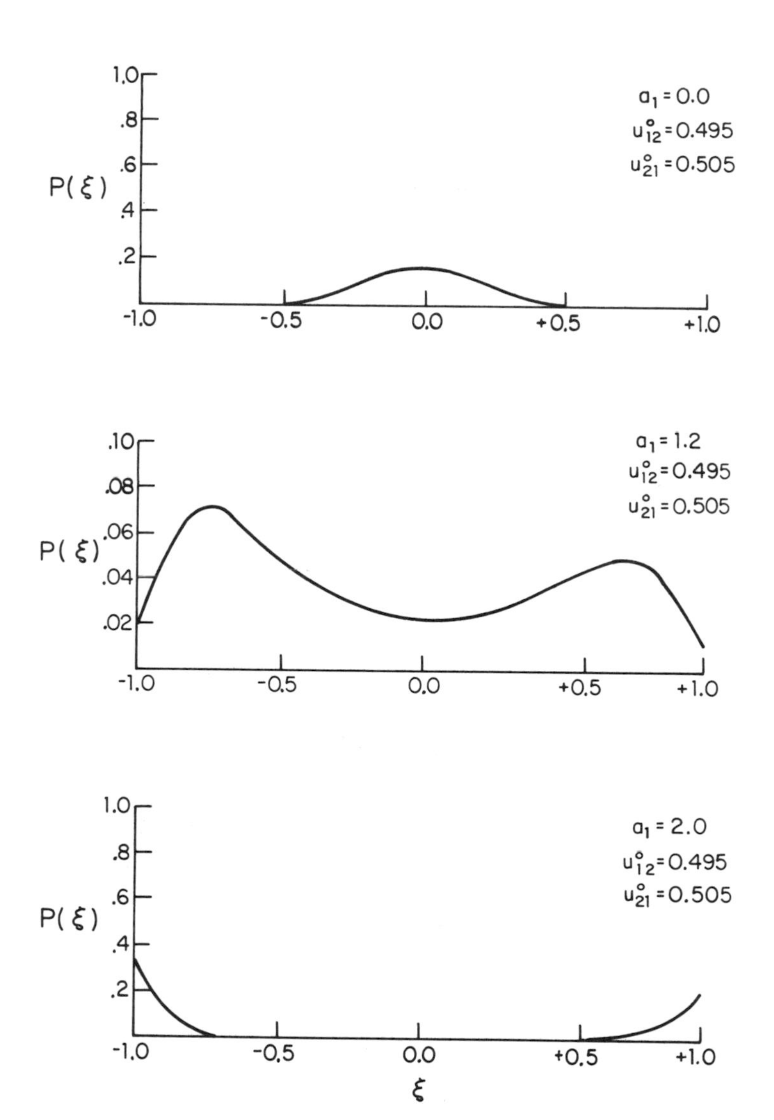

choices by other persons) at zero intrinsic bias ϕ, that is, in the case where $\phi \triangleq u^0_{12} - u^0_{21} = 0$ and $\mathscr{R}_0 = 1$. The $P(\xi)$ curve remains symmetric around $\xi = 0$. *Right panels:* The $P(\xi)$ series shows the qualitative effect of a very small ϕ ($= -0.01$) and an $\mathscr{R}_0 = 1.02$ both below and above the transition point. The ethnographic curves were calculated from Eq. (4-22) in the case of a group of twenty-five individuals. (Modified from Lumsden and Wilson, 1980a.)

where $\langle \cdot | \cdot \rangle$ is the usual inner product

$$\langle \mathbf{x} | \mathbf{y} \rangle \triangleq \sum_{k=1}^{K} x_k y_k . \tag{4-45}$$

The standard definition supplied in Eq. (4-44) merely generalizes the notions of Euclidean distance, familiar from two and three dimensions, to real vector spaces of arbitrary but finite dimension. When the epigenetic rules are shifted through a distance

$$d(\mathbf{a}, \mathbf{a}') = \langle \mathbf{a} - \mathbf{a}' | \mathbf{a} - \mathbf{a}' \rangle^{1/2} = \left[\sum_m (a_m - a'_m)^2 \right]^{1/2} , \tag{4-46a}$$

the ethnographic distribution moves a distance

$$d(\mathbf{P}, \mathbf{P}') = \langle \mathbf{P} - \mathbf{P}' | \mathbf{P} - \mathbf{P}' \rangle^{1/2} = \left[\sum_{n_1} (P(n_1) - P'(n_1))^2 \right]^{1/2} \tag{4-46b}$$

The analog for an ethnographic curve $P(\xi)$ with continuous ξ is

$$d(P, P') = \left[\int_{-1}^{1} [P(\xi) - P'(\xi)]^2 d\xi \right]^{1/2} \tag{4-46c}$$

Unlike A, $d(\mathbf{P}, \mathbf{P}')$ is not naturally suited to the structure of $P[\xi(n_1)]$, and concise analytic results are not available. However, the distance functions are readily evaluated numerically so it is possible to map out the Euclidean distance moved by an ethnographic curve when its epigenetic rules, specified by $\mathbf{a}$, are shifted by a given amount. Rules having many parameters $a_1, \ldots, a_M$, $M \gg 1$ require an epigenetic space of high dimension and produce a cumbersome mapping problem. Here we will use examples with two or at most three parameters. The findings are illustrated in Figure 4–24. For a hunter-gatherer group of size twenty-five, the ethnographic distribution space is twenty-six dimensional.

On the whole the values for Eq. (4-46b) appear more modest than those for the amplification factor A. This is to be expected since $d(\mathbf{P}, \mathbf{P}')$ is composed of arithmetic differences between quantities $P(n_1)$ and $P'(n_1)$, which are necessarily bounded between zero and unity. The amplification factor A, in contrast, can be arbitrarily large or small depending on the ratio $\mathcal{R}_0$ of the innate biases. The distance shifts documented in Figure 4–24 are placed in perspective by noting that the distance between the two most extreme conceivable forms of ethnographic curve, which we

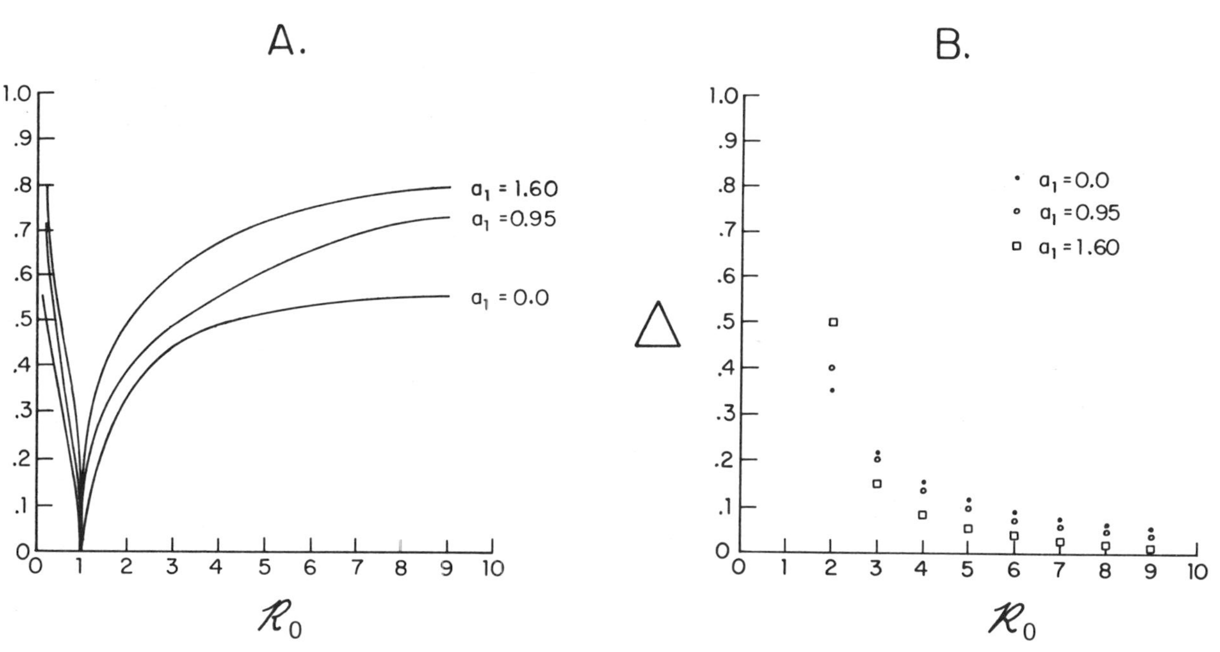

Figure 4–24 Euclidean distances in a space containing the ethnographic distributions, (Eq. 4-22). The calculations are illustrated for exponential trend watchers with $v_{ij} = u_{ij}^0 e^{\pm a_1 \xi}$. Note that the value $a_1 = 0$ corresponds to the ξ-independent epigenetic rules v_{ij} = constant. *A* shows the distance d of ethnographic distributions, each determined by a specific value of $\mathcal{R}_0 = u_{21}^0/u_{12}^0$, from the reference distribution with $\mathcal{R}_0 = 1.0$ (tabula rasa state). The absolute rate of displacement $|\partial d/\partial \mathcal{R}|$ is very high initially, but tapers off as the distributions approach the vertices of the unit simplex (see also Figure 4–25). *B* shows the distances between adjacent distributions on the representative orbit $u_{12}^0 = 0.1$, $u_{21}^0 = 0.1(1 + k)$, $k = 1, \ldots, 9$. Thus Δ is the distance between the distributions with $u_{21}^0 = 0.1(1 + k)$ and $u_{21}^0 = 0.1k$. The results for the other models of Figure 4–20 are similar.

call $P^*(n_1)$ and $P^{**}(n_1)$, is only

$$d(\mathbf{P}^*, \mathbf{P}^{**}) = \sqrt{2} \sim 1.414 \tag{4-47}$$

where

$$P^*(n_1) = \begin{cases} 1, & n_1 = 0 \\ 0, & n_1 = 1, 2, \ldots, N \end{cases} \tag{4-48a}$$

and

$$P^{**}(n_1) = \begin{cases} 0, & n_1 = 0, \ldots, N - 1 \\ 1, & n_1 = N \end{cases} \tag{4-48b}$$

(Figure 4–25A). The set of realizable curves forms a fairly tight cluster in its space $\mathbf{R}^{N+1}$, occupying the diagonal face of the unit simplex (Figure 4–25B). Thus the distances of 0.05 and larger accompanying small absolute changes in the $v_{ij}(\xi|\mathbf{a})$ cover a significant amount of the "territory" accessible to $\mathbf{P}$ in $\mathbf{R}^{N+1}$. These results confirm again the sensitivity of the ethnographic distribution to the structure of the epigenetic rules.

Although the values attained by the amplification and distance factors are remarkable, the estimates provided by Eqs. (4-42) and (4-46) are likely to be conservative. In real cognitive systems the network of epigenetic rules is expected to be reticulate (recall Figure 4–1) more often than it is univalent as idealized here. Changes in any single v_{ij}^0 then affect the operation of not just one, but many different epigenetic subsystems. Furthermore, learning and long-term semantic memory, guided by a reinforcement processor tuned to selective advantage (see Chapter 3), will amplify the raw biases v_{ij}^0 themselves over time, leading to increasingly large amplifications and to significant changes in the ethnographic curves as the pull of the genetic leash translates upward to the cultural level. Rigorous theoretical study of these more complex effects in gene-culture translation for the time being remains an open problem.

In summary, *a barely detectable amount of selectivity in an epigenetic rule operating during the behavior of individuals can strongly alter social patterns*. Even when the assimilation functions are relatively insensitive to social patterns, small differences in the $v_{ij}(\xi)$ still have a substantial qualitative impact on the ethnographic curves. Many of the disparities depicted in the illustrations are large enough to be readily apparent in thorough ethnographic surveys. They demonstrate the powerful but hitherto largely unappreciated linkage we believe to exist between the behavior patterns that have been the principal concern of the experimental psychologists and the social patterns that are traditionally studied by anthropologists, sociologists, and economists.

Are the human epigenetic rules sufficiently specific to create such

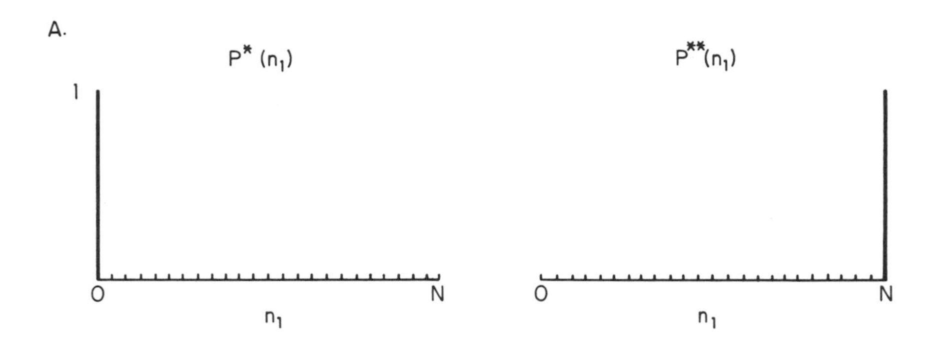

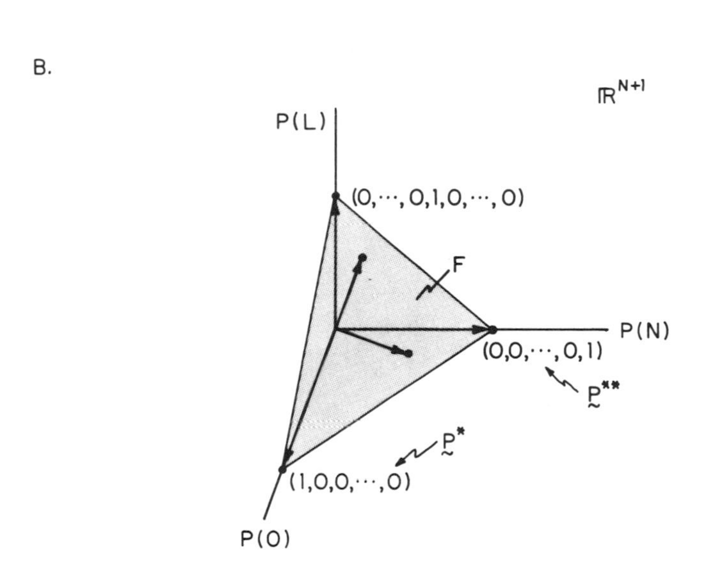

Figure 4–25 *A:* The two extreme conceivable ethnographic distributions. Under $P^*(n_1)$ all cultures are 100-percent c_2-users with unit probability. The opposite is the case for $P^{**}(n_1)$. *B*: Because $\Sigma_{n_1} P(n_1) \triangleq 1$, the vectors $\mathbf{P}$ of the ethnographic distributions in $\mathbb{R}^{N+1}$ are restricted to the diagonal face F of a tetrahedron with apex on the origin and sides of unit length. This body is the unit simplex in $\mathbb{R}^{N+1}$. The distributions $P^*(n_1)$ and $P^{**}(n_1)$ correspond to points $\mathbf{P}^*$ and $\mathbf{P}^{**}$ on vertices of the simplex.

marked constraints on the cultural patterns? The answer is that they often are; in some cases they exceed the marginal levels just demonstrated by one or two orders of magnitude. In Table 4–1 we have summarized the cases known to us in which approximate values of relative assimilation probabilities can be estimated from experimental data. These examples are limited principally to initial enculturation, but it is reasonable to suppose that the innate biases carry over into later transitions between the

Table 4–1 Estimates of innate preferences by young children. The preferred culturgen is arbitrarily designated c_1, and the estimated probability of choice of this culturgen over c_2 is denoted u_{21}. (Modified from Lumsden and Wilson, 1980a.)

Alternative responses	Age at first response	u_{21}/u_{12}	Later social effects	Source
Preference for sugar (sucrose, fructose, glucose, lactose)	Newborn	0.6/0.4 to 0.8/0.2 according to sugar and at 0.2–0.3 M	Sugar preference extends at least into childhood and influences adult cuisine	Maller and Desor (1974), Chiva (1979)
Discrete four-color classification over continuous or other discrete classifications	4 months	Approaches 1.0/0 in fully color-sighted persons	Linguistic color classifications across cultures can be mapped from one-to-one to three-to-one onto the four categories	Ratliff (1976), Bornstein (1979), Wattenwyl and Zollinger (1979)
Preference for schematic pattern of human face over similar designs	Newborn	≥0.51/0.49 to ≥0.6/0.4 according to design	Long term focusing on face, especially eyes; facilitation of parent-offspring bonding, and perhaps later forms of interpersonal bonding	Freedman (1974)
Preference for intermediate complexity in visual design: approximately 10 turns in figures as opposed to 5 or 20 turns	Newborn	≥0.55/0.45	Followed by comparable degree of preference in school-age children; adults also prefer intermediate complexity, with a redundancy of about 20 percent	Hershenson et al. (1965), Young (1978)
Fear response to strangers	8 months	>0.75/0.25	Possibly contributes to early group distinctions and hostility to strangers and out-groups by children and adults	Hess (1973), Argyle and Cook (1976), Eibl-Eibesfeldt (1979)
Infant holding: women carry babies on left side; men carry them at random on left or right	Adult; possibly traceable to sex differences in object-carrying behavior that extends back to preadolescence	0.6/0.4 to 0.7/0.3 according to age of infant	Proximity to heartbeat soothes infant, possibly facilitates mother-infant bonding	Salk (1973), Lockard et al. (1979)

culturgens. This persistence of bias is in fact well marked in the cases of sugar preference, color classification, and infant holding.

There have been relatively few such developmental studies, and the responses have not been investigated with reference to their dependence on the behavior of the rest of the society. However, the examples cited are not likely to be very flexible in this regard, because they are either primary in nature, entailing relatively fundamental sensory discrimination, or else occur in socially invariant circumstances such as early physical contact between mother and infant. In fact, the values that have been ascertained may prove to be among the highest that actually exist, for the reason that extremely inflexible and selective rules are those most likely to catch the attention of experimental psychologists. Yet as demonstrated by the projections of Figures 4–21 to 4–24, far less selective rules can still create powerful canalizing effects. For this reason we expect the essential qualitative results of the translation model to prove robust.

Because virtually no psychological or ethological studies have been conducted with gene-culture translation in mind, few data relevant to the assimilation functions appear to exist beyond those listed in Table 4–1. Similarly, almost no ethnographic curves are available that can be linked directly to assimilation data. Nevertheless, certain categories of behavior are sufficiently well known and appear tractable enough to encourage further analysis. We have selected three examples of such categories to illustrate the great potential diversity of gene-culture structures. The first is brother-sister incest avoidance, which is a strongly "biological" case possessing a large but covert adaptive value (homozygosity reduction) and underlain by a relatively simple but powerful epigenetic rule. Next, the fissioning pattern of Yanomamö villages is treated as a complex social phenomenon, which nonetheless appears to stem from relatively simple and quantifiable epigenetic rules. The phenomenon is typical of a class of great concern to anthropologists who explore the ecological basis of social patterns. Finally, variation in women's dress is an ultimately "cultural" phenomenon that might seem at first to lie beyond the reach of biological rules. We shall show that in fact it can be analyzed in a novel manner by recourse to gene-culture theory.

Brother-Sister Incest Avoidance

In searching for illustrations of the process of gene-culture translation, we begin with the case of the nearly universal avoidance of marriage and full sexual relations between brothers and sisters. The epigenetic rule ap-

pears well established: a deep sexual inhibition develops between people who live in close domestic contact during the first six years of life. The evidence suggests that the process is of a relatively simple form and one especially tractable to analysis. In particular, the epigenetic rule strongly biases individual development but is unresponsive, or at most weakly responsive, to choices made by other members of the society. In this section we construct a model with a flat assimilation curve, which yields an ethnographic distribution in the form of a binomial density function (Eq. 4-49). We then employ a normal distribution approximation for the ethnographic curve in large societies. Although data accumulated by social scientists are too few to draw the actual ethnographic curve, the available information indicates that the mean value is close to that predicted by the model. Also, an illustration is provided for the principle that even rigid epigenetic rules can give rise not only to amplification effects, but also to substantial cultural diversity, with the diversity nevertheless conforming overall to a particular probability distribution (the ethnographic curve).

Many human societies tolerate and even encourage marriages between first cousins, but nearly all forbid it between siblings and half-siblings. A very few societies, including the Incas, Hawaiians, ancient Egyptians, Buganda, and Bunyoro, institutionalized brother-sister incest in the case of royalty or other groups of high status. The taboo was abrogated only amidst ritual and mythic justification. Furthermore, van den Berghe and Mesher (1980) note that the incestuous males were polygynous, behavior that resulted in outbreeding and higher personal genetic fitness. Because of the general human trend toward hypergamy, especially in such patrilineal societies, high-ranking women were less likely to marry downward in rank and hence were more susceptible to matches with their brothers.

The avoidance of sibling incest is generated by a strong epigenetic rule in which close domestic familiarity during the first six years of life neutralizes sexual attraction (see Chapter 3). This rule has been verified by studies in Israel and Taiwan that are separate from the mass of largely anecdotal cross-cultural accounts just cited, and they can be used as independent information in estimating the process of gene-culture translation. In modeling the phenomenon, we recognize two culturgens, outbreeding and brother-sister incest. The exact level of incest demarcation then must be specified. It can entail marriage or equivalent long term bonding, full sexual activity, and the procreation of children, or it can be defined at any lower level of intensity down to and including transient, casual sexual contact. We consider here the more extreme forms, including the practice of full sexual intercourse with or without marriage-like bonding.

From studies of the interactions of unrelated children raised together in Israeli kibbutzim (Shepher, 1971) and Taiwanese villages (Wolf, 1966, 1968, 1970; Wolf and Huang, 1980), we can tentatively assign a value of the transition probability u_{21} in favor of outbreeding (culturgen c_1) close to unity and u_{12} in favor of brother-sister incest (culturgen c_2) close to zero. Our reasoning is as follows. The young people in the large Israeli sample never chose heterosexual intercourse or marriage with unrelated children with whom they had been raised, even in a permissive environment that encouraged matches within the kibbutzim. Kaffman (1977) reports that this pattern of preference has been maintained during the subsequent period of "sexual revolution" in the Israeli society at large, although he reports anecdotes of some amount of heterosexual activity (of an unspecified nature) among adolescents of the same kibbutz. Hence, when given free rein, the young people express a virtually absolute preference for marriage outside their domestic group, a response that matches the general aversion to brother-sister mating existing in other, more conventionally organized societies. Put another way, the likelihood of transition to outbreeding approaches unity in a society that is uniformly outbreeding but otherwise offers free individual choice.

The Taiwanese studies permit an evaluation of preference in a society that has an intermediate level of culturgen usage. Wolf considered the effects of "minor marriages," in which unrelated infant girls are adopted by families, raised with the biological sons of the family in a typical brother-sister relationship, and later married to the sons. Even in the region of Taiwan where such arrangements were common practice, comprising approximately half of all virilocal marriages, they were generally resisted by the young people who participated.

If the nineteen families analyzed by Wolf's 1966 report, for example, the young couples refused to go ahead with the match in fifteen of the cases. In two cases one member of the pair died in childhood, while the two remaining couples married. Couples in the larger samples subsequently analyzed showed a higher level of acquiescence, but resistance was still extremely strong and the results far less successful than the major marriages contracted in the same region. Twenty-four percent of the minor marriages ended in divorce, whereas only 1.2 percent of the major marriages did. In 33.1 percent of the minor marriages the woman was reported to have committed adultery, as opposed to 11.3 percent of the major marriages. During the first twenty-five years after the wedding, minor marriages produced 30 percent fewer children than major marriages, and many of these were reported to be products of the adulterous relations. There were some reports of parents having to coerce young people to consummate minor marriages, even to the point of threatening physical punishment, but no accounts of coercion in major marriages. Thus when the culturgen usage was approximately 50 percent ($\xi \sim 0$), the preference was still strongly in favor of outbreeding. The preference may well be as close to absolute as in the Israeli kibbutzim, because the minor marriages of Taiwan were arranged and compelled by the parents. This assessment is supported by the fact that following the Japanese occupation of Taiwan, and a consequent weakening of parental authority, minor marriages fell from about 50 percent to 10 percent in virilocal arrangements (Wolf, 1970).

The phenomenon of sexual inhibition due to early propinquity appears to exist in other societies, as indicated especially by anecdotal accounts of the Trobrianders, Tallensi of Ghana, and Tikopians (Fox, 1980).

It is of considerable interest that a parallel epigenetic rule has been discovered in free-living chimpanzees (Pusey, 1980). At the Gombe National Park, Tanzania, females abruptly dropped their association with their brothers and other males closest to them when they entered their full estrous cycles. As a result, sexual activity was rare between siblings and between mothers and sons. In addition, most or all adolescent females transferred to other chimpanzee communities, sometimes permanently, as a result of greater attraction to unfamiliar males during occasional visits to the other groups. In general, inbreeding appeared to be avoided as a consequence of reduced sexual attraction between individuals who were familiar with one another during immaturity. Because the chimpanzee is genetically the closest species to man, the epigenetic rule inhibiting incest in both species may be truly homologous, in other words based on a genetic prescription that has persisted from a common ancestor.

The ethnographic curves for the case of brother-sister incest indicated by the Israeli and Taiwanese data have a simple but revealing structure. Since the $v_{ij}(\xi)$ appear to be insensitive to ξ (in spite of the fact that young people can be coerced to act against their own personal preferences), these transition rates can be approximated as positive real constants and the formulas (4-22) and (4-28) for the ethnographic curves can be solved exactly. The epigenetic rules are represented graphically in Figure 4–26.

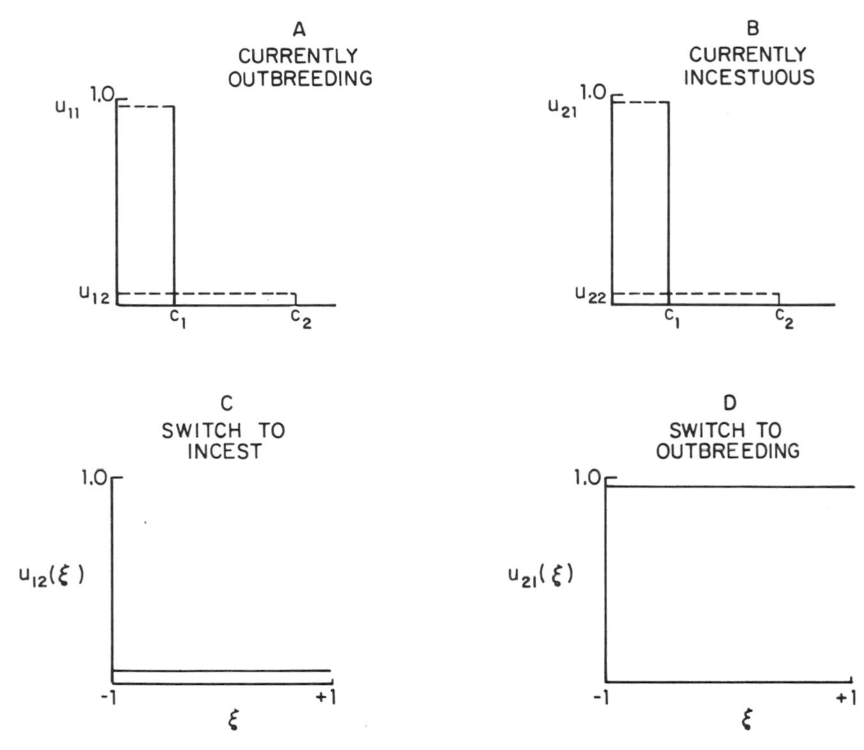

Figure 4–26 The epigenetic rules inferred for brother-sister incest. The culturgen c_1 is incest avoidance and c_2 is the incestuous relationship. *A*: Individuals currently avoiding incest have a strong intrinsic preference to continue doing so. *B*: Individuals engaging in incest have a strong intrinsic tendency to switch to outbreeding. These epigenetic rules are considered to be largely independent of the observed practice of others, as indicated in the flat $u_{ij}(\xi)$ curves of *C* and *D*.

For the discrete-ξ representation applicable to all group sizes (Eq. 4-22), the ethnographic curve is, in terms of the number of outbreeding individuals n_1, the binomial density

$$P(n_1) = \binom{N}{n_1} \rho^{n_1} (1 - \rho)^{N-n_1}, \qquad n_1 = 0, 1, 2, \ldots, N \quad (4\text{-}49)$$

with probability parameter ρ set by the assimilation functions:

$$\rho \triangleq v_{21}/(v_{12} + v_{21}). \quad (4\text{-}50)$$

Recalling that for $i \neq j$, $v_{ij} = r_i u_{ij}$, we find

$$\rho = (1 + \tau_2 u_{12}/\tau_1 u_{21})^{-1} = (1 + \mathcal{R}^{-1})^{-1}, \tag{4-51}$$

where in the second step we have used Eq. (4-32).

Ethnographic distribution (4-49) has mean $N\rho$ and variance $N\rho(1 - \rho)$; thus if *population sample estimates* are available for these two moments, then the relative values of v_{12} and v_{21} *for the individual epigenetic rules* can be inferred. The most probable number of outbreeders in the group is n_1^* such that

$$(N + 1)\rho - 1 < n_1^* \leq (N + 1)\rho \tag{4-52}$$

(Feller, 1958:140). When in a group of size N the transition rates are such that

$$v_{21} > Nv_{12} \tag{4-53}$$

pancultural avoidance of incest, corresponding to a value $n_1^* = N$, becomes the most probable state and the mode of $P(n_1)$ is located at $\xi = -1$. The value of $P(n_1)$ at n_1^* is for $N \gtrsim 25$ roughly proportional to the inverse of its standard deviation:

$$P(n_1^*) \sim [2\pi N\rho(1 - \rho)]^{-1/2}. \tag{4-54}$$

Equation (4-54) illustrates an important organizing principle of gene-culture translation stated in general terms earlier, namely that even rigid and highly selective epigenetic rules are compatible with cultural diversity. This is reflected in the values of $P(n_1^*)$: even for an extreme $\rho = 0.99$, a band of size twenty-five has $P(25) \sim 0.8$ and one of size seventy-five has $P(75) \sim 0.46$. In the first case 20 percent of bands observed would not show pancultural incest avoidance, while in the second more than half would have some degree of incest activity. It is the probabilistic pattern of cultural diversity across many societies and across the histories of individual societies that is determined by the epigenetic rules.

Figure 4–27 presents a series of ethnographic curves based on Eq. (4-49). In this illustration we have taken the transition values to be independent of the culturgen possessed by the individual, in other words set $\tau_1 = \tau_2$, $u_{12} = u_{22}$ and $u_{21} = u_{11}$. The mean number of outbreeders is then given by the linear relation

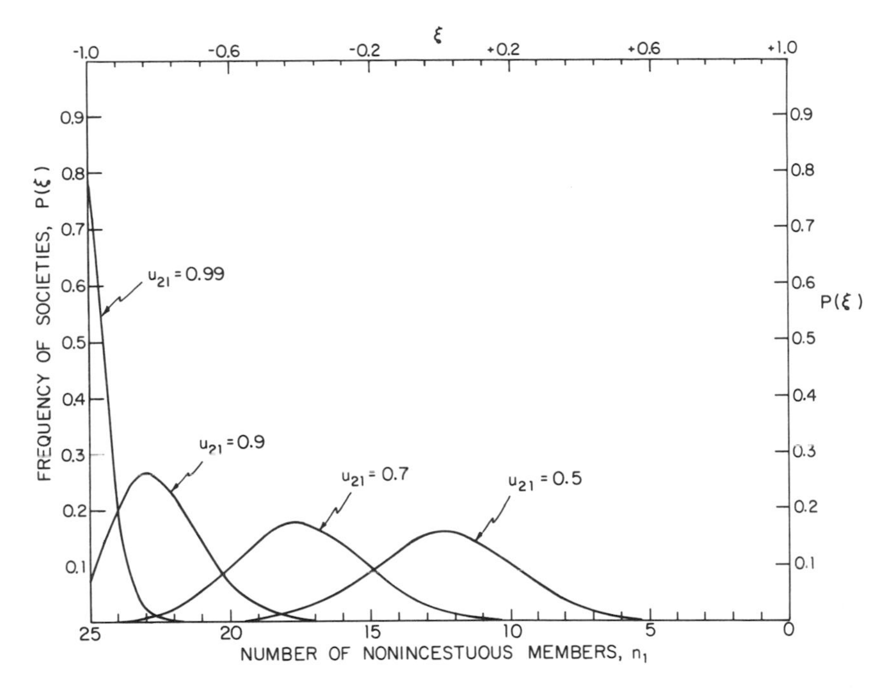

Figure 4–27 Ethnographic curves derived from the brother-sister incest model. The size of the group is $N = 25$. $P(\xi)$ for $n_1 = 0, 1, \ldots, 25$ was calculated using Eq. (4-22). Only ξ values corresponding to these n_1 values are realizable. (From Lumsden and Wilson, 1980b.)

$$\bar{n}_1 = u_{21}N \qquad (4\text{-}55)$$

(Figure 4–28). The strong aversion to the incestuous relationship when the social arrangements are such as to make the liaison all but a fait accompli suggests that this special case is a useful approximation to the real epigenetic rules. It is possible that a better approximation to the holding times is $0 < \tau_2 \ll 1 \ll \tau_1$, so that once individuals have commenced outbreeding they reevaluate that preference only occasionally. Furthermore, such an epigenetic rule system accelerates once an incest error is made and increases the transition rate $r_2 u_{21}$ from incest to outbreeding. In this "satisfaction limit" $r_2 \gg r_1$ it follows that

$$\rho \sim u_{21}/u_{21} = 1. \qquad (4\text{-}56)$$

The corresponding ethnographic distribution is concentrated around $\bar{n}_1 \sim N$.

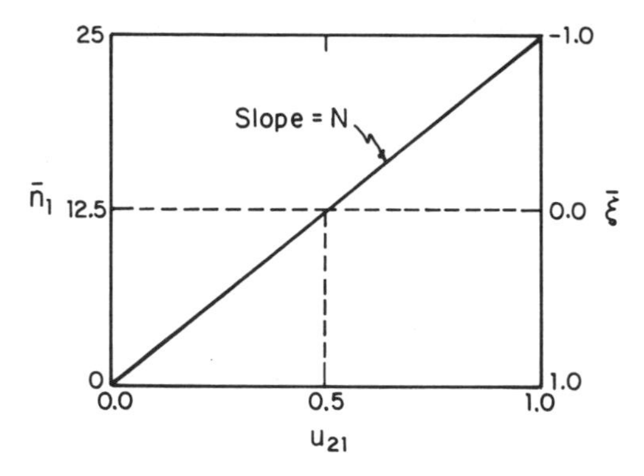

Figure 4–28 The incest avoidance case. Displacement of the mean of the ethnographic curve $P(\xi)$ with changes in the assimilation probability u_{21}. Drawn for a group with $N = 25$ members (see Eq. 4-55).

In the large-N limit $N \gtrsim 25$, $P(\xi)$ is approximated by the continuous-ξ equation (4-28). Evaluating (4-28) by direct integration,

$$P(\xi) \sim \frac{N^{1/2}}{(2\pi)^{1/2}} e^{-N\xi^2/2} \tag{4-57}$$

if $v_{12} = v_{21}$. If v_{12} and v_{21} are unequal but not too different, then the ethnographic distribution (4-49) is again very accurately approximated by a normal probability density with mean $\mu = N\rho$, variance $\sigma^2 = N\rho(1 - \rho)$ (see, for example, Feller, 1958:168; Hamburg, 1977:206). For the incest case, $v_{12} \ll v_{21}$, so that v_{12} and v_{21} are very different in absolute value. The ethnographic curve is skewed far toward $\xi = -1$ and an approximation using the normal density is less accurate.[5] The exact solution of the translation equation (4-28) in the general case of $v_{12} \neq v_{21}$ is not restricted to $|v_{21} - v_{12}| \ll 1$ and gives the ethnographic distribution

$$P(\xi) = C(\beta - \phi\xi)^{N\Lambda-1} e^{N\beta\xi/\phi}. \tag{4-58}$$

In terms of the function $Q(\xi)$,

$$P(\xi) = C(N/2)^{N\Lambda-1}[Q(\xi)]^{N\Lambda-1} e^{N\beta\xi/\phi}. \tag{4-59}$$

5. A popular rule of thumb holds that the normal approximation works well when $N\rho \geq 5$ and $N(1 - \rho) \geq 5$ (Hamburg, 1977:206). For the case of $u_{12} = 0.1$ and $u_{21} = 0.9$ in Figure 4–27, $N\rho = 22.5$ but $N(1 - \rho) = 2.5$ and the conditions are not met.

The quantity C is a normalization constant given by

$$C^{-1} = \frac{e^{N\beta^2/\phi^2}}{|\phi|}\left[\frac{\phi^2}{N\beta}\right]^{N\Lambda} \epsilon \left[\gamma\left(N\Lambda, N\beta\frac{(\beta-\phi)}{\phi^2}\right) - \gamma\left(N\Lambda, N\beta\frac{(\beta+\phi)}{\phi^2}\right)\right] \tag{4-60}$$

such that

$$\phi = v_{12} - v_{21}$$
$$\beta = v_{12} + v_{21} \tag{4-61}$$
$$\Lambda = \frac{\beta^2 - \phi^2}{\phi^2}$$

The symbol ϵ is defined by

$$\epsilon = \begin{cases} 1 & \text{if} \quad v_{12} < v_{21} \\ -1 & \text{if} \quad v_{12} > v_{21} \end{cases} \tag{4-62}$$

and the $\gamma(\cdot,\cdot)$ are incomplete gamma functions of the first kind (Abramowitz and Stegun, 1965:260). The principal merit of Eq. (4-58) is that for large N the simple threshold condition (Figure 4–29)

$$N\Lambda - 1 = 0 \tag{4-63}$$

gives an estimate to within several percent of Eq. (4-53) those values of the assimilation functions necessary to make complete avoidance of incest the most probable ethnographic state.

Although few exact ethnographic data are available, anecdotal accounts of a wide range of societies (Murdock, 1949; Berelson and Steiner, 1964; van den Berghe and Mesher, 1980) suggest that the true curve is closest to that generated by $u_{21} = 0.99$ than to the others displayed in Figure 4–27. This is the qualitative result to be expected from the strength of the epigenetic rule of incest avoidance revealed by developmental studies (Shepher, 1971) and makes pancultural incest avoidance the most probable ethnographic state in culture groups up to size approximately 100.

Although the ethnographic data support a ξ-independent system of

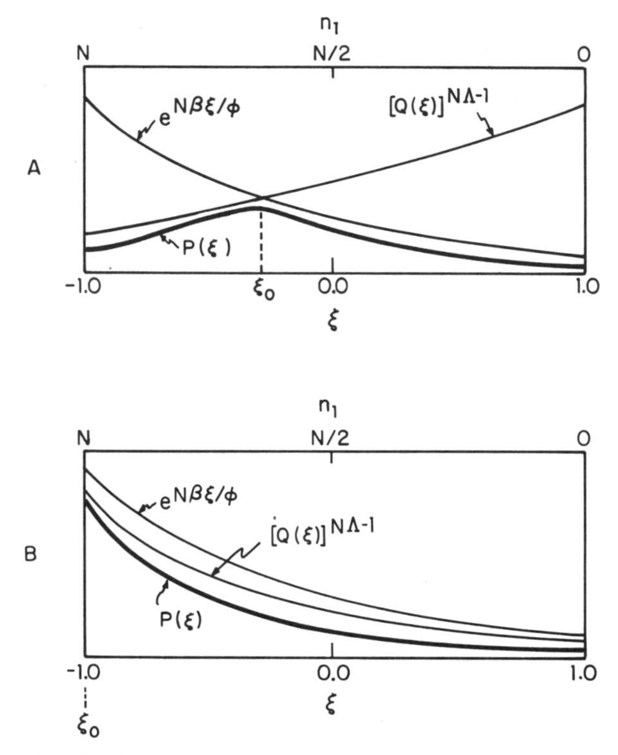

Figure 4–29 Qualitative behavior of the ethnographic curve for brother-sister incest. $Q(\xi)$ is monotone increasing on $(-1, 1)$ and $e^{N\beta\xi/\Phi}$ is monotone decreasing. For $N\Lambda - 1 > 0$, there is an interior mode (*A*); when $N\Lambda - 1 < 0$, the mode is pancultural avoidance of incest (*B*). Magnitudes of the three curves have been drawn to show the mode position clearly.

epigenetic rules as a first approximation, a more accurate model incorporates gradual, monotonic shifts upward in v_{12} and downward in v_{21} as ξ increases. The relative inflexibility of the epigenetic rules implies that a first-order Taylor series fitted to $v_{ij}(\xi)$,

$$v_{ij}(\xi) \sim v_{ij}(\xi^*) + (\xi - \xi^*)dv_{ij}(\xi^*)/d\xi \tag{4-64}$$

where

$$-1 \leq \xi \leq 1, \qquad \xi^* \in [-1, 1],$$

is a good approximation. These epigenetic rules are linear functions of ξ. Discrete variable analogs of the relations (4-64) are thus

$$v_{21}(n_1) = v_{21}(0) + n_1 a_{21}$$

and

$$v_{12}(n_1) = v_{12}(0) + (N - n_1)a_{12} \tag{4-65}$$

with slope parameters a_{ij}. The magnitudes of the a_{ij} are related to the $v_{ij}(0)$ by the condition of small relative change over $n_1 = 0,1, \ldots, N$:

$$a_{12}/[v_{12}(0) + v_{21}(0)] \ll 1$$

and

$$a_{21}/[v_{12}(0) + v_{21}(0)] \ll 1. \tag{4-66}$$

In the special case $a_{12} = a_{21}$ of equal but opposite rates of change in the $v_{ij}(n_1)$, the general translation equation (4-22) has a concise solution. It is the contagious binomial density (Coleman, 1964):

$$P(n_1) = \binom{N}{n_1} \prod_{k=0}^{n_1-1} (\rho_0 + k\zeta) \prod_{k=0}^{N-n_1-1} (1 - \rho_0 + k\zeta) \Big/ \prod_{k=0}^{N-1} (1 + k\zeta) \tag{4-67}$$

with

$$\rho_0 \triangleq v_{21}(0)/[v_{12}(0) + v_{21}(0)] \tag{4-68}$$

$$\zeta \triangleq a_{12}/[v_{12}(0) + v_{21}(0)]. \tag{4-69}$$

For incest $\zeta \ll 1$ and the behavior of this $P(n_1)$ is essentially that of the binomial ethnographic distributions treated earlier.

In conclusion, brother-sister incest provides a relatively direct entrée to the gene-culture amplification problem because of the robustness of its epigenetic rule and the binary nature of the cultural choice. Additional rules of sexual preference can be sought in the following gradients adduced by Murdock (1949) from ethnographic data:

(1) The influence of ethnocentrism, with close nonsibling members of an individual's culture being most preferred and members of very alien cultures and different physical races being least preferred.

(2) The influence of exogamy, with a decline in preference in going from nonrelatives to relatives.

(3) The tendency to prefer heterosexual over homosexual relationships.

(4) The influence of age, with females one generation or more older

than the male being less preferred by the males, and females in the same age group or younger being more preferred.

A linkage between quantitative developmental and ethnographic studies might be achieved with reference to any one of these rules of sexual preference and would result in an additional testing and clarification of the process of gene-culture coevolution.

Fission in Yanomamö Villages

For the second example of gene-culture amplification we take the more complicated behavior leading to village fission and emigration by the Yanomamö, a South American tribal group. The basic epigenetic rules of individual behavior underlying the response are not known, although they can be inferred to entail at least in part the more basic forms of binary group recognition and bonding (see Chapter 3). We utilize an intermediate ansatz motivated by the data: beyond a critical village size, aggression and strife become unbearable to a sufficient number of village members to induce emigration by part of the population. In the model we suggest, which appears compatible with the data, individuals can become very sensitive to the decision of others on whether to stay or depart. However, this pattern of responsiveness (in other words, the assimilation function) depends on the size of the group. Thus the epigenetic rules are context dependent. Assimilation functions are postulated to be of a step-function or steep logistic form, such that decisions on the part of family groups to stay or leave can be changed by either alterations in village size or shifts in the day-by-day aggregate of individual choices. The resulting ethnographic curves are unimodal or multimodal and can change from one such form to another as a consequence of relatively small shifts in the parameters of the epigenetic rules. This example also illustrates how future studies of developmental psychology might be designed to illuminate even the more subtle and complex forms of group behavior described in the literature of cultural anthropology.

The approximately fifteen thousand Yanomamö of southern Venezuela and adjacent portions of Brazil are organized into villages containing from 40 to 250 people. As shown by Chagnon (1976, 1977), the members of each village are tightly linked by intricate bonds and rituals of kinship, and they are all in close daily contact. The Yanomamö are exceptionally aggressive. Their frequent wars are almost always over women, since the acquisition of wives by barter, raiding, and seduction is a goal fueled by a cultural emphasis on polygyny. When Chagnon asked the Yanomamö men why they fought, they answered approximately as follows: "Don't ask me such stupid questions! Women! Women! Women! That's what started it! We fought over women!" These emic reports match the etic evaluations carried out by Chagnon. There is also a high level of internecine squabbling, often over women but also involving both sexes in matters of status, food, and various domestic issues.

Because the Yanomamö population is growing and expanding its range, there has been a long history of village multiplication by fission (Figure 4–30). Fission results from accumulated strife and tension, which increase disproportionately as the population grows in size. When the vil-

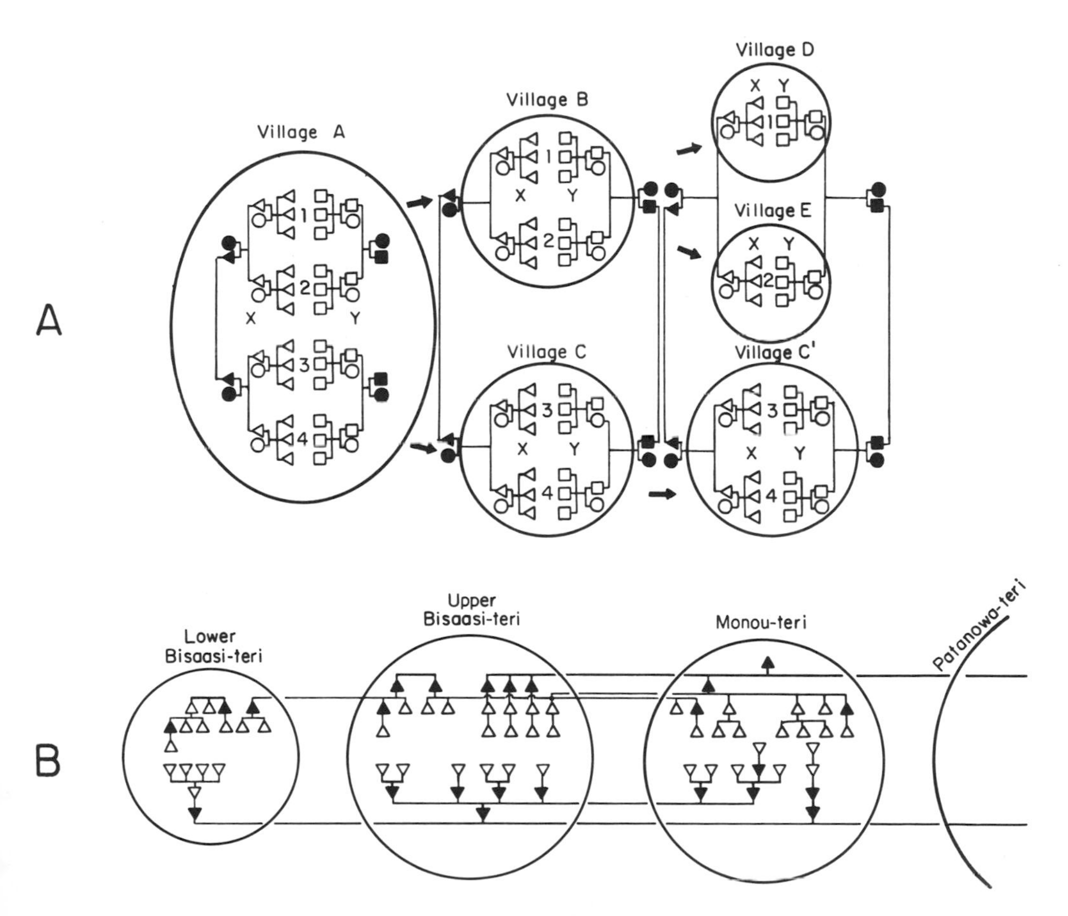

Figure 4–30 The fissioning pattern of Yanomamö Indian villages in Venezuela. *A*, idealized; *B*, observed. (After Chagnon, 1977.)

lage can no longer be held together by the bonds of kinship, marriage arrangements, and the relatively weak authority of the headman, a single dispute can trigger the departure of a family group. This critical level is often reached when the population exceeds eighty to a hundred individuals. Fissioning is rare in villages with less than eighty members, regardless of the level of internal strife. The reason is that a village must have at least ten able-bodied men to engage in raiding and defense. With an ordinary age-sex distribution, this cohort of adult males will be found only in groups with a total of forty to sixty members. Consequently a village must be composed of at least eighty members in order to produce two self-sufficient villages when it divides (Figure 4–31).

In terms of gene-culture theory, it is practical to treat this behavior of the Yanomamö as a binary decision: to stay together or to separate. The assimilation functions with reference to this pair of alternatives vary according to the size of the group. There is an overall high transition rate from "depart" to "remain" when the village contains about one hundred people or less, gradually giving way to the strong reverse tendency as the population size approaches two hundred individuals. Chagnon notes that villages expand until they reach an apparent "critical mass." When the group is small, squabbles die down quickly and individuals are relatively unresponsive to isolated instances of strife. But beyond a certain population size, small confrontations inflame rapidly and spread throughout the entire group.

These effects can be approximated by the threshold decision logic shown in Figure 4–32. Culturgen c_1 is the "remain" culturgen and c_2 is

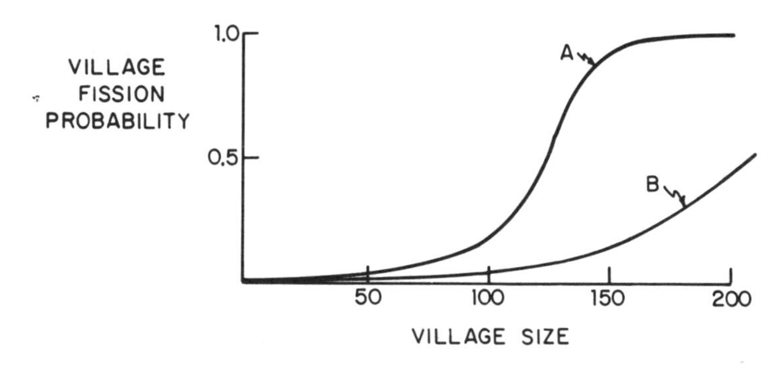

Figure 4–31 Quantitative models of fission probabilities for Yanomamö villages, based on the accounts of Chagnon (1976, 1977). *A*, low external pressure from warfare and low internal kinship; *B*, high external pressure from warfare and high internal kinship.

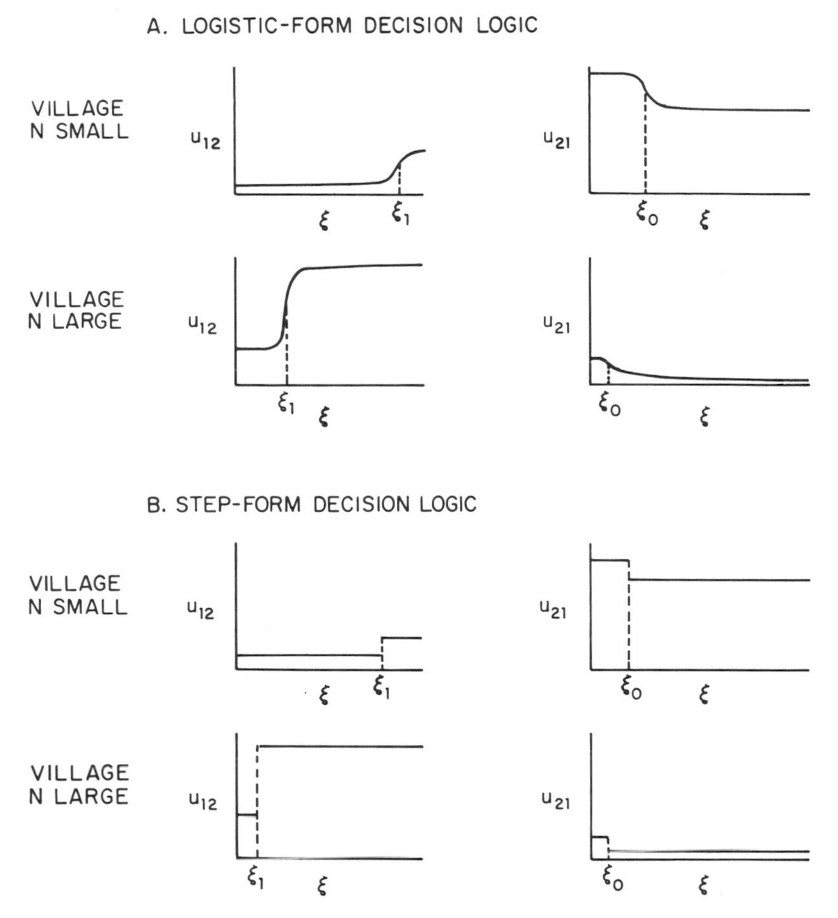

Figure 4–32 Threshold structure postulated for the Yanomamö decision rules u_{ij}. *A*, smooth, logistic-type model; *B*, step-function model. The behavior of the v_{ij} is similar.

the "depart" culturgen. The decision rules $u_{ij}(\xi)$ are responsive to patterns of social conflict, as represented by the frequency of individuals who have engaged in a confrontation and prefer the "depart" culturgen c_2. These individuals act both as spreading centers for c_2 and as foci of social pressure for the conversion of others from c_1 to c_2. The ξ-dependence exhibits a threshold effect: for any group size there is a ξ_1 below which pressure to depart is largely ignored and above which it is highly effective. In the more general case, the possibility of neophobia (or status quo maintenance) must be accommodated. It shows up in the approximation being used as a second threshold, ξ_0, below which ξ must fall in order to promote a significant likelihood for a "departer" to again become a "stayer."

The position of the thresholds and the values of $u_{ij}(\xi)$ are N-dependent as shown in Figure 4–32. Although real systems will not follow exactly the step-function approximation represented in the B series of Figure 4–32, this Yanomamö model can be analyzed quite completely. Simulations made using smooth approximations to the logistic form of the assimilation functions, such as the saturable trend-watcher model of Figure 4–20, reveal that even for moderate slopes of the $v_{ij}(\xi)$ at ξ_0, ξ_1 little accuracy is lost with the step functions. Let us therefore consider in some detail the impact of threshold decision rules on stationary ethnographic curves.

Figure 4–33 sets out the notation for the step-function approximation to the assimilation curves. Below ξ_0 advocacy for village unity is sufficient to

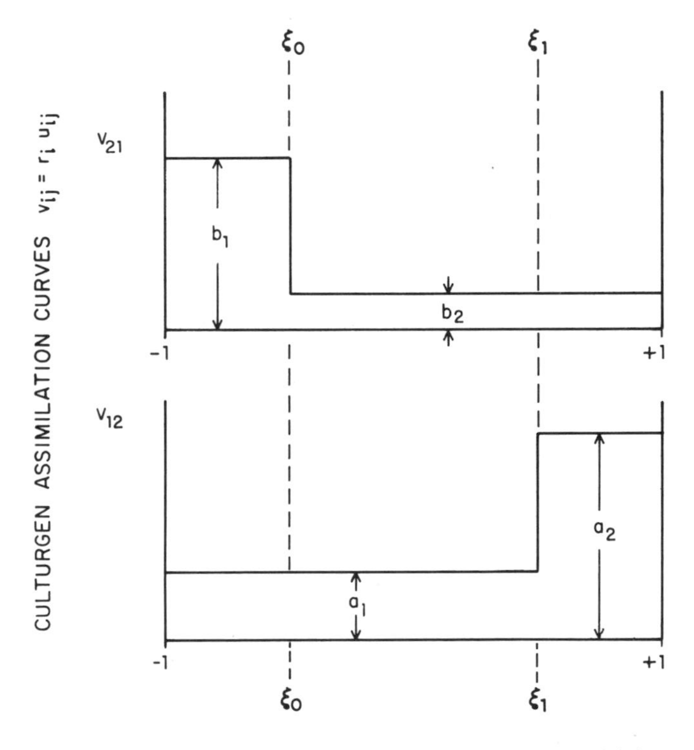

Figure 4–33 Threshold model of the assimilation functions, where the v_{ij} is the probability per unit time of transitions between the "depart" and "remain" culturgens.

induce an increased v_{21} value, b_1, that a member advocating fission will become a stayer. This v_{21} value falls to b_2 when the advocacy for staying together weakens and ξ becomes greater than ξ_0. When the proportion of villagers advocating fission grows beyond ξ_1, even those previously in favor of staying together jump to the v_{12} value a_2 that the assimilation rules will switch to the alternative option of splitting the village. Below ξ_1 this v_{12} value is a_1. The epigenetic rules guiding the decisions are therefore envisaged as containing two semantic triggers.

From Eq. (4-27)

$$X(\xi) = (1 - \xi)v_{12}(\xi) - (1 + \xi)v_{21}(\xi)$$

$$= [v_{12}(\xi) - v_{21}(\xi)] - [v_{12}(\xi) + v_{21}(\xi)]\xi. \qquad (4\text{-}70)$$

This equation has the form of a straight line

$$X = m\xi + b \qquad (4\text{-}71)$$

with ξ-dependent slope

$$m = -[v_{12}(\xi) + v_{21}(\xi)] \qquad (4\text{-}72)$$

and ξ-dependent intercept

$$b = v_{12}(\xi) - v_{21}(\xi). \qquad (4\text{-}73)$$

Since both v_{12} and v_{21} are always ≥ 0, it follows that $m \leq 0$. Reading from Figure 4–33 the various values of $v_{12}(\xi)$ and $v_{21}(\xi)$ and employing Eqs. (4-71) to (4-73), we obtain the following explicit equations for $X(\xi)$ in the model:

$$\text{for } -1 < \xi < \xi_0: \quad X(\xi) = (a_1 - b_1) - (a_1 + b_1)\xi$$

$$\text{for } \xi_0 < \xi < \xi_1: \quad X(\xi) = (a_1 - b_2) - (a_1 + b_2)\xi \qquad (4\text{-}74)$$

$$\text{for } \xi_1 < \xi < 1: \quad X(\xi) = (a_2 - b_2) - (a_2 + b_2)\xi.$$

This is a felicitous result, because $X(\xi)$ is shown to generate a distinct straight-line segment of negative slope in each of the three regions. Similarly, for the function $Q(\xi)$ we obtain the following:

$$\text{for } -1 < \xi < \xi_0: \qquad Q(\xi) = \frac{2}{N}[(a_1 + b_1) - (a_1 - b_1)\xi]$$

$$\text{for } \xi_0 < \xi < \xi_1: \qquad Q(\xi) = \frac{2}{N}[(a_1 + b_2) - (a_1 - b_1)\xi] \qquad (4\text{-}75)$$

$$\text{for } \xi_1 < \xi < 1: \qquad Q(\xi) = \frac{2}{N}[(a_2 + b_2) - (a_2 - b_2)\xi],$$

which is again a system of straight-line segments. The slopes of these segments are positive, zero, or negative depending on the values of the a_j and the b_j.

Figure 4–34 shows the effects of increasing the postthreshold assimilation values a_2 and b_1 on the shape of the ethnographic curve for the Yanomamö population, in the case where there is no prethreshold preference for one culturgen over the other. The phenomenon of greatest interest in this sequence is the existence of transition thresholds at which $P(\xi)$ splits from unimodal to multimodal. The change occurs when $X(\xi)$ acquires multiple zeros and creates one or more corresponding lateral peaks in the ethnographic distribution. This is the circumstance illustrated in the sequence of panels shown in Figures 4–35 and 4–36.

It can be seen that even a mild threshold response exercises a profound effect on the ethnographic distribution. This responsiveness itself is postulated to be genetically canalized; it can follow a rigid surface rule regardless of the remainder of the environment and history of the society, or it can be relatively flexible at the surface, adapting the individual to particular conditions by means of a relatively firm core strategy. In the phenomenon of village fissioning, the Yanomamö might be exhibiting assimilation rules about group size, composition, and complexity that are quite inflexible overall. The rules could in fact be a general primate trait, as Chagnon (1977) has suggested. It is clear from Chagnon's data, however, that information on the particular details of intravillage relatedness and external warfare is weighted into decisions concerning fissioning. The pattern of responsiveness can therefore vary from one culturgen category to another through strict genetic programming; in some instances it can vary from one environment and history to another within the same culturgen category. The important point, once again, is that even under the regime of a rigidly fixed response curve and a genetically canalized culturgen preference, there can be substantial cultural diversification. This counterintuitive result is another reason for stressing that the mechanisms

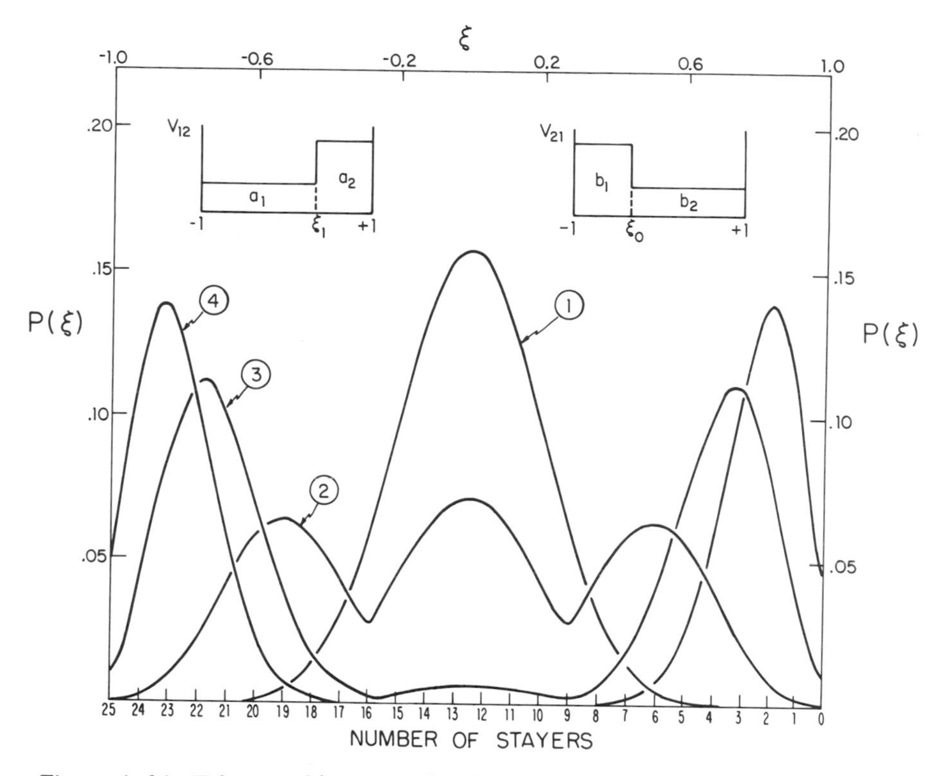

Figure 4–34 Ethnographic curves for the threshold assimilation rules approximating Yanomamö village fissioning (see Figure 4–33). The mean times τ_1 and τ_2 between social events that force culturgen reevaluation are taken to be unity for both c_i states. The curves show the effect of increasing assimilation responses b_1 and a_2 beyond the thresholds ξ_0 and ξ_1 as indicated in the legends. Curve 1 is below the transition thresholds, the rest are above. For all $P(\xi)$, $\xi_1 = -\xi_0 = 0.25$; $a_1 = b_2 = 0.1$. *Curve 1:* $a_2 = b_1 = 0.1$. *Curve 2:* $a_2 = b_1 = 0.3$. *Curve 3:* $a_2 = b_1 = 0.6$. *Curve 4:* $a_2 = b_1 = 0.99$. The large impact of small absolute changes in the epigenetic rules is apparent. For all four cases the group size is twenty-five and Eq. (4-22) is used directly. $P(\xi)$ is trimodal above the transition threshold in this sequence, although for a_2, $b_1 \gtrsim 0.6$ the central mode has negligible value.

underlying cultural diversity can be understood fully and in a predictive manner only with reference to the epigenetic rules. In most cases they await clarification by appropriate empirical studies of cognitive and behavioral development.

In the fissioning process the prethreshold and postthreshold assimilation probabilities are biased by events that accompany changes in village

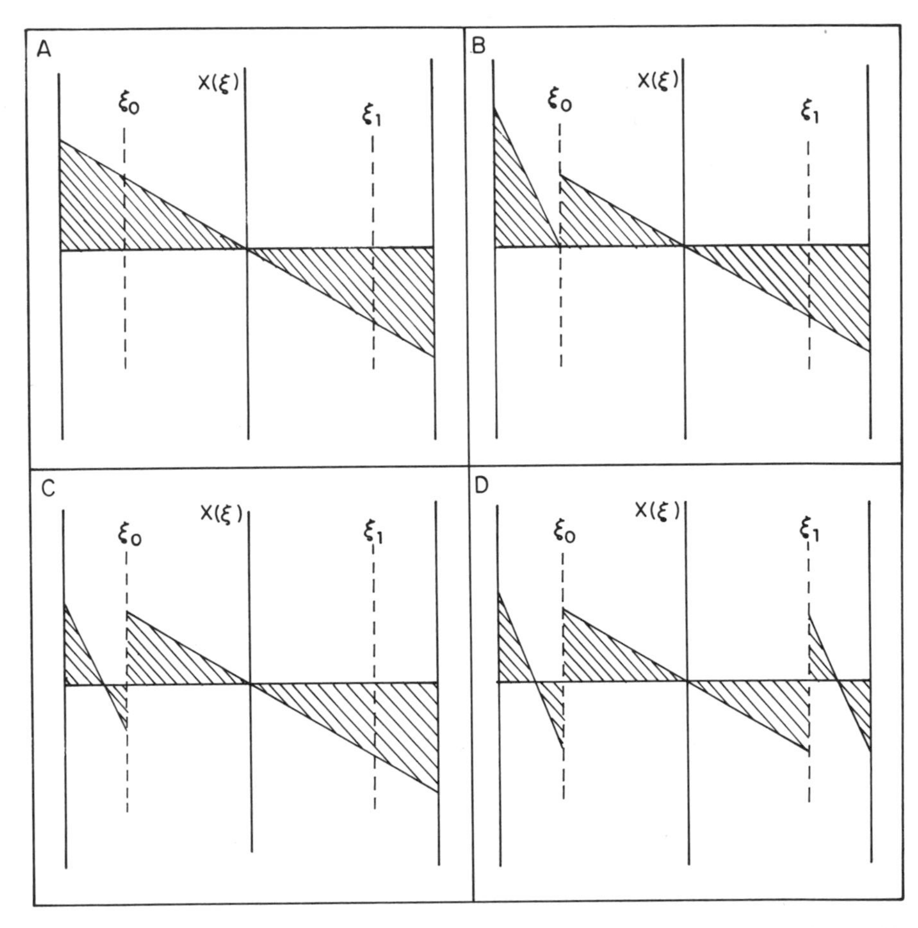

Figure 4–35 The effects on $X(\xi)$ of thresholds and of increasing postthreshold response to patterns of depart/stay advocacy. *A*: Thresholds are absent and the v_{ij} are independent of ξ. The reference state shown is for $v_{12} = v_{21}$. $X(\xi)$ is antisymmetric about $\xi = 0$. *B* and *C*: The threshold in $v_{21}(\xi)$ is at $\xi = \xi_0$ with $b_1 > b_2$ (recall Figure 4–33). As b_1 increases, the slope of $X(\xi)$ on $(-1, \xi_0)$ grows increasingly negative, until, at $b_1^* = a_1(1 - \xi_0)/(1 + \xi_0)$, $X(\xi)$ finally intersects the ξ-axis at ξ_0 (in *B*). For $b_1 > b_1^*$, $X(\xi)$ cuts partly below the ξ-axis, inducing a lateral peak in $P(\xi)$. (See also Figure 4–36.) *D*: The same as *B* and *C*, but with a threshold at $\xi = \xi_1$ in the decision rule $u_{12}(\xi)$ and the assimilation function $v_{12}(\xi)$.

size (recall Figure 4–30). In small villages ($N < 100$–150) a "depart" decision is unlikely and those making it will probably switch back quickly. In large villages ($N > 200$–250) confrontations increase in frequency and duration, and the assimilation biasing reverses; a "depart" decision be-

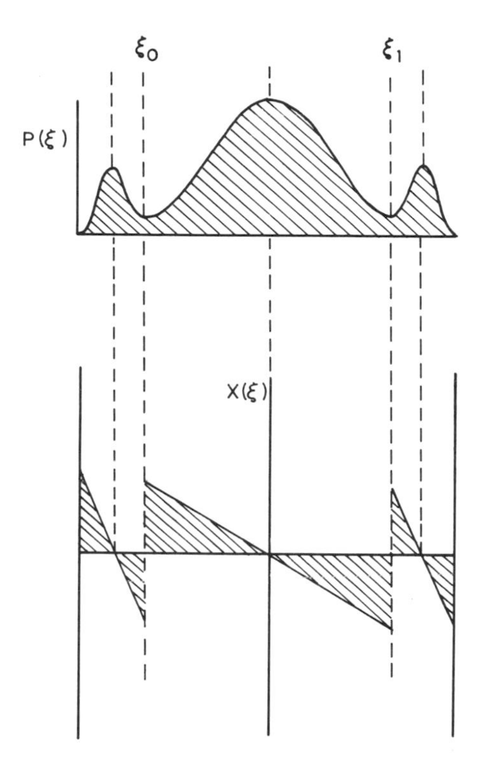

Figure 4–36 The consequences of the transition threshold applied to Yanomamö village fissioning on $X(\xi)$ and the ethnographic curve, which is defined by $P(\xi)$.

comes increasingly likely and more sensitive to advocacy ($\xi_1 \rightarrow 0$). A villager in the "depart" state is also less likely to reverse his decision.

The impact of this size-dependence on the ethnographic curve is deduced in Figure 4–37. The function $X(\xi)$ is negative for individuals in small villages over much of $(-1, +1)$, reaching zero at $\xi < 0$. The ethnographic curve has a single peak close to $\xi = -1$, the point where there is a consensus to stay. In large villages $X(\xi)$ is positive over much of $(-1, +1)$, reaching zero at $\xi > 0$. The ethnographic curve peaks near the consensus to depart ($\xi = +1$), and the anthropologist doing repeated samples on villages in this state will find a $P(\xi)$ skewed far to the right. Figure 4–38 makes these predictions quantitative with $P(\xi)$ curves calculated from Eqs. (4-22) and (4-28) for specific values of ξ_0, ξ_1, a_1, a_2, b_1, and b_2. Although vastly oversimplified, the gene-culture translation models in the two-culturgen Markov decision approximation account for significant features of the village fissioning process.

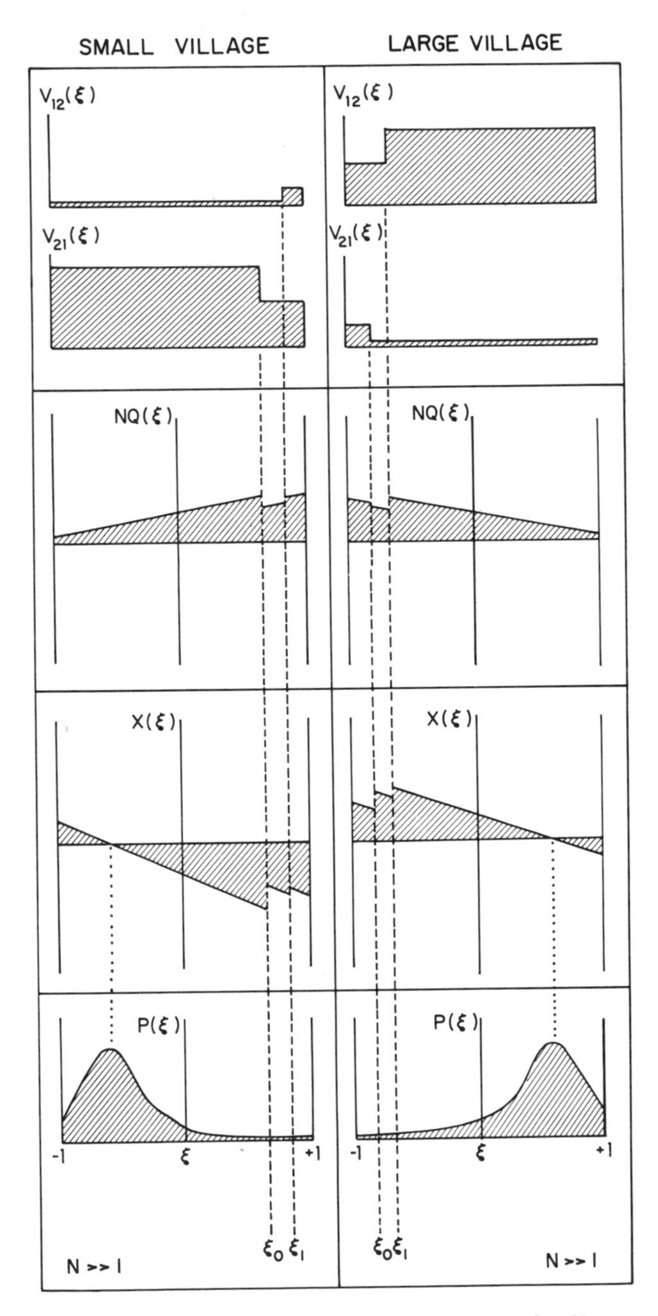

Figure 4–37 Derivation of the ethnographic curve $P(\xi)$ for Yanomamö village fissioning. $Q(\xi)$ is on the order of N^{-1} and is converted graphically into $NQ(\xi)$ to make it visible on the scale of $X(\xi)$.

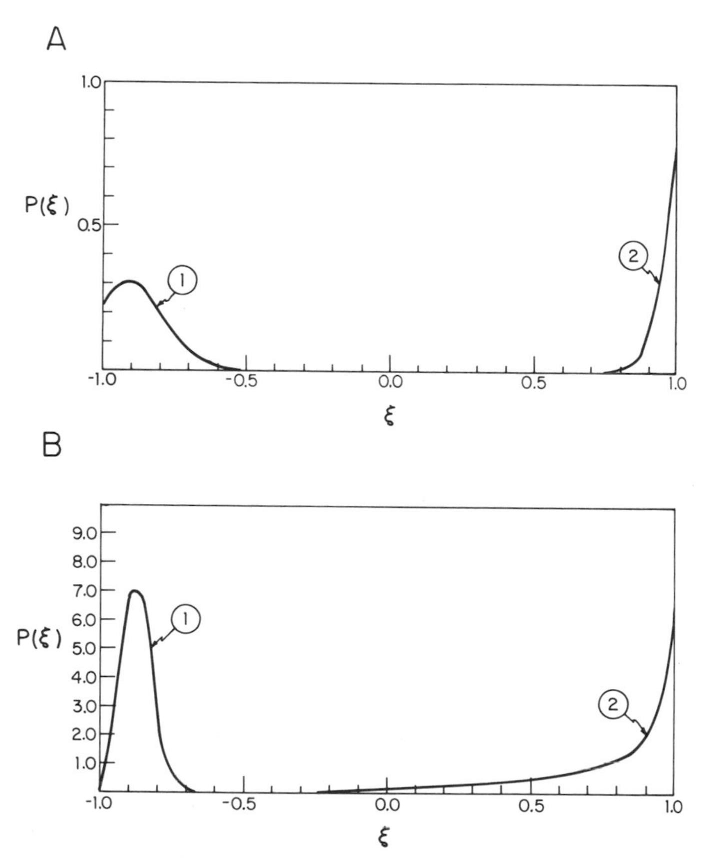

Figure 4–38 Ethnographic curves for the Yanomamö assimilation rules of Figure 4–33. *A, curve 1:* The small-N pattern of assimilation rules holds, with $\xi_1 = 0.9$, $a_1 = 0.05$, $a_2 = 0.1$, $\xi_0 = -0.9$, $b_1 = 0.9$, $b_2 = 0.7$ in the notation of Figure 4–33. *Curve 2:* large-N pattern of assimilation rules, with $\xi_1 = -0.85$, $a_1 = 0.8$, $a_2 = 0.95$, $\xi_0 = -0.95$, $b_1 = 0.05$, $b_2 = 0.01$. For ease of comparison, both patterns were calculated for a village with a population of twenty-five. $P(\xi)$ was derived from Eq. (4-22), for which only values of ξ such that $(N/2)(1 - \xi) =$ integer are meaningful. *B*: The same as *A*, except that the Fokker-Planck equation (4-28) is used to approximate $P(\xi)$ for large groups of actual size 100 (*curve 1*) and 150 (*curve 2*). In *B*, ξ is a continuous variable and all values on $(-1, 1)$ are meaningful.

Fashion in Women's Dress

In this final case study we consider ethnographic evolution that is cyclical and thus never converges to a steady state in the manner of incest avoidance and village fissioning. The data we use concern fashion in women's clothing over a 350-year period. They reveal wide fluctuations in such features as waist height and décolletage, some of which appear to follow a cycle of about 100 years. Even though we do not know the basic epigenetic rules underlying the style preferences, they entail what intuitively seem to be two interacting forms of competition: among couturiers to lead in innovation, and among women to gain status. In addition, the fashions swing back and forth across central values that conform to the true body shape, such as a waistline positioned around the actual waist. In our translation model we add a 100-year time cycle to the assimilation function. The transition rates at which individuals switch from one style to another depend not only on the choices made by others but also on the position of the society as a whole in the 100-year cycle. A simulation of the data is generated. Although individual behavior is here based on oversimplified epigenetic rules, we are convinced that an improved connection can be made with the aid of future studies in developmental psychology, and we expect that translation models of fashion change of the kind proposed here will be correspondingly refined.

In dialect, paralinguistic gestures, dress, and a few other forms of social behavior, people select among a broad but far from unlimited array of culturgens. The members of a society adopt only one or a relatively few at any one time, then shift to new choices with the passage of months, years, or generations. The phenomenon has been documented in fine detail in the study of women's formal dress by Richardson and Kroeber (1940), who took measurements from European and American paintings and fashion magazines from 1605 on. The principal features were claimed to oscillate back and forth through periods lasting about 100 years. During each century, for example, waist height went from high, near the bustline, to short, hugging the hips, and then back up to high again. Similar excursions occurred in dress length and décolletage (Figure 4–39). There appears to be an ideal although largely unappreciated pattern within these fluctuations toward which fashion gravitates: a long full skirt, the waist as slender as possible and in its true anatomical position, and much of the shoulders, arms, and upper portions of the bosom exposed.

Episodes of unsettlement occur within the long runs of dress data, during which there is more aggressive experimentation and overall variation in style. These developments often take place during periods of social and political upheaval such as the Napoleonic era and the modern world wars. But they also occur when the dress pattern evolves too far in one direction or another—toward, say, too short a skirt length or too high a

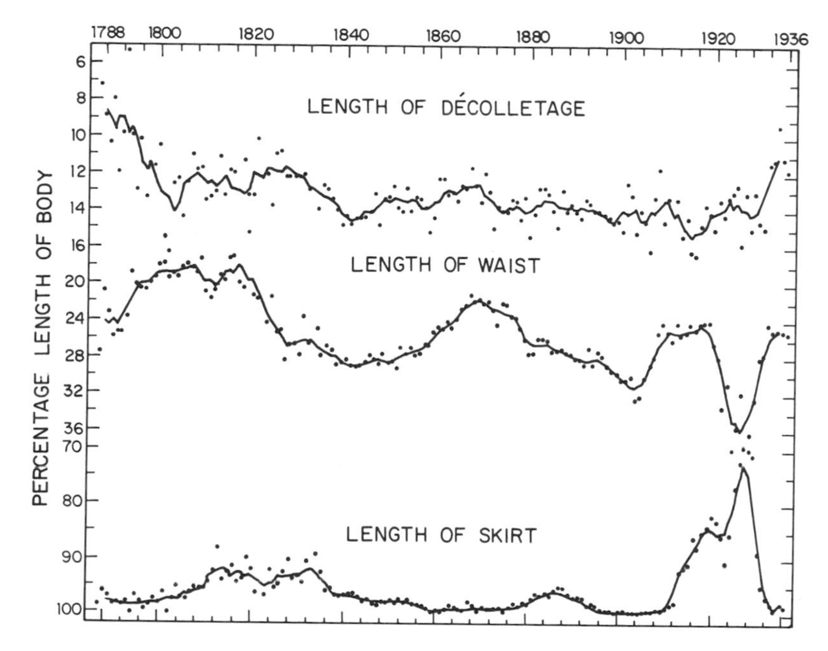

Figure 4–39 The Richardson-Kroeber data on several dimensions of women's dress in European and American fashion for the period 1788–1936. The numbers are the percentage of the line drawn from the mouth to the feet. The lines give the five-year running averages and the points are the yearly means. (Modified from Richardson and Kroeber, 1940.)

waistline. Although styles fluctuate around the "ideal" pattern, they do not converge toward a steady state in its vicinity. Innovation is built into the system. Competition among couturiers initiates experimentation and inevitably triggers departure from the contemporary mode.

These observations can be built into an elementary translation model. There they illuminate the fact that the ethnographic distributions (4-22) and (4-28), originally conceived as steady states, can also be applied to certain types of time-dependent historical processes. A given style component or set of components can be divided arbitrarily into two categories, for example high and low for waist height. The historical evidence suggests that in the case of waist height there is no innate preference for either of these culturgens, provided the dividing line between them is made at approximately the anatomically natural midwaist level. Because of unrelenting innovativeness, there is no relative frequency of the two styles, ξ, at which the values of the assimilation functions $v_{21}(\xi)$ and

$v_{12}(\xi)$ remain fixed. We can further characterize the $v_{ij}(\xi)$ by noting that as the styles approach either of the extremes, that is toward very high or very low waistlines, the transition rate in style choice shifts to favor the opposite style.

A consideration of human sociobiology suggests the motive forces: status competition among adult females and subsistence competition among couturiers, each reinforcing the other. For centuries novelty in adornment has provided positive status tokens for women. Furthermore, to lead one's peers in fashion rather than merely to stand apart is the goal. The prestige symbolized by successful innovation increases access to information, social dominance, and attention from males.

Couturiers are engaged in a different form of competition, the outcome of which determines both their subsistence and their social position. Their only strategy is culturgen innovation and propagation. The goal of most is to move fashion away from the accepted norm, either by embellishment or, in the case of most haute couture, radical redesign. Wide variation is inevitably generated, but when an attribute such as waist height exceeds limits set by anatomy or transgresses nudity taboos and other group norms, a rebound will occur. This redirection of innovation can be sudden or gradual, and it will tend to be most abrupt in societies through which cultural information flows rapidly and widely.

In their original study Richardson and Kroeber anticipated key elements of this mechanism and cited earlier precedents, but assessed them all pessimistically. The required cognitive mechanisms of valuation, decision, and communication seemed closed to objective study. The gap between individual and culture also appeared impassable. This is no longer the case. Earlier, in Chapters 2 and 3, we showed that cognitive psychology has become empirical and sophisticated enough to make the processes underlying human judgment and decision accessible to measurement. Furthermore, the models of gene-culture translation we have developed show that no formal, in-principle barrier exists between the patterns of human thought and those of culture.

Thus a consideration of what appear to be elementary traits of human behavior leads to the expectation that the cultural dynamics of fashion will show some aspects of temporal pattern, with styles replacing one another in dynastic succession. To incorporate this process within the present model, consider two style culturgens c_1 and c_2. For example, c_1 can be "low waist" and c_2 "high waist"; hence $v_{12}(\xi)$ is the assimilation function for a low- to high-waist transition and $v_{21}(\xi)$ the function for a high- to low-waist transition. This is the type of binary decision that faces all trend watchers—either to stay with current usage or to adopt a novel style.

Prevailing usage is now time dependent, with $\xi = \xi(t)$ and both $v_{12} = v_{12}(\xi,t)$ and $v_{21} = v_{21}(\xi,t)$, functions of time. The steady environmental conditions suitable for brother-sister incest and the Yanomamö village fissioning no longer exist. For the first time we encounter an ethnographic curve $P(\xi)$ with permanently dynamic properties.

The Richardson-Kroeber data imply, however, that on the time scale of human decision making and communication the style changes are slow. Their cycle lengths are on the order of 100 years. Rather than solve the complete dynamic problem, we shall introduce an *adiabatic approximation* that is accurate under the conditions just cited. In this approximation the $v_{ij}(\xi,t)$ change in time, but only slowly, and at each moment the group is at or close to the steady state $P(\xi)$ that would apply to the current $v_{ij}(\xi,t)$. This is made possible by the relative speed of decision and communication, whereby the group as a whole "relaxes" rapidly and adjusts itself to the prevailing conditions. Of course when relaxation is not rapid and the $v_{ij}(\xi,t)$ accelerate, which has occurred in the past during revolutionary eras and wartime, the adiabatic approximation is no longer accurate and the full dynamic problem must be solved. But for initial models of the Richardson-Kroeber process, adiabaticity is sufficient.

Although the adiabatic approximation can be used on $v_{ij}(\xi,t)$ of general form, to accord with the basic Richardson-Kroeber model we make the assimilation functions cyclic processes $v_{ij} = v_{ij}(t) \in \mathbb{R}$ with period 100 years (Figure 4–40A). A generalization of the model to incorporate complex ξ-dependence, overtones, and randomness is feasible (see for example Wang and Uhlenbeck, 1945), but it will require careful attention to the statistical properties of the time-series data as well as to the specific mechanisms of prestige competition.

In the present case the ethnographic curve has the time-dependent structure

$$P(n_1,t) = \binom{N}{n_1} \rho(t)^{n_1} (1 - \rho(t))^{N-n_1} \tag{4-76}$$

where $\rho(t)$ is the time-dependent analog of Eq. (4-50), namely

$$\rho(t) = v_{21}(t)/[v_{12}(t) + v_{21}(t)] \tag{4-77a}$$

$$\sim u_{21}(t). \tag{4-77b}$$

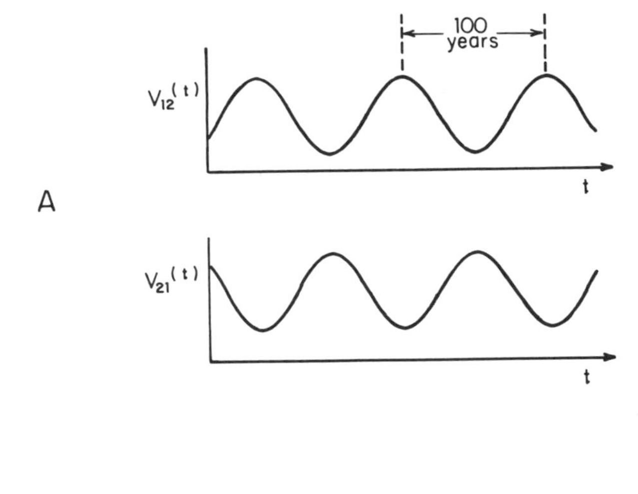

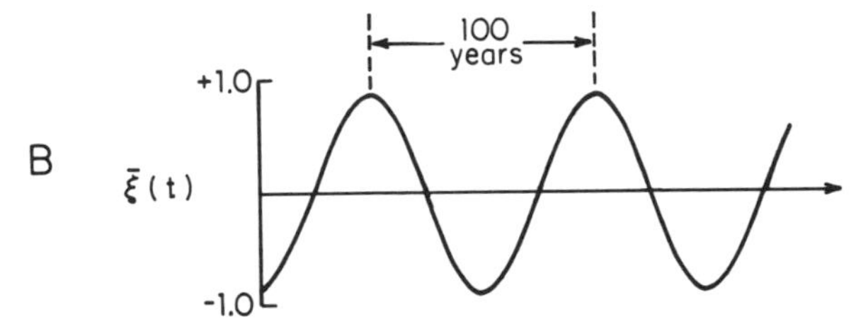

Figure 4–40 Gene-culture amplification for the Richardson-Kroeber model of fashion change. *A*, individual assimilation rules; *B*, dynamics of mean style usage.

The second step, (4-77b), gives the form of $\rho(t)$ when the assumptions underlying the limiting form, (4-55), are obeyed. The mean style usage pattern at time t is

$$\bar{\xi}(t) = 1 - 2\rho(t) \qquad \text{(4-78a)}$$

$$\sim 1 - 2u_{21}(t). \qquad \text{(4-78b)}$$

The quantity $\bar{\xi}(t)$ is the theoretical construct corresponding roughly to the Richardson-Kroeber cycle. In the simplest case (4-78b) $\bar{\xi}(t)$ has a period of 100 years. The behavior of this $\bar{\xi}(t)$ is sketched in Figure 4–40B, and the corresponding motion of the ethnographic curve is shown in Figure 4–41.

The dynamics of the transition probabilities $v_{ij}(t)$ have been modeled here as a purely phenomenologic process. The gene-culture mechanisms

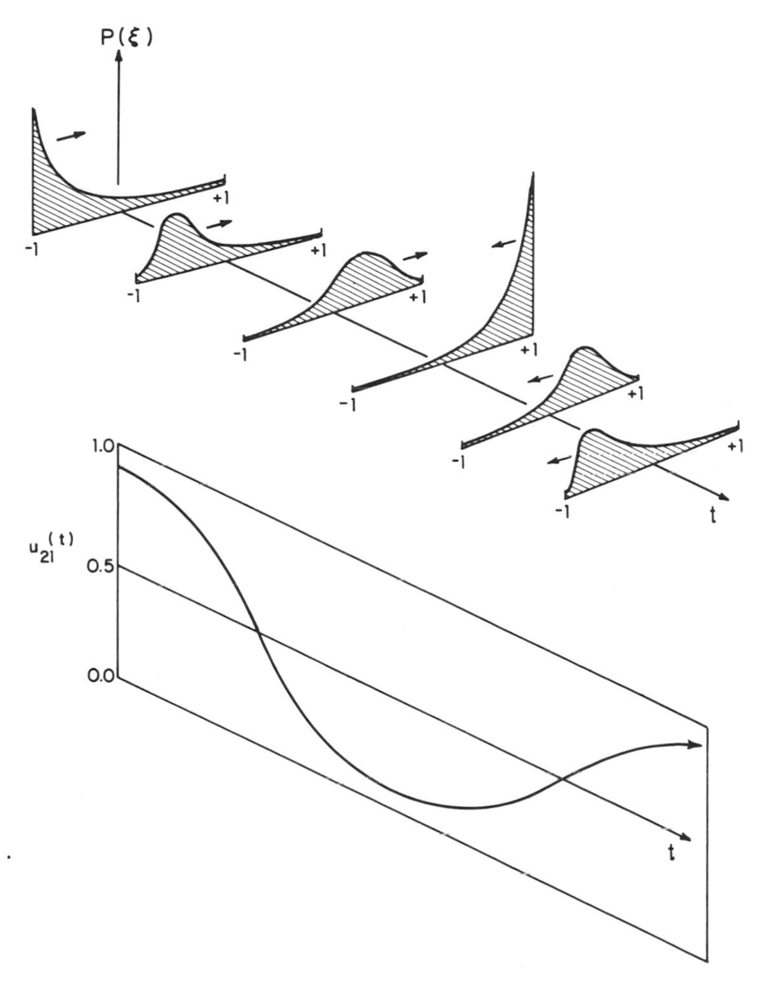

Figure 4–41 The ethnographic cycle for the Richardson-Kroeber process of fashion change.

underlying the dynamics of culturgen-culturgen competition will be treated in Chapters 6 and 7.

While the regime of quantitative style competition models can be expected to include some periodic systems, our adiabatic amplification has focused on periodicity in order to reflect the essential Richardson-Kroeber hypothesis. The extent to which the fashion data warrant this conjecture about periodicity is a related but for the most part separate question. In view of the wide attention the work by Richardson and Kroeber has received, however, their original published data (1940) deserve to be scrutinized using modern techniques not employed by the original investigators. From this perspective one notes that a statistically significant statement regarding cycles on the order of a century generally requires at least 400 years of sequential data (Box and Jenkins, 1970), roughly three times the data used by Richardson and Kroeber. Our own calculations of the periodograms and power spectra corresponding to the dynamics of these fashions display features suggesting cycles about one century long, but their statistical significance is dubious. We therefore feel that until more evidence is gathered, the classic fashion cycles remain largely hypothetical and may ultimately prove to be an artifact of a more complex, perhaps chaotic time-series process.

Can Culture Have a Life of Its Own?

We have perceived culture as the product of a myriad of personal cognitive acts that are channeled by the innate epigenetic rules. The "invisible hand" in this marketplace of culturgens has been made visible by characterizing the epigenetic rules at the level of the person and translating them upward to the social level through the procedures of statistical mechanics. Gene-culture coevolutionary analysis runs counter to the organicist conception of many social scientists, which views culture as a virtually independent entity that grows, proliferates, and bends the members of the society to its own imperatives.

Cultural determinism has been given many shades of meaning by social anthropologists over the years, as noted in the reviews by Marvin Harris (1968, 1979), Hatch (1973), and Leaf (1979). The extreme form dismisses biological feedforward altogether. According to White, for example, "Culture exerts a powerful and overriding influence upon the biological organisms of *Homo sapiens,* submerging the neurological, anatomical, sensory, glandular, muscular, etc., differences among them to the point of insignificance" (1963:116). Hence, he claims, culture possesses a life of its own: "Flowing down the ages, it embraces the members of each generation at birth and molds them into human beings, equipping them with their beliefs, patterns of behavior, sentiments, and attitudes. Human behavior is but the response of a primate who can symbol to this extrasomatic continuum called culture" (1949a:379). As a result, culture must be *sui generis,* comprehensible only through an autonomous discipline of culturology, because it "changes and develops in accordance with laws of its own, not in obedience to man's desire or will. A science of culture would disclose the nature and direction of the culture process, but would not put into man's hands the power to control or direct its course" (1948:213).

Evidence accumulated during the past ten years concerning cognitive epigenesis, which we analyzed in Chapters 2 and 3, negates this extreme interpretation. Culture is in fact the product of vast numbers of choices by individual members of the society. Their decisions are constrained and biased in every principal category of cognition and behavior thus far subjected to developmental analysis. In the present chapter we have shown that even small propensities to behave in one manner in preference to another tend to be exponentially amplified into strongly distinctive cultural patterns. Furthermore, and of considerable importance to social theory, measurement of the amplification is technically feasible. Gene-culture theory leads to the inference that laws governing culture qua culture must exist, but they can be synthesized from the principles governing the mind. The derivation of social pattern from biologically grounded individual cognition is not just logical; it appears to provide the only method for gaining knowledge of the organic mechanisms underlying such principles.

What form do cultural laws take? The gene-culture translation model developed in the previous sections can lead to a partial answer. It was shown that a key measure of cultural pattern is the culturgen proportion ξ, and Eqs. (4-21) and (4-26) are the dynamical laws governing its probability. These formulations yield still other equations that define the history of ξ experienced in particular cultures. For example, during periods of history when the ethnographic curve $P(\xi, t)$ is sharply peaked around a single mode, ξ is never far from its mean $\bar{\xi}(t)$, and $\bar{\xi}(t)$ obeys a law of cultural evolution

$$\frac{d}{dt}\bar{\xi}(t) \sim X[\bar{\xi}(t)]$$

where the function X is given by Eq. (4-27a). When $P(\xi, t)$ is broad or has multiple sharp peaks, one can either resort to the full ethnographic curve, to coupled equations of motion for the moments $\langle \xi^k \rangle$, $k = 1, 2, 3, \ldots$ that follow directly from $P(\xi, t)$, or to laws for the stochastic motion of ξ itself (see for example Mortensen, 1969, and Goel and Richter-Dyn, 1974). Furthermore, in a chapter that gives attention to the simplest models of gene-culture translation, we should note that the formal tools are not limited to largely unstructured social systems containing just two culturgens and a single cultural pattern variable ξ. Although comparatively less well understood, methods are available for handling many cul-

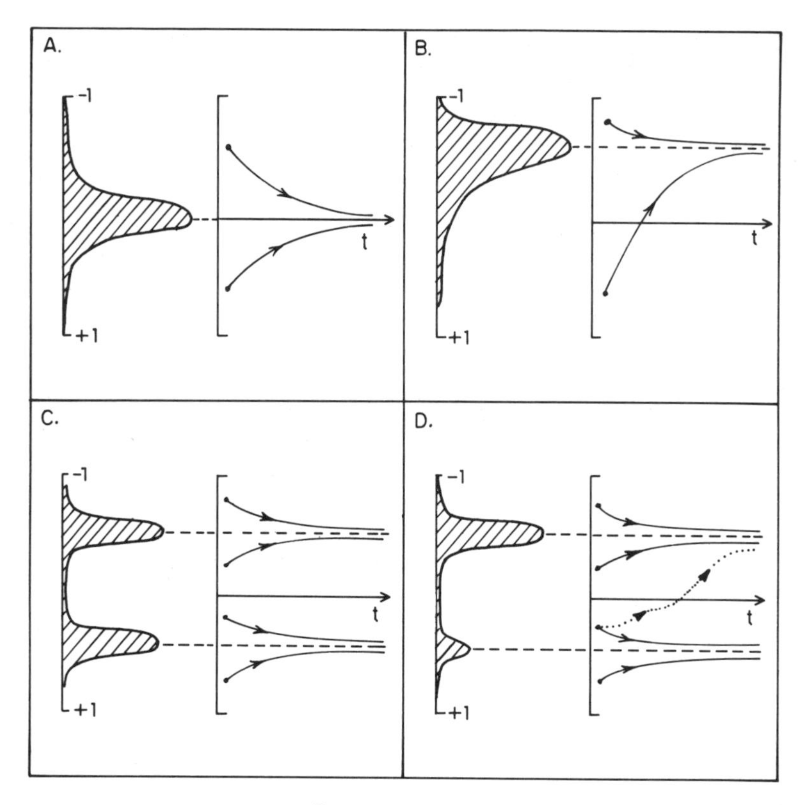

Figure 4–42 The dynamics of $\bar{\xi}(t)$, the average value of ξ, in the steady ethnographic state. The modes of $P(\xi)$ represent local attractors for the motion of $\bar{\xi}(t)$. Individual societies change stochastically, with the result that they are able not only to move against the overall trend but also, in the bimodal case, to cross over between stationary subdistributions (the dotted line in *D*).

turgens at once and for synthesizing the covering laws for their diverse measures of structure and process (for background see Appendix 4–3; Haken, 1977; Nicolis and Prigogine, 1977; Penrose, 1979). These techniques and their background hypotheses will require attention in increasingly sophisticated applications of gene-culture theory connecting genes through epigenetic rules to the cultural laws they underlie.

One nevertheless might conceive of the possible emergence of institutions and customs so powerful, so able to command resources through heterarchic feedback, that they grow and proliferate even while contravening the epigenetic rules and lowering genetic fitness. In such a case then, surely we can speak of culture's acquiring a life of its own, utterly independent of individual concerns. But no, not at all; we have merely returned to the leash principle of Chapter 1. It is possible to demonstrate that no cultural juggernaut will persist indefinitely under such ill-fitting conditions. The cognitive core rules might permit a changing of the assimilation function in reaction to the new cultural milieu, generating a new set of individual developmental probabilities (see Figure 3–3). The evidence shows that such flexibility exists but is very limited against the scale of conceivable biological possibilities. If epigenetic rules are contravened, they can be expected to exert a steady pressure until the culture is realigned into a more congenial form. Only by changing the genetic basis of the epigenetic rules or of the more fundamental cognitive core rules themselves could a previously maladaptive culture be preserved indefinitely.

Finally, it should not be forgotten that the cultures obedient en masse to the stationary ethnographic distributions are not individually stationary. Single societies can and do shift in the proportions of the alternative culturgens assimilated by their members. As suggested in Figure 4–42, the form of the ethnographic curve affects not only the likelihood of the occurrence of the culturgen usage pattern in individual societies but also, via the covering laws for ξ, the probability of the trajectory they will follow during cultural evolution within both the stationary and the time-dependent distributions. In the case of single-peaked distributions, illustrated in Figure 4–42A and B, states closer to the mean are more likely to occur than states farther away, so that $\bar{\xi}$, the expected value of ξ (the proportion of culturgens) is attracted to the most probable state. A similar circumstance obtains for the neighborhood of the two peaks in the case of the bimodal distribution (Figure 4–42C and D). An intermediate state (here $\xi = 0$) serves as a barrier between the two modes. However, the time course of each system is stochastic, so that actual trajectories are not smooth but erratic, permitting the occurrence of chance crossovers between the two subdistributions.

Summary

Gene-culture translation is the biological feedforward through the epigenetic rules of individual development to the formation of social patterns. The general properties of human socialization and enculturation are favorable to the reconstruction of translation by means of statistical models. In the great majority of societies culture is systematically passed along not just by the nuclear family (the "ordinary" circumstance in some industrialized Western societies) but by a much broader array of relatives and parent surrogates. This condition ensures a relatively uniform exposure of younger people to most culturgens. Even though later exchanges of information are channeled to some degree through guilds and other preference groups, the overall transmission occurs rapidly and evenly enough in many instances for culturgen adoption to be accurately described by relatively simple diffusion models. The dissemination patterns of many kinds of culturgens studied by social scientists during the past twenty years can be derived from elementary assumptions about information exchange, imitation, and evaluation. These processes in turn are causally linked to the epigenetic rules and core decision procedures of individual development.

As a result, gene-culture translation is not exclusively a feedforward of individual decisions to the level of social patterning. Often the proportion of other members of the society utilizing a given culturgen, together with additional macrocultural properties, affect the probabilities that an individual will adopt that culturgen. We have made this relationship explicit in the assimilation function $v_{ij}(\xi)$, where the transition rate from use of culturgen i to use of culturgen j is a function of ξ, defined so that $\xi = -1$ when no one possesses culturgen j and $+1$ when all possess it (see Figure 4–13). The influence of the group is enhanced by the effects of reification, through which the mind treats organizational properties and group norms as though they were discrete entities with an independent existence. This mental operation simplifies the decision process of the individual and converts the culture into a heterarchy, in which different levels of organization feed back upon one another. Assimilation functions, like epigenetic rules that generate them, are considered to be biological traits. Our fragmentary information indicates that they vary greatly among behavioral categories, and their effects amplify differences in the dependent social patterns (Figures 4–17 and 4–21 through 4–24).

In order to model gene-culture translation, we have explored the individual decision-making process. The important determining element in quantifying social patterns is the transition probability per unit time among competing culturgens. This rate can be related exactly to the probability of culturgen adoption and the waiting time between decision events in a way that makes its estimation practicable. The society is envisaged as a system in flux, so that the proportions of members carrying alternate culturgens composes a probability density distribution. This distribution is called the ethnographic curve; it gives the frequencies of societies utilizing each proportion of the competing culturgens (for example in 0.4 of the societies, 0.5 of the members utilize c_1 and 0.5 utilize c_2; in 0.05 of the societies 0.9 of the members utilize c_1 and 0.1 utilize c_2; and so on through all culturgen frequencies; see Figure 4–12). As the proportions change in the course of cultural evolution, so do the transition probabilities, in accordance with the assimilation functions that characterize the behavioral category. We have provided the master equation for the translation of the assimilation function into the ethnographic curve, as well as a Fokker-Planck equation that permits a slightly less exact but sometimes more convenient evaluation of culturgen dynamics. A general formula for the ethnographic steady state is also applied.

Two findings of general significance have emerged from the translation models. The first is that even small differences in the epigenetic rules, as reflected in the assimilation function, are magnified in the resulting ethnographic patterns. Even differences as low as 0.02 in the intrinsic bias toward competing culturgens, possibly below the detectable level in standard studies of behavioral development, can generate differences 1.5 times greater or more in the corresponding modes of the ethnographic curves. An amplification law has been derived that provides an exact and simple relation between the magnitude of the intrinsic bias, the values of ξ selected for analysis, and the amplification observed in the ethnographic curves (see Eq. 4-42). Where bias has been measured in a variety of developmental studies, the values are mostly an order of magnitude greater than those required to produce easily detectable effects in ethnographic data (Table 4–1).

The second finding is a corollary of the first: even when the underlying epigenetic rules and assimilation functions are rigidly constrained by genetic prescription, they can generate wide cultural diversity (see for example Figure 4–23). Fine tuning in these innate parameters can create large shifts in the dependent social patterns. Additional variation arises from the probabilistic nature of ethnographic distributions due to continuing flux in the decisions of individuals.

Although suitable data concerning the epigenetic rules exist in some developmental studies and can be used to construct rough ethnographic curves, we know of no case in which information on both individual development and ethnography are sufficient to examine the full linkage between the two levels. Perhaps the closest approach is encountered in the category of brother-sister incest. It is possible to make an approximation of the assimilation function, which on the basis of developmental studies is considered to be relatively insensitive to social context. We have derived the relatively narrow range of ethnographic curves in which brother-sister incest should lie. The available ethnographic measures, which are largely anecdotal, appear to fall within the predicted range. Another case chosen for analysis is village fissioning in the Yanomamö Indians, where the assimilation function seems to approach a step form based on the tension between kinship ties and intravillage strife. A third example is fashion in women's dress, which fluctuates through a period of about a hundred years. The rebound apparently arises from the conflict between competition for status through innovation and exclusiveness and the tendency to adhere to natural body form. These three case histories demonstrate some of the great variety to be found in the assimilation functions and their deduced ethnographic effects. They also illustrate the nature of the data that will be required to link developmental psychology and human sociobiology to the remainder of the social sciences by means of quantitative theory.

Appendix 4–1

Cognitive Dynamics and Algorithmic Languages

In the search for a formalism that expresses naturally and concisely the component processes of cognition, much attention has been focused recently on algorithmic languages, or *information processing languages* (IPLs). These systems deal explicitly with information processing and symbol manipulation (see Newell and Simon, 1972; Schank and Colby, 1973; Norman and Rumelhart, 1975; Lindsay and Norman, 1977; Colby, 1978). Although principally employed in psychology and machine intelligence research, they are also suitable for many applications in cognitive anthropology and in ethnosemantics (Abelson, 1973; Colby, 1973, 1978; Simon, 1979).

When written out in full, hypotheticodeductive models expressed in IPL dialects resemble programs written for electronic computers, and they have the same inherent advantages and limitations. It is our view that formal IPLs, because they are both precise and embedded in theories of symbol manipulation, can be very useful for organizing models and for clarifying in a cross-cultural perspective otherwise vague ideas about cognitive mechanisms.

Hypotheticodeductive models of cognition can also be expressed in the language of dynamics. The objective of this form of model is to capture the properties of the mind that are best described by numbers. Thus throughout Chapter 4 we have used the equations that govern the dynamics of Markov processes in order to predict how many people in a society use a particular culturgen at any given time. But the IPL and the equation of motion approaches are not disjunct. While the formal statements of an IPL model specify each step followed by a cognitive mechanism, equations of motion show the mechanism in action. As a consequence there is an intimate relationship, only partially understood, between these two basic ways of treating the mind. Here we wish to illustrate this deep connection by using a particular case from ethnography. We shall present it in the context of a model based on the theorems of Gregg and Simon (1967).

The Warao Indians of Venezuela are a water people dependent upon dugout canoes for transportation and fishing. Over a long period extending through adolescence and young manhood, the Warao male learns and perfects the skill of canoe building for which his people have been celebrated through history. But as Wilbert (1976) has shown, the attainment of maturity as a canoe builder does not finish the Warao male's socialization experience. Many aspire to the status of *moyotu,* or master craftsman, where the myths of the canoe craft are acquired and communication with the canoe spirits is established through the dream state. The appropriate chants and rituals for this esoteric communion are learned. The yeoman builder-shaman brings his mystic experiences under the guidance of an already established master craftsman. The master's role is relatively passive; little overt instruction is provided and the initiate is left to discover and to experience the hypotheses behind the mythworld for himself. Hallucinogenic smoke inhalation and physical isolation help to induce the dream states. Specific dream experiences, viewed as communication with spirits, are attained by the initiate and reported to the master. The master acts as judge. When the initiate reaches the state when the hypothetical mythworld has been constructed in the initiate's mind and experienced correctly according to social norms, the master pronounces the period of apprenticeship over. The initiate enters the group of master craftsmen, and his prestige and power within the society attain new heights. His own welfare and that of his relatives is manifestly improved. "It is the genius of Warao culture," Wilbert states, "that enculturation along informal and nonformal channels succeeds so well in motivating the developing craftsman by holding out new goals each time a change in personal identity becomes desirable."

On the basis of the ethnography it is possible to write a simple IPL model for the gradual structuring of the initiate's thoughts under the master's guidance. It has the following form, with M the master and I the initiate.

M1: Do M3; then do M4.
I1: If current hypothesis ∈ dream,
  then response ← "Have experienced mythworla."
  else response ← "Have not experienced mythworld."
M2: If response = correct response,
  then reinforcement ← "Master says 'Yes.'"
  else reinforcement ← "Master says 'No.'"
I2: If reinforcement = "No,"
  then current hypothesis ← I5.
M3: Generate dream in initiate via dream state
  (sampling randomly from each pair of possible outcomes on
    attribute structure of dreams) (⇒ generated dream)
  dream ← possible dream.
M4: If correct hypothesis ∈ dream,
  then correct response ← "Have experienced mythworld."
  else correct response ← "Have not experienced myth-
    world."
I5: Generate hypothesis (sampling randomly from possible
    hypotheses) (⇒ new hypotheses)
  current hypothesis ← new hypothesis.
M0: Do M1, do I1, do M2.
  If reinforcement = "Yes,"
    then tally of initiate's progress = tally + 1,
    else tally ← 0,
  If tally = K = full set of experiences,
    then halt,
    else do I2, then repeat M0.

This model of Warao enculturation is a direct adaptation of a concept-learning model created by Gregg and Simon (1967), written in a somewhat informal mode appropriate to the present example. It illustrates the form that cognitive models of the Wilbert data could take if expressed in algorithmic, symbol-manipulative form. Such information-processing or algorithmic explanations are intentionally mechanistic and well suited to implementation on computers. Gregg and Simon demonstrated that this model of discovery obeys an equation of motion that can be mapped exactly onto a Markov process. They number the hypotheses or possible hypotheses from 1 to $2N$, assigning 1 to correct hypothesis and 2 to its complement on the attribute structure of the stimulus (in our example, the dream). If the current hypothesis is i, and the initiate has made a correct interpretation, his cognitive system is said to be in state iR. If a wrong response is made, it is in state iW. There are $4N - 2$ accessible states, two for any possible hypothesis chosen, except for the first two. Their matrix of the transition probabilities for the Markov process then has the structure shown in Figure 4–43A. Furthermore, take AR as an aggregate state such that the initiate is in AR if he is in iR, $i \neq 1$; W an aggregate state equivalent to iW, $i = 2, \ldots, N$; and R equivalent

A. To state

$$
\text{From state}\quad
\begin{array}{c}
1R \\ \vdots \\ iR \\ \vdots \\ jR \\ \vdots \\ 2W \\ \vdots \\ iW \\ \vdots \\ jW \\ \vdots
\end{array}
\overset{\begin{array}{ccccccccccccc} 1R & \cdots & iR & \cdots & jR & \cdots & 2W & \cdots & iW & \cdots & jW & \cdots \end{array}}{
\begin{bmatrix}
1 & \cdots & 0 & \cdots & 0 & \cdots & 0 & \cdots & 0 & \cdots & 0 & \cdots \\
\cdot & \cdots & \cdot & \cdots & \cdot & \cdots & \cdot & \cdots & \cdot & \cdots & \cdot & \cdots \\
0 & \cdots & \frac{1}{2} & \cdots & 0 & \cdots & 0 & \cdots & \frac{1}{2} & \cdots & 0 & \cdots \\
\cdot & \cdots & \cdot & \cdots & \cdot & \cdots & \cdot & \cdots & \cdot & \cdots & \cdot & \cdots \\
0 & \cdots & 0 & \cdots & \frac{1}{2} & \cdots & 0 & \cdots & 0 & \cdots & \frac{1}{2} & \cdots \\
\cdot & \cdots & \cdot & \cdots & \cdot & \cdots & \cdot & \cdots & \cdot & \cdots & \cdot & \cdots \\
\frac{1}{2N} & \cdots & \frac{1}{4N} & \cdots & \frac{1}{4N} & \cdots & \frac{1}{2N} & \cdots & \frac{1}{4N} & \cdots & \frac{1}{4N} & \cdots \\
\cdot & \cdots & \cdot & \cdots & \cdot & \cdots & \cdot & \cdots & \cdot & \cdots & \cdot & \cdots \\
\frac{1}{2N} & \cdots & \frac{1}{4N} & \cdots & \frac{1}{4N} & \cdots & \frac{1}{2N} & \cdots & \frac{1}{4N} & \cdots & \frac{1}{4N} & \cdots \\
\cdot & \cdots & \cdot & \cdots & \cdot & \cdots & \cdot & \cdots & \cdot & \cdots & \cdot & \cdots \\
\frac{1}{2N} & \cdots & \frac{1}{4N} & \cdots & \frac{1}{4N} & \cdots & \frac{1}{2N} & \cdots & \frac{1}{4N} & \cdots & \frac{1}{4N} & \cdots \\
 & & & & & \vdots & & & & & &
\end{bmatrix}}
$$

B. To state

$$
\text{From state}\quad
\begin{array}{c} 1R \\ AR \\ W \end{array}
\overset{\begin{array}{ccc} 1R & AR & W \end{array}}{
\begin{bmatrix}
1 & 0 & 0 \\
0 & \frac{1}{2} & \frac{1}{2} \\
\frac{1}{2N} & \frac{(N-1)}{2N} & \frac{1}{2}
\end{bmatrix}}
$$

Figure 4–43 The transition probabilities for the Gregg-Simon concept-learning model. *A*, complete transition matrix; *B*, aggregated matrix.

to $1R$. Then it is possible to show that the initiate's cognitive dynamics is equivalent to the much simpler, three-state, aggregate Markov process shown in Figure 4–43B.

These and similar results illustrate the rich connections between the IPL and equation-of-motion approaches to learning and cognition. The epigenetic rules emphasized in our development of gene-culture theory are the cognitive mechanisms that ultimately produce the transition probabilities within the state matrix; in other words, they are part of the information and symbol manipulation strategies that give the IP algorithm its specific structure.

Appendix 4–2

Deriving the Motion Equation of Culturgens

Use Eq. (4-24) to rewrite (4-21a) as

$$\begin{aligned}\partial_t P(N\nu_1, N\nu_2, t) = {} & N(\nu_1 + \Delta\nu)v_{12}[N(\nu_1 + \Delta\nu), N(\nu_2 - \Delta\nu)] \\ & \cdot P[N(\nu_1 + \Delta\nu), N(\nu_2 - \Delta\nu), t] \\ & + N(\nu_2 + \Delta\nu)v_{21}[N(\nu_1 - \Delta\nu), N(\nu_2 + \Delta\nu)] \\ & \cdot P[N(\nu_1 - \Delta\nu), N(\nu_2 + \Delta\nu), t] \\ & - N\{\nu_1 v_{12}[N\nu_1, N\nu_2] + \nu_2 v_{21}[N\nu_1, N\nu_2]\} \\ & \cdot P[N\nu_1, N\nu_2, t]. \end{aligned} \tag{4-A1}$$

Defining $P'(\nu_1, \nu_2, t) \triangleq P(N\nu_1, N\nu_2, t)$ and $v'_{jk}(\nu_1, \nu_2) \triangleq v_{jk}(N\nu_1, N\nu_2)$ and dropping the primes, we can write Eq. (4-A1) as

$$\begin{aligned}\partial_t P(\nu_1, \nu_2, t) = {} & N(\nu_1 + \Delta\nu)v_{12}(\nu_1 + \Delta\nu, \nu_2 - \Delta\nu)P(\nu_1 + \Delta\nu, \nu_2 - \Delta\nu, t) \\ & + N(\nu_2 + \Delta\nu)v_{21}(\nu_1 - \Delta\nu, \nu_2 + \Delta\nu)P(\nu_1 - \Delta\nu, \nu_2 + \Delta\nu, t) \\ & - N[\nu_1 v_{12}(\nu_1, \nu_2) + \nu_2 v_{21}(\nu_1, \nu_2)]P(\nu_1, \nu_2, t). \end{aligned} \tag{4-A2}$$

Expanding the right-hand side of (4-A2) to second order in $(\Delta\nu)$, we find that

$$\partial_t P(\nu_1, \nu_2, t) \sim$$

$$N\left[1 + \Delta\nu\left(\frac{\partial}{\partial\nu_1} - \frac{\partial}{\partial\nu_2}\right) + \frac{(\Delta\nu)^2}{2}\left(\frac{\partial^2}{\partial\nu_1^2} - \frac{2\partial^2}{\partial\nu_1\partial\nu_2} + \frac{\partial^2}{\partial\nu_2^2}\right)\right]$$

$$\cdot\ \nu_1 v_{12}(\nu_1, \nu_2) P(\nu_1, \nu_2, t)$$

$$+ N\left[1 - \Delta\nu\left(\frac{\partial}{\partial\nu_1} - \frac{\partial}{\partial\nu_2}\right) + \frac{(\Delta\nu)^2}{2}\left(\frac{\partial^2}{\partial\nu_1^2} - \frac{2\partial^2}{\partial\nu_1\partial\nu_2} + \frac{\partial^2}{\partial\nu_2^2}\right)\right]$$

$$\nu_2 v_{21}(\nu_1, \nu_2) P(\nu_1, \nu_2, t)$$

$$- N[\nu_1 v_{12}(\nu_1, \nu_2) + \nu_2 v_{21}(\nu_1, \nu_2)] P(\nu_1, \nu_2, t). \qquad \text{(4-A3)}$$

But $\Delta\nu = 1/N$ by definition, with the result that

$$\partial_t P(\nu_1, \nu_2, t) \sim \left(\frac{\partial}{\partial\nu_1} - \frac{\partial}{\partial\nu_2}\right)[\nu_1 v_{12}(\nu_1, \nu_2) P(\nu_1, \nu_2, t)]$$

$$+ \left(\frac{-\partial}{\partial\nu_1} + \frac{\partial}{\partial\nu_2}\right)[\nu_2 v_{21}(\nu_1, \nu_2) P(\nu_1, \nu_2, t)]$$

$$+ \frac{1}{2N}\left(\frac{\partial}{\partial\nu_1} - \frac{\partial}{\partial\nu_2}\right)^2 [\nu_1 v_{12}(\nu_1, \nu_2) P(\nu_1, \nu_2, t)]$$

$$+ \frac{1}{2N}\left(\frac{-\partial}{\partial\nu_1} + \frac{\partial}{\partial\nu_2}\right)^2 [\nu_2 v_{21}(\nu_1, \nu_2) P(\nu_1, \nu_2, t)] \qquad \text{(4-A4)}$$

after collecting terms in (4-A3).

Collecting terms in (4-A4) gives

$$\partial_t P(\nu_1, \nu_2, t) \sim$$

$$\left(\frac{\partial}{\partial\nu_1} - \frac{\partial}{\partial\nu_2}\right)[\nu_1 v_{12}(\nu_1, \nu_2) - \nu_2 v_{21}(\nu_1, \nu_2)] P(\nu_1, \nu_2, t)$$

$$+ \frac{1}{2N}\left(\frac{\partial}{\partial\nu_1} - \frac{\partial}{\partial\nu_2}\right)^2 [\nu_1 v_{12}(\nu_1, \nu_2) + \nu_2 v_{21}(\nu_1, \nu_2)] P(\nu_1, \nu_2, t).$$

$$\text{(4-A5)}$$

The variable ξ contains all of the information expressed by the two culturgen frequencies ν_1 and ν_2, and it follows from

$$\xi = \nu_2 - \nu_1 \quad \text{and} \quad \nu_1 + \nu_2 = 1 \tag{4-A6}$$

that

$$\nu_2 = \frac{1}{2}(1 + \xi) \quad \text{and} \quad \nu_1 = \frac{1}{2}(1 - \xi). \tag{4-A7}$$

By the chain rule for a function $f = f(\nu_1, \nu_2) = f[\nu_1(\xi), \nu_2(\xi)]$,

$$\frac{\partial f}{\partial \xi} = \frac{\partial f}{\partial \nu_1}\frac{\partial \nu_1}{\partial \xi} + \frac{\partial f}{\partial \nu_2}\frac{\partial \nu_2}{\partial \xi} = -\frac{1}{2}\frac{\partial f}{\partial \nu_1} + \frac{1}{2}\frac{\partial f}{\partial \nu_2}. \tag{4-A8}$$

Hence

$$2\frac{\partial}{\partial \xi} = -\frac{\partial}{\partial \nu_1} + \frac{\partial}{\partial \nu_2}. \tag{4-A9}$$

Using Eqs. (4-A7) to (4-A9) in (4-A5), we obtain

$$\partial_t P(\xi,t) \sim -2\frac{\partial}{\partial \xi}\left[\frac{1}{2}(1 - \xi)v_{12}(\xi) - \frac{1}{2}(1 + \xi)v_{21}(\xi)\right] P(\xi,t)$$

$$+ \frac{2}{N}\frac{\partial^2}{\partial \xi^2}\left[\frac{1}{2}(1 - \xi)v_{12}(\xi) + \frac{1}{2}(1 + \xi)v_{21}(\xi)\right] P(\xi,t). \tag{4-A10}$$

Hence we obtain the approximate equation of motion

$$\partial_t P(\xi,t) = -\frac{\partial}{\partial \xi}[X(\xi)P(\xi,t)] + \frac{1}{2}\frac{\partial^2}{\partial \xi^2}[Q(\xi)P(\xi,t)] \tag{4-A11}$$

where

$$X(\xi) = (1 - \xi)v_{12}(\xi) - (1 + \xi)v_{21}(\xi) \tag{4-A12}$$

and

$$Q(\xi) = \frac{2}{N}(1 - \xi)v_{12}(\xi) + \frac{2}{N}(1 + \xi)v_{21}(\xi). \tag{4-A13}$$

The generation of partial differential equations from master equations such as (4-A1) is sometimes called the Kramers-Moyal technique. For reviews and further discussion see Haken (1975), Görtz (1976), Horsthemke and Brenig (1977), and Nicolis and Prigogine (1977).

Appendix 4–3

An M-Culturgen Model

Let M culturgens, $c_1, \ldots, c_M$, be accessible to each society member. When $M > 2$, the amplification equation can be derived by a direct generalization of the two-culturgen case.

The allowed transitions are of the form

$$(n_1, \ldots, n_j + 1, \ldots, n_k - 1, \ldots, n_M) \rightleftarrows (n_1, \ldots, n_j, \ldots, n_k, \ldots, n_M) \quad \text{(4-A14)}$$

for which the master equation is

$$\begin{aligned} \frac{d}{dt} P(n_1, \ldots, n_M, t) = & \\ & \sum_{\substack{j=1 \\ j \neq k}}^{M} \sum_{k=1}^{M} (n_j + 1) v_{jk}(n_1, \ldots, n_j + 1, \ldots, n_k - 1, \ldots, n_M) \\ & \quad \cdot P(n_1, \ldots, n_j + 1, \ldots, n_k - 1, \ldots, n_M, t) \\ & - \sum_{\substack{j=1 \\ j \neq k}}^{M} \sum_{k=1}^{M} n_j v_{jk}(n_1, \ldots, n_j, \ldots, n_k, \ldots, n_M) \\ & \quad \cdot P(n_1, \ldots, n_j, \ldots, n_k, \ldots, n_M, t). \end{aligned} \quad \text{(4-A15)}$$

Defining the culturgen frequencies $\nu_k = n_k/N, k = 1, M$, one obtains for $0 < \nu_k < 1$ the Fokker-Planck equation of motion

$$\partial_t P(\boldsymbol{\nu}, t) = - \sum_{j=1}^{M} \frac{\partial}{\partial \nu_j} [X_j(\boldsymbol{\nu}) P(\boldsymbol{\nu}, t)] + \frac{1}{2} \sum_{j=1}^{M} \sum_{k=1}^{M} \frac{\partial^2}{\partial \nu_j \partial \nu_k} [Q_{jk}(\boldsymbol{\nu}) P(\boldsymbol{\nu}, t)] \quad \text{(4-A16)}$$

where

$$X_j(\boldsymbol{\nu}) = \sum_{\substack{k=1 \\ k\neq j}}^{M} [\nu_k v_{kj}(\boldsymbol{\nu}) - \nu_j v_{jk}(\boldsymbol{\nu})] \tag{4-A17}$$

$$\text{and } Q_{jk}(\boldsymbol{\nu}) = \begin{cases} N^{-1} \sum_{\substack{k'=1 \\ k'\neq j}}^{N} [\nu_j v_{jk'}(\boldsymbol{\nu}) + \nu_{k'} v_{k'j}(\boldsymbol{\nu})] & \text{when } j = k, \\ - N^{-1}[\nu_j v_{jk}(\boldsymbol{\nu}) + \nu_k v_{kj}(\boldsymbol{\nu})] & \text{when } j \neq k. \end{cases} \tag{4-A18}$$

In contrast to the two-culturgen case, there is no easy procedure for getting at the stationary distribution unless the X_j and Q_{jk} have very simple properties. The difficulty can often be circumvented by clustering the culturgens into two sets, and proceeding with the analysis for the two-culturgen case.

Appendix 4–4

The Stationary Solution

From Eq. (4-26),

$$\partial_t P(\xi,t) = -\frac{\partial}{\partial\xi}[X(\xi)P(\xi,t)] + \frac{1}{2}\frac{\partial^2}{\partial\xi^2}[Q(\xi)P(\xi,t)] \tag{4-A19}$$

we seek $P(\xi)$ such that

$$\partial_t P(\xi) = 0 = -\frac{\partial}{\partial\xi}[X(\xi)P(\xi)] + \frac{1}{2}\frac{\partial^2}{\partial\xi^2}[Q(\xi)P(\xi)]$$

$$= \frac{\partial}{\partial\xi} J(\xi) \tag{4-A20}$$

where

$$J(\xi) \triangleq -X(\xi)P(\xi) + \frac{1}{2}\frac{\partial}{\partial\xi}[Q(\xi)P(\xi)], \tag{4-A21}$$

the so-called probability flux. It is evident that $\partial_t P(\xi) = 0$ when $J(\xi) =$ constant.

What is the value of this constant? We know that the society is of constant size N, so that system state transitions that take N to any different value are forbidden. In terms of ξ, this means that ξ is bounded between -1 and $+1$. Consequently no probability density $P(\xi)$ has a chance to progress beyond the interval $[-1, +1]$, and $\xi = -1$ and $\xi = +1$ are reflecting boundaries for the stochastic process $\xi(t)$. But since $J(\xi = -1)$ and $J(\xi = +1)$ are simply the loss or gain of amounts of $P(\xi)$ at the boundaries, whatever enters the reflecting boundaries is certain to be reflected back, and hence the net flow $J(\xi = -1)$ and $J(\xi = +1)$ must be zero.

However, $J(\xi)$ is constant in the stationary state everywhere on $[-1, +1]$, and it can be concluded that for the boundary condition

$$\partial_t P(\xi) = 0 \text{ if and only if } J(\xi) \equiv 0, \qquad -1 \leq \xi \leq 1 \tag{4-A22}$$

Then, from Eq. (4-A21),

$$-X(\xi)P(\xi) + \frac{1}{2}\frac{d}{d\xi}[Q(\xi)P(\xi)] = 0, \tag{4-A23}$$

or

$$\left[-X(\xi) + \frac{1}{2}\frac{dQ}{d\xi}(\xi)\right]P(\xi) + \frac{1}{2}Q(\xi)\frac{dP}{d\xi}(\xi) = 0, \tag{4-A24}$$

or

$$\int_{P(-1)}^{P(\xi)} dP(\xi) = \int_{-1}^{\xi} 2\frac{X(\xi)}{Q(\xi)}\,d\xi - \int_{Q(-1)}^{Q(\xi)} \frac{dQ(\xi)}{Q(\xi)}, \tag{4-A25}$$

or

$$\ln P(\xi) = 2\int_{-1}^{\xi} \frac{X(\xi)}{Q(\xi)}\,d\xi - \ln Q(\xi) + \text{constant}, \tag{4-A26}$$

which gives the form

$$P(\xi) = \frac{C}{Q(\xi)} \exp\left[2\int_{-1}^{\xi} \frac{X(\xi)}{Q(\xi)}\,d\xi\right] \tag{4-A27}$$

where C is a normalization constant.

CHAPTER FIVE

The Gene-Culture Adaptive Landscape

Our treatment of gene-culture amplification makes it clear that the interaction between genes and culture is especially simple when the epigenetic rules function independently of the cultural surround or any learning experiences. We have reviewed the evidence that several important cases, including brother-sister incest, patterns of color naming, and mother-infant bonding appear to approximate this condition closely. Such epigenetic rules guide the assembly of behavioral programs in a manner little influenced by the existence or nature of other culturgens. The relative simplicity of such rules makes them the natural point from which to obtain our first concrete visualization of the coupling between genetic and cultural evolution. Our purpose in this enterprise is to formulate coevolutionary pictures, in mathematical form, that incorporate the reciprocity of the two modes of change. We wish to draw conclusions from the pictures about the evolutionary mechanisms that lead to the prevalence of specific types or designs of epigenetic rules.

Whenever possible the equations and their antecedent arguments will be written in a form parallel to that used in conventional population genetics, in order to emphasize both the similarities and the differences between pure genetic evolution and gene-culture coevolution. By this means we shall show that it is possible to marshal a substantial number of existing concepts from theoretical population genetics and ecology to analyze certain aspects of gene-culture coevolution, particularly the existence of adaptive peaks for the epigenetic rules, the effects of environmental heterogeneity, and the role of purely phenotypic variation within populations.

A straightforward model of a social population with epigenetic rules independent of usage patterns is formulated in the early sections of this chapter. It is a simplified version of the models developed in Chapter 4 and is suitable for application to evolutionary time scales. Its epigenetic rules are shaped by two alleles at a single locus, and the population reproduces across partially overlapping generations. Since it is clear that single-locus models are useful for theory, but by no means obvious that they are relevant to real cognitive systems, we first gather the evidence on this question. Current findings about trait control by polygenes and by major genes suggest that properly applied single-locus models will find useful application. We split our treatment of this point between the present chapter and the one following. In the next section we examine the status of population genetics in general, reserving for Chapter 6 a more detailed treatment of the profound effects that mutations at a single locus can have on the structure of neural circuits, brain ontogeny, and the dynamics of cognition and behavior.

Since our objective is to draw conclusions about the effects of selection forces on the evolutionary trajectories of epigenetic rules, we begin with a time-independent ecology and constant selection pressures. The model is then sufficiently simple to demonstrate the existence of an adaptive landscape for the gene frequency dynamics; this landscape is a novel entity built from components of both the epigenetic rules and the fitness coefficients. Its principal merit is the clarity with which it exposes the tight coupling between selection forces and epigenetic rules in the coevolutionary process.

Next we turn to environments that vary in time or space. By applying both fitness set arguments and the Haldane-Jayakar method, we obtain a first classification of the type of epigenetic rules selected in heterogeneous environments. In the final section we treat the role of developmental noise and phenotypic variability in the coevolution, deducing that the major conclusions achieved earlier carry over unchanged and that variance in the epigenetic rules plays a potentially important secondary role in shaping the evolutionary trajectory.

In order to begin the treatment of adaptive peaks it is useful to envisage, in the manner of Wright (1932, 1970), local arrays of genotypes deployed onto the surface of a landscape, with elevation symbolizing their adaptive values. Think of a two-dimensional grid on which all possible gametes or diploid recombinants of a population are mapped. Beyond it stretches the surface of all possible gametes or recombinants that can be generated by mutation and the immigration of new allelomorphs from nearby populations. A literal, point-by-point visualization of the adaptive landscape is difficult, if not impossible, because the ultimate potential genotypic diversity is astronomically large.[1] Furthermore, for n polymorphic loci every combination and its fitness would require a diagram in $(n + 1)$-dimensional space. But a rough geometric conception of the landscape has often proved valuable. It is an analog that allows a rapid initial grasp of the process of evolution, and it has led to more precise formulations for restricted cases in the theory of population genetics, as reviewed for example by Turner (1970) and Edwards (1977).

The topography contains hills and mountains of high genetic fitness and valleys of low genetic fitness. Often the prominences are clustered into ranges, representing major adaptive categories such as the ungulate type among mammals, the finch type among birds, and the caste-prolific forms of social insects. Each population occupies a patch on the surface of the landscape. More precisely, the population is made up of various numbers of individuals possessing the genotypes represented by points on the surface. Evolution proceeds as natural selection multiplies the genotypes on the upper edge of the slope and eliminates those on the lower edge. The rate of progression is a function of the relative frequencies of the alleles and the fitnesses of their genotypic combinations. In the diallelic case, for example, where p and q are the gene frequencies and $\overline{W}$ is the mean fitness,

$$\overline{W} = p^2 W_{AA} + 2pqW_{Aa} + q^2 W_{aa}, \qquad (5\text{-}1)$$

and the rate of gene frequency change is

$$\Delta p = \frac{pq}{2\overline{W}} \frac{d\overline{W}}{dp} \quad \text{per generation.} \qquad (5\text{-}2)$$

Thus the frequency of individuals located on points of the surface containing an a allele increases as a function of the frequency of a and the mean fitness of the diploid genotypes generated by A and a. Such an ideal evolving population can achieve a steady state when most or all individuals possess the genotype that occupies the nearest adaptive peak. Or it can settle indefinitely around genotypes located on a saddle point or on an intermediate point of a knife-edge crest, where each departure is more

1. Fifty diallelic loci alone can generate 7×10^{23} diploid combinations, or approximately the number of molecules in a gram molecular weight, while one thousand diallelic loci, an ensemble actually exceeded by many species of higher plants and animals, can generate more combinations than there are atoms in the visible universe.

likely to move it down a steep slope than farther up along the crest. The population can also be deflected from its upward course by genetic drift or by the immigration of disadvantageous gene combinations from other populations. Either of these events can have sufficient impact to carry the population to another slope and start its progression toward a second, previously unattainable adaptive peak.

A more concise conception of the adaptive landscape is generated by the use of points on the surface to represent the genotype frequencies of the entire population. Thus one point P_1 is the location of a population with a particular set of frequencies of *AA*, *Aa*, and *aa* respectively. The elevation of the landscape is $\overline{W}$, the mean fitness of the population defined in Eq. (5-1). Populations tend to evolve in such a way as to gain in mean fitness; that is, they move from P_1 to other frequency combinations, say P_2 and P_3, that possess higher mean fitnesses. This is the visualization we shall apply later in the chapter in an extension of the theory to gene-culture coevolution.

The models of classic population genetics typically treat the adaptive topography as if it were static through evolutionary time. But the landscape is more accurately described as a seascape of thick syrup. Through evolutionary time, peaks slowly subside while nearby valleys fold upward into ridges and hills. The resident populations move through zigzagging trajectories; most are trapped in troughs and die out. Some regions of the surface can become even more fluid. To be explicit, important changes can occur frequently during the course of a single life cycle: forest fires strike, prey species die out, famine deepens, new predators invade, but then a rich nutfall is discovered, and so on.

The evolutionary trajectories seldom if ever lead directly toward global optima, that is, toward the highest but invariably distant mountain ranges. Conceivably there is an ideal ant—with a huge brain and steel jaws, perhaps—but it is not within reach of the ten thousand contemporary species of Formicidae during the remaining time allotted to them or to any other form of life on earth. A great variety of other only slightly inferior designs can be imagined, but they too represent virtually unattainable adaptive peaks. For this reason long-range evolution is excessively difficult if not impossible to predict, and extremum principles have proved to have limited utility in evolutionary biology (Oster and Wilson, 1978).

Yet the conception of the adaptive landscape can illuminate the short-term histories of species and help to estimate the accuracy and precision of the adaptedness of populations to their environment. For example, the caste of the leaf-cutter ant *Atta sexdens* that gathers fresh vegetation to serve as the substrate for growth of symbiotic fungus consists in workers with head widths 2.2–2.4 mm. This is precisely the group, falling within a total colony head-width range of 0.6–5.4 mm, that experiments have proved to be optimum according to the criterion of energetic efficiency (Wilson, 1980b). When models were constructed that projected additional short-range behavioral evolution, the leaf-cutting caste was found to be within 10 percent of the theoretical optimum caste (head width 2.6–2.8 mm). Thus in its crucial foraging activity *Atta sexdens* can be said to be both optimal and nearly optimized; it is located very near the top of a local adaptive peak.

The study of adaptation is most likely to progress through such small steps. Like soft landers scanning a few square kilometers of the surfaces of unknown planets, optimality techniques can be applied to experimentally tractable species to provide close views of the process of short-term evolution. Crude maps of local adaptive topography can be drawn, making it possible eventually to reconstruct evolutionary theory on a firmer empirical base.

The Genetics of Populations

Population genetics, the discipline charged with analyzing the dynamics of gene frequency change and the adaptive landscapes, has created a sophisticated theory based on the facts of Mendelian and molecular genetics. But it remains weak in its connections to the rest of biology, in particular to the study of epigenesis and behavior and to ecology, the field most concerned with the pressures of natural selection. This disparity is partly the result of historical accident. Much of the experimental research on population genetics has been devoted to *Drosophila* fruit flies, whose natural ecology is exceptionally difficult to study, and to *Homo sapiens,* which has the most complex and least tractable development and ecological relationships of any organism. It is important to note some of the difficulties in population genetic analysis and the constraints they place on immediate future studies of gene-culture coevolution.

Application of the basic theory to real systems is hampered by the potentially great complexity of interaction of gene loci above the two-locus case (Lewontin, 1974; Barker, 1979). First of all there is epistasis, defined as nonadditive interactions in the effects of alleles located on different loci. In some cases epistasis consists in the complete masking of the phenotype of alleles on one locus by those on another. However, it also includes more subtle phenomena, such as intermediate degrees of suppression across loci and multiplicative contributions by the loci to a common phenotype.

When alleles from more than one locus contribute to a single character, they are called polygenes, and the inheritance is said to be polygenic or multifactorial. Multifactorial inheritance defined in the broadest sense is all but universal; the variation of few if any traits is totally under the control of a single locus. Variation affected by polygenes can be expressed in any one of three ways (Hartl, 1980). It can be deployed into discrete classes, such as the number of bristles on the thorax of a fruit fly and the number of petals in a flower. Or the variation can incorporate a threshold: when a sufficient number of polygenes of a particular kind is present in a specified environment, the trait is expressed; below this number (in the same environment), it is absent. Examples in human beings include diabetes and schizophrenia. Finally, in what is probably the prevailing mode of expression, polygenes influence continuous variation among a vast array of traits in most categories of anatomical features, physiological processes, and behavior. Some polygenes affect the activity of others and hence are epistatic. Other kinds of polygenes simply contribute to the phenotype in an additive fashion.

Another major form of interlocus interaction is linkage disequilibrium, the departure from random association of alleles on different loci that occurs when the loci are located on the same chromosome. If alleles a_1 and b_2, representing two loci, are found mostly on one chromosome and a_2 and b_1 are found mostly on the second, homologous chromosome, the combinations a_1b_2 and a_2b_1 are likely to be more frequent among the gametes of at least the next few generations than would be expected from chance alone. Sometimes this linkage disequilibrium results simply from random genetic drift or the mixing of previously isolated and genetically different populations. The more interesting case, however, is the epistatic interaction of alleles. If alleles reinforce one another in such a way as to gain higher fitness than the fitness they possess when alone, they can attain a stable linkage disequilibrium. In a changing environment, or in the face of repeated invasions from other, genetically different populations, the linked alleles will appear to be a "supergene" prescribing a set of phenotypes different from those prescribed by alternative genes on the same loci. If a supergene heterozygote possesses a higher fitness than its respective homozygotes, the disequilibrium can be stable even in a steady-state, closed population. Examples of such assemblages that appear to have been stabilized by the coadaptiveness of their member alleles, with or without heterozygote superiority, have been well documented in field studies. They include the genes that control shell color and banding in the land snail *Cepaea* (Jones et al., 1977) and some of the constituent alleles of the chromosome inversions of *Drosophila* (Dobzhansky, 1970; Dobzhansky et al., 1977).

The effects of linkage on microevolution can be complex. Theoretical studies indicate that genes can be included in equilibrium ensembles by the hitchhiking effect, and under various conditions multiple gene frequency equilibria can arise, some in stable linkage disequilibrium and others in linkage equilibrium (Maynard Smith and Haigh, 1974).

The effect that is the inverse of interaction among loci is pleiotropy, the multiple phenotypic effects of single alleles. Although one component of fitness might be increased by the presence of a given allele, a second, unsuspected component might be diminished. Pleiotropic effects often cut across anatomy, physiology, and behavior (Futuyma, 1979). In the Norway rat the gene *achondroplasia* causes an overall abnormal growth of cartilage early in embryonic development, which in turn creates an inability to suckle, an occlusion of the incisors, and deficient pulmonary circulation. Pleiotropic effects can also be epistatic. For example, the mutation *su-pr* both suppresses the *purple* eye mutation in *Drosophila* and enhances the expression of *hairy wing,* which is yet a third mutation responsible for the growth of excess bristles.

Because of their nonlinearity and often surprising effects, interlocus interactions and pleiotropy militate against exact genetic analysis of all but the simplest biochemical traits. Yet population and developmental genetics are today moving swiftly toward solution of less tractable problems of the kind encountered regularly in the study of complex anatomy and behavior. It is relatively easy to get an order-of-magnitude estimate of the number of loci involved in polygenic control (Milkman, 1979; Hartl, 1980). Substantial progress has been made in the chromosomal mapping of polygenes in *Drosophila* (Thoday, 1979; Thompson, 1979; Thompson and Kaiser, 1979). Rough estimates of the number of polygenes have been made in the case of maze learning, morphine sensitivity, and dominance behavior in mice (Oliverio, 1979). Where explicit information cannot be immediately obtained about the genes themselves, a large armamentarium of techniques has been developed to estimate the heritability of traits. These permit close approximations of the relative contributions to phenotypic variation of the genotype, environment, and genotype-environment interaction (Falconer, 1960; Jinks, 1979; Cloninger et al., 1979a,b; Hartl, 1980; Karlin, 1980a,b).

Some authors have nevertheless remained pessimistic about the prospects of analyzing polygenic systems at the level of the population. If the human genome comprises on the order of a hundred thousand genes (as indicated for example by McKusick and Ruddle, 1977), and if a sizable fraction of these influence the variation of particular traits, the permutations of phenotype and fitness are astronomically large and the traditional dynamical equations of population genetics based on one or two loci can no longer be put into service (Lewontin, 1974). However, current research suggests that the problem may not be nearly so formidable as the more naive arithmetic exercises suggest. Thompson and Thoday (1979) have expressed the matter as follows: "There is little evidence that the number of polygenes affecting any particular trait is exceptionally large. Indeed, in those instances in which gene number has been studied in detail, a very limited number of loci account for the great majority of the genetic variation." To take one of the more carefully analyzed examples, Spickett (1963) characterized three polygenes that account for most of the known variation in sternopleural bristle number of *Drosophila melanogaster*. In the realm of behavior, Kovach's (1980) study of color preference in quail chicks has implicated four to eight segregating units of inheritance.

Moreover, many systems observed in natural populations are not under the control of polygenes as defined in the narrow sense, with multiple loci making roughly equal contributions to phenotypic variation. There is instead a single major gene (or "switch gene") whose effects may be altered to some degree by modifiers. During the past ten years numerous single-locus mutants have been identified in *Drosophila* that alter behavior and information processing through the modification of individual steps in virtually every conceivable episode of cell assembly, histogenesis, and functioning of the nervous system (Hall and Greenspan, 1979). Their various effects touch most of the sensory modalities and range in magnitude from very minor to massive and lethal. Some even alter particular forms of learning ability, including the *x*-chromosome *amnesiac,* which reduces memory and apparently nothing else.

Examples of relatively simple forms of control in human heredity include a major gene affecting certain forms of spatial ability (Ashton et al., 1979) and what is probably a dominant autosomal gene that confers on adults the ability to produce lactase, and hence to digest milk. The frequency of the latter trait shows a large amount of variation among human populations, closely correlated with the use of milk and milk products in the diet (Kretchmer, 1972). When ensembles of loci are involved, their alleles are sometimes stably linked into segregating units by inversions, translocations, or strong epistatic interactions and as such can be treated by the more elementary models of population genetics based on few-locus systems (Dobzhansky, 1970).

Microevolutionary studies that focus on major genes and blocks of genes have been conducted on a diversity of species of butterflies, moths, *Drosophila* fruit flies, flowering plants, and other organisms. It is not uncommon to observe selective advantages of 20 percent or higher under natural conditions (Ford, 1971). We see no reason to doubt that similar information at the genic level can be obtained for directional microevolution in human populations, such as the increase in brachycephaly in Central European populations during the past seven hundred years (Bielicki and Welon, 1971), the shortening of stature and lightening of skin pigment in equatorial forest pygmies and pygmoids during the past twenty thousand years (Hiernaux, 1977), and the presumed increase in endurance and resistance to pain of Papuans during a still unknown period of time (Gajdusek, 1970). The prospect is brightened further by rapid progress in human genetics as a result of the electrophoretic separation of isoalleles, the full chemical characterization of enzymes, and other molecular techniques. By 1977, according to McKusick and Ruddle, 1,200 human genes had been identified, of which 210 were mapped onto a particular chromosome. At least one gene had been located on each of the twenty-three pairs of chromosomes. Although most of these genes affect primarily anatomical and biochemical traits, many alter behavior as well. Of equal significance, the first human genes have now been isolated and chemically characterized (Derynck et al., 1980).

We suggest that an efficient strategy of research on the genetic basis of brain and behavioral microevolution will be to concentrate on cases in which the variation of relatively well defined epigenetic patterns is controlled by a major gene, or at most a relatively small number of polygenes. The evidence accumulated thus far indicates that these simpler systems are common and will be increasingly accessible with the aid of biochemical techniques. The single-locus and few-locus models arising in application of gene-culture theory to these systems are also relatively simple and will therefore generate a rapid increase in our understanding of both the organisms and the basic theory.

The Existence of the Coevolutionary Landscape

If we proceed on the assumption that the genetic analysis of social behavior is feasible, an exact theory of coevolution must start by asking two questions of basic importance. Does a gene-culture adaptive topography exist that is in any sense comparable to the pure genetic topography? And if so, can the theory of population genetics and ecology be extended to characterize the gene-culture topography in a manner that enhances the analysis of coevolutionary strategies? We have been able to answer both of these questions affirmatively for the classic case of one locus. There exists a function of gene and culturgen frequencies for specified survival rates, reproductive rates, and epigenetic rules that increases or at least remains constant in each successive generation. As a result, equations of motion can be written for gene-culture coevolution that extend conventional population genetic theory into a new domain.

The generations in the idealized populations that we shall consider overlap in time except for their respective periods of reproduction. The generations are made to be quasi-discrete in a way that allows for a transparent accounting of gene frequencies. The life cycle of an organism engaged in gene-culture coevolution within this population proceeds in successive steps:

(1) Zygotes are formed and the young grow and develop.

(2) The young pass through an enculturation period. Biased to a greater or lesser extent by the particular epigenetic rules they have inherited genetically, the young acquire a specific repertory of cultural schemata. The process of culturgen innovation plays a key role in the continuing coevolution. When the frequency of a culturgen is low, a few deaths or emigrations can result in its abrupt loss from the population and hence a drop in cultural diversity. As a necessary condition for the indefinite maintenance of both culturgens, we suppose that the members are innovators. As a result culturgens c_1 and c_2 are continuously rediscovered and are thus always accessible to the juveniles.

(3) As the elders die off, the young survive and reproduce at rates set by the culturgens they have assimilated. Although the alleles shape cognitive epigenesis, it is the culturgens expressed in actual behavior that determine fitness. For these organisms culture has become a crucial part of the adaptive repertory. The young adults mate at random with respect to the genes and culturgens.

(4) The offspring of the young start the next life cycle.

Each of the principal conditions in these steps can be eased to a substantial degree without dire effects on the qualitative conclusions now to be derived from the more rigid elementary version.

The pattern of enculturation is consistent with the known rules of socialization in cultural animals (Bonner, 1980) and the great majority of economically more primitive human societies (Williams, 1972a,b). In the case of human beings, the young are taught by a substantial portion of the adult members of the group both within and outside the family, so that in many categories of culturgens there is a relative homogeneity in the teaching and learning processes. Furthermore, hunter-gatherer groups, in which most gene-culture coevolution has taken place, have throughout human history contained only about fifteen to seventy-five individuals (Wobst, 1974; Hage, 1976; Buys and Larson, 1979). Thus it is realistic to posit a case in which there are uniform enculturation practices throughout the society within any given generation. (See Chapter 4 for a fuller discussion of enculturation.)

In the simplest case, the coevolution of two alleles (A,a) and two culturgens (c_1, c_2) is traced by following the population repetitively through its life cycle:

Zygotes. In a population of N_t individuals at time t, the numbers of the diploid genotypes among the zygotes are given by the Hardy-Weinberg distribution:

$$p_t^2 N_t + 2p_t q_t N_t + q_t^2 N_t = N_t \tag{5-3}$$

where p_t and q_t are the frequencies of the two alleles A and a, and $p_t + q_t = 1$.

Enculturation. Most protocultures and eucultures are likely to fall somewhere between the two extreme strategies of enculturation shown in Figure 5–1A. For human beings the assimilation of culturgen alternatives and their retention in long term memory has been firmly demonstrated for categories of experience involving both observational and vicarious learning (for a detailed review see Rosenthal and Zimmerman, 1978), in accordance with everyday introspection. The limiting case of saturation by a single culturgen during enculturation is therefore of greatest interest in those cases where individuals do not switch culturgens following an initial episode of evaluation and decision. This pattern is expected to occur in categories of thought shaped by the primary epigenetic rules and the subset of relatively selective and inflexible secondary epigenetic rules, such as those governing fears and phobias. It might also characterize epigenetic rules that are flexible and thus entertain multiple culturgens, yet summon feelings of extreme neophobia once the individual has made an initial decision. This condition appears to typify the activation of specific codes of moral conduct and religious ideology, systems of thought that can become extremely persistent once installed in the mind.

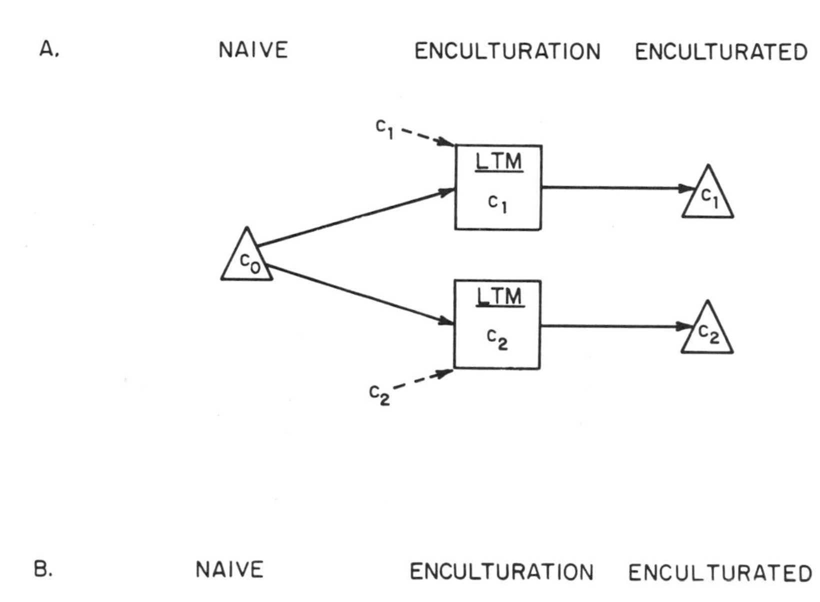

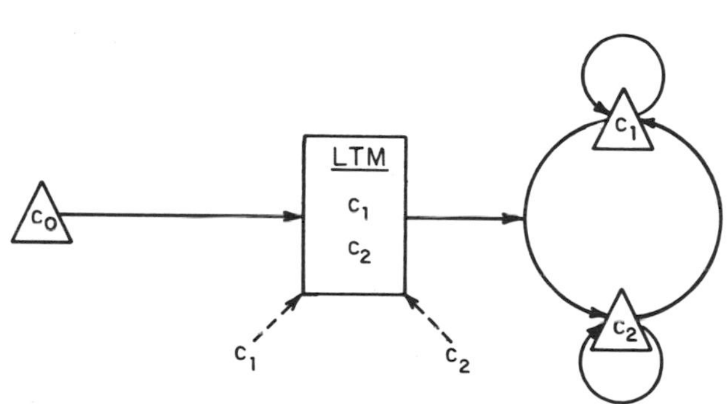

Figure 5–1 The full decision process in the life cycle, which entails enculturation (passage from c_0, or no culturgens) as well as later decisions to retain the same active culturgen or to shift to the alternative choice. *A*: Enculturation in which exactly one culturgen from each culturgen category is used. Thus the epigenetic rules shut down the learning and choice process as soon as one member of the category is learned, or following the first choice between learned alternatives. *B*: More realistic enculturation strategy with multiple culturgens learned, followed by continued decision making. LTM = long term memory.

Furthermore, as cultures grow in complexity and there is an accompanying increase in the number of culturgens circulating within a social group, strategies devoted to the acquisition of all or even a large fraction of the set of existing culturgens will cease to be efficient unless a macroevolutionary reorganization occurs in the brain systems that serve long term memory. The reason is the relatively lengthy times needed to write a culturgen into long term memory (approximately 10 seconds per each old chunk incorporated within the new chunk; see Chapter 3). The organism simply does not have the time to "know everything" and must cede to others certain rights to information. But for our present purposes it is sufficient to note that a full-assimilation strategy will again be most relevant for protocultural groups, in which relatively few culturgens circulate, and for the eucultures of small hunter-gatherer bands, in which jack-of-all-trades enculturation is still practiced.

In Chapter 4 we dealt in depth with both the epigenetic mechanisms of decision making and their consequences for cultural structure and change. In this chapter we propose to treat in detail the consequences of the epigenetic strategy displayed in Figure 5–1A for gene frequency changes on evolutionary time scales. Thus each juvenile comes to rely on a single culturgen. The epigenetic rule that biases individuals toward c_1 or c_2 is inflexible; in other words, we shall in this chapter direct our attention toward ξ-independent assimilation curves like those of brother-sister incest. Chapter 6 will then introduce a unified mathematical theory in which the interaction of genes and culture is based upon the far more general models of the coevolutionary process shown in Figure 5–1B.

As described in the previous chapters, the epigenetic rules generate probability distributions for the acquisition and utilization of various culturgens within a given culturgen category. For convenience these distributions will again be called *usage bias curves* (or, in the interest of brevity, just *bias curves*). The quantity $u(c_k|G_iG_j)$ will be the probability that following enculturation the individual will have chosen to use the culturgen c_k, on which it relies for the rest of its life (see Figure 5–2). When the context is clear, the symbol $u(c_k|G_iG_j)$ will be abbreviated to u_k.

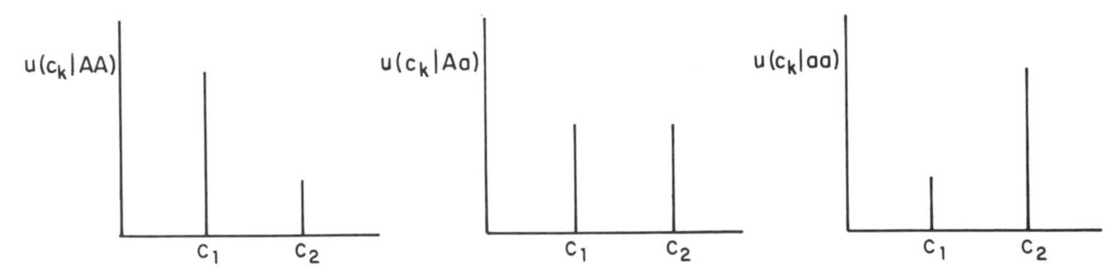

Figure 5–2 An example of usage bias curves in which there are two culturgens (c_1 and c_2) and two alleles (A and a). The curves result from the operation of epigenetic rules during development, which in turn are under the control of the alleles.

Selection. The fitnesses $W_{ij}(c_k)$ will represent conventional absolute selective values, with $2W_{ij}(c_k)$ being the total number of gametes produced by an organism that has used c_k throughout the prereproductive period T.

We now define an *absolute fitness value vector* for genotype ij as

$$\mathbf{W}_{ij} \triangleq [W_{ij}(c_1),\ W_{ij}(c_2)]. \tag{5-4}$$

Similarly, a *usage bias vector* for genotype ij is defined:

$$\mathbf{L}_{ij} \triangleq [u(c_1|ij), u(c_2|ij)]. \tag{5-5}$$

The components of $\mathbf{W}_{ij}$ are simply the conventional absolute selective values, one for each culturgen. Similarly, the components of $\mathbf{L}_{ij}$ are the predisposition values of the bias curves, which are derived from the genetically determined epigenetic rules. Let $\circ$ denote the dot product operation between vectors.

The frequency of allele A after one generation of *gene-culture transmission* is

$$p_{t+1} = \frac{p_t^2\mathbf{W}_{AA} \circ \mathbf{L}_{AA} + p_t q_t \mathbf{W}_{Aa} \circ \mathbf{L}_{Aa}}{p_t^2\mathbf{W}_{AA} \circ \mathbf{L}_{AA} + 2p_t q_t \mathbf{W}_{Aa} \circ \mathbf{L}_{Aa} + q_t^2 \mathbf{W}_{aa} \circ \mathbf{L}_{aa}}. \tag{5-6}$$

This expression may be usefully compared with the equation for gene frequency change in the case of ordinary *genetic transmission:*

$$p_{t+1} = \frac{p_t^2 W_{AA} + p_t q_t W_{Aa}}{p_t^2 W_{AA} + 2p_t q_t W_{Aa} + q_t^2 W_{aa}}. \tag{5-7}$$

Similarly, population growth dependent on gene-culture transmission is

$$N_{t+1} = N_t(p_t^2\mathbf{W}_{AA} \circ \mathbf{L}_{AA} + 2p_t q_t \mathbf{W}_{Aa} \circ \mathbf{L}_{Aa} + q_t^2 \mathbf{W}_{aa} \circ \mathbf{L}_{aa}), \tag{5-8}$$

which compares with population growth dependent on conventional genetic transmission:

$$N_{t+1} = N_t(p_t^2 W_{AA} + 2p_t q_t W_{Aa} + q_t^2 W_{aa}). \tag{5-9}$$

Thus ordinary fitness W_{ij} is replaced by gene-culture fitness $\mathbf{W}_{ij} \circ \mathbf{L}_{ij}$. The epigenetic rules transform the genetic landscape by creating bias curves that "fold into" the topography based on genotypes alone (Figure 5-3).

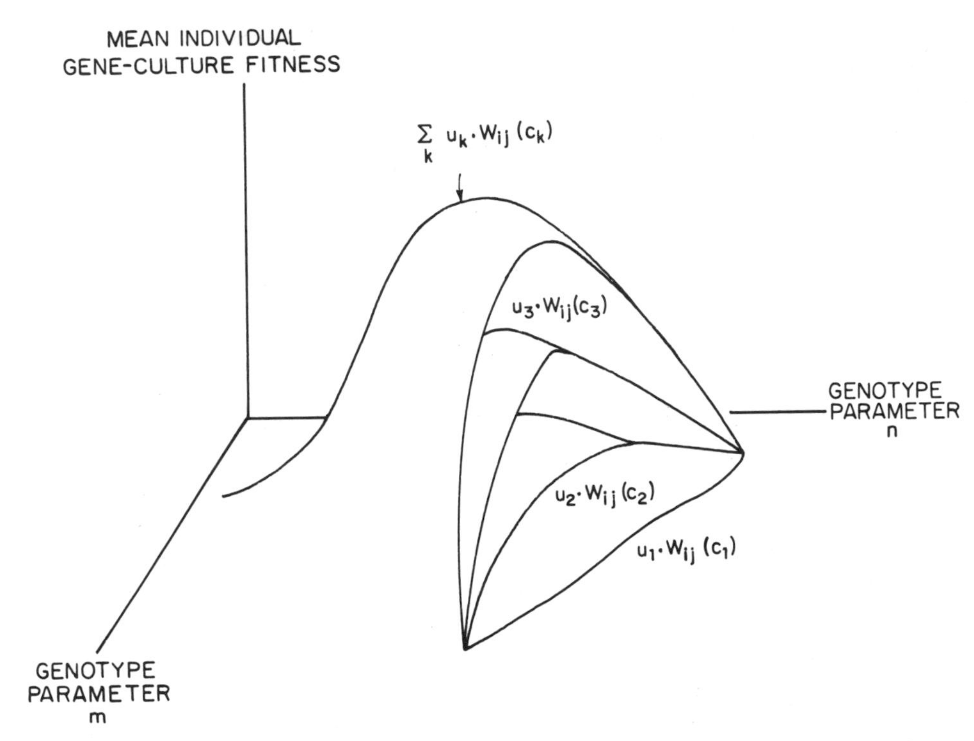

Figure 5–3 The individual gene-culture fitness landscape for a genotype G_iG_j and culturgen series c_k. The mean fitness is a linear combination of selective values $W_{ij}(c_k)$ over the culturgens c_k. The weighting factors are the epigenetic rule values $u_k \triangleq u(c_k|G_iG_j)$. Insofar as the epigenetic rules and the fitness values are expressions of the activity of the underlying genotype, there will be parameters, including the epigenetic rules themselves, that determine the location of a specific genotype on this individual fitness landscape.

The topography can now be characterized more explicitly. We define the *population-average gene-culture fitness* $\langle \mathbf{WL} \rangle$ by the following equation:

$$\langle \mathbf{WL} \rangle = p_t^2 \mathbf{W}_{AA} \circ L_{AA} + 2p_t q_t \mathbf{W}_{Aa} \circ \mathbf{L}_{Aa} + q_t^2 \mathbf{W}_{aa} \circ \mathbf{L}_{aa}. \qquad (5\text{-}10)$$

The gene frequency dynamics expressed in Eq. (5-6) is equivalent to a dynamics

$$\Delta p = \frac{1}{2} p_t q_t \frac{d}{dp_t} \ln \langle \mathbf{WL} \rangle \qquad (5\text{-}11)$$

on the $\langle \mathbf{WL} \rangle$ surface, where $\Delta p = p_{t+1} - p_t$ is the frequency change in a generation. In the expected manner of an entity moving upward on a landscape, the population evolves so as to maximize its mean gene-culture fitness, $\langle \mathbf{WL} \rangle$. It finds equilibria at the zero-slope points of the landscape surface, and the standard theorems concerning single-locus trajectories then apply.

The abundance of the culturgens can also be derived. Let $l_{ij}(c_k)$ be the probability that a newly enculturated young organism survives the prereproductive period T to become a fertile adult. The number of c_1 culturgens in the time t adult phase is given by

$$n_{1,t} = l_{AA}(c_1)u(c_1|AA)p_t^2N_t + l_{Aa}(c_1)u(c_1|Aa)2p_tq_tN_t$$

$$+ l_{aa}(c_1)u(c_1|aa)q_t^2N_t. \tag{5-12}$$

The number of c_2 culturgens is obtained by replacing c_1 with c_2 everywhere. The culturgen frequencies are then

$$\nu_1 = n_1/(n_1 + n_2) \quad \text{and} \quad \nu_2 = n_2/(n_1 + n_2) \tag{5-13}$$

and the parameter ξ of cultural organization is given by

$$\xi = \nu_2 - \nu_1. \tag{5-14}$$

It can readily be seen that if gene-culture coevolution favors an all-AA population on a static landscape, the developing young in the population will saturate on the AA bias curve, and the culturgen frequencies among the adults will be determined by $l_{AA}(c_1)u_1$ and $l_{AA}(c_2)u_2$. Similarly, an all-aa population will saturate on the aa bias curve. If a stable polymorphism is attained, by heterozygote superiority or some other special selection regime, the associated culturgen frequencies are a function of all three genotypes and are given by Eq. (5-13).

The determination of the ethnographic distribution $P(\xi)$ for this culture is now straightforward. Since there is no opportunity for the society members to reevaluate their initial choice of culturgen, they cannot switch between c_1 and c_2. The usage pattern is fixed at the value of ξ given by Eq. (5-14), and $P(\xi)$ itself is a Dirac delta function centered on this ξ. Hence for the present class of epigenetic rules the corresponding ethnographic curves are easily worked out.

This basic model captures gene-culture transmission in the case where genetic fitness of individuals is determined by their personal genotypes and social behavior. A more complete model will eventually incorporate a full measure of inclusive fitness, in which the effects of personal behavior on the fitnesses of kin are tracked explicitly; increments to the fitness of kin are devalued by the coefficient of relationship ($r = \frac{1}{2}$ for siblings, $\frac{1}{8}$ for first cousins, and so forth) and then added to the base level fitness of the individual. In the original formulation by Hamilton (1964) the relationship was conceived as additive and can be expressed most concisely as follows:

$$w = 1 + \delta w + e,$$

where 1 is the basal reference level, δw is the individual's contribution, e is the kin effect, and w is the total fitness. More general formulations will allow for nonlinear effects between ego and its relatives and among the relatives, a circumstance especially likely to exist in complex eucultural systems; for example

$$w = 1 + f(e) + \delta w(e).$$

Heterogeneous Environments

In evolutionary time, spanning in most cases ten generations or more, few adaptive topographies are static. The environment of the population is inevitably heterogeneous in time. Qualities important to the survival and reproduction of its members fluctuate, often violently. Even if for some period the physical environment remains approximately constant the biotic environment, composed of predators, parasites, competitors, symbionts, and food species, is certain to change. In the human environment the socioeconomic structure itself can be altered drastically, with or without accompanying ecological change (see Figure 5-4).

Figure 5–4, opposite Examples of environmental fluctuations of importance in gene-culture coevolution. These variations can be ecological or social, and they can occupy vastly different time spans. Here we see seasonal changes in the principal food plants of the ≠Kade San of the central Kalahari Desert; annual rainfall at Visakhapatnam, India, over an 85-year period; and Sorokin's estimates of major social disturbances in Europe, including social, political, and religious upheavals, during a 1,400-year period. (Modified from Tanaka, 1976; Thomas, 1971; and Sorokin, 1957.)

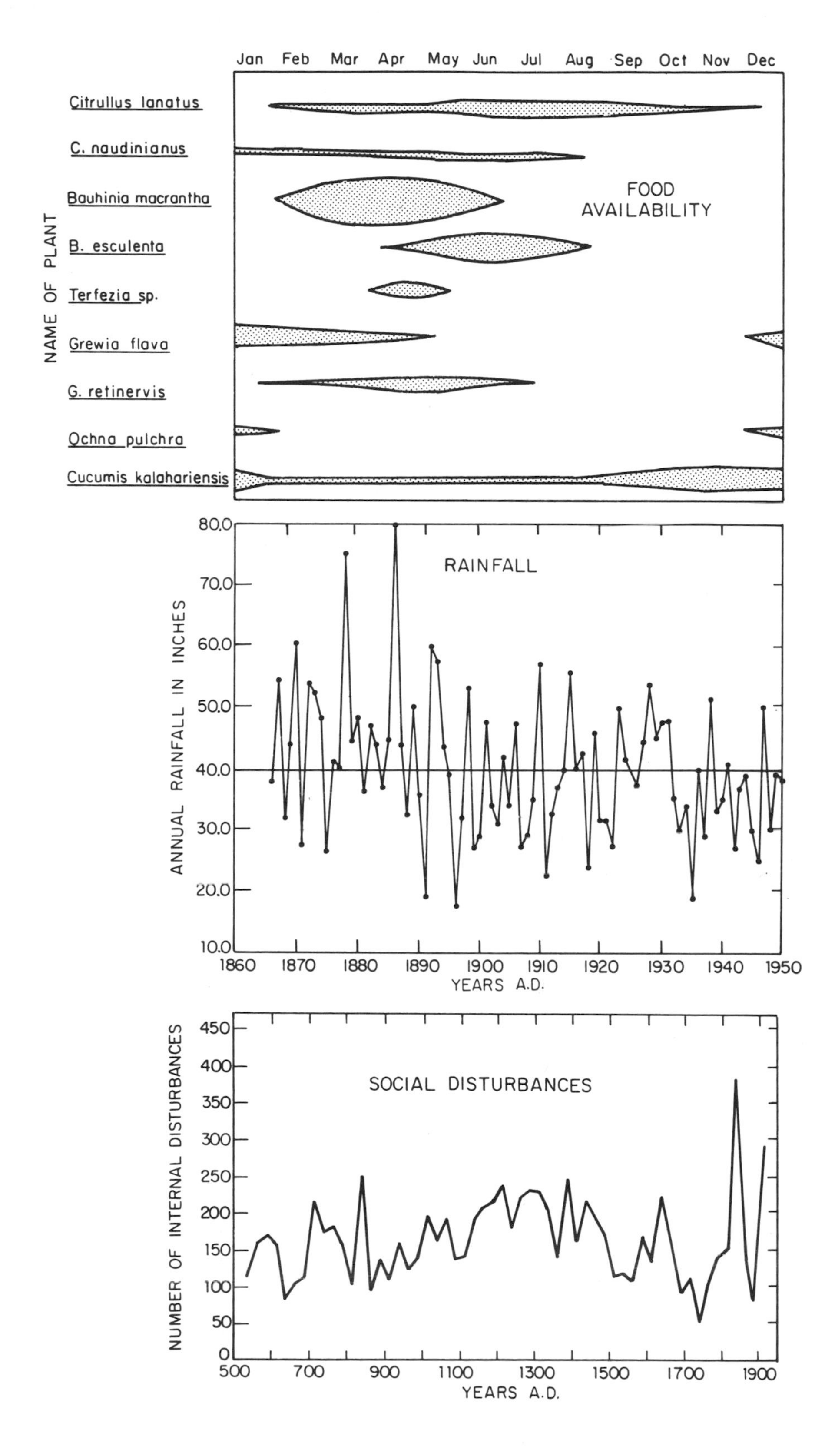

Jan Feb Mar Apr May Jun Jul Aug Sep Oct Nov Dec
NAME OF PLANT
Citrullus lanatus
C. naudinianus
Bauhinia macrantha
B. esculenta
Terfezia sp.
Grewia flava
G. retinervis
Ochna pulchra
Cucumis kalahariensis
FOOD AVAILABILITY
RAINFALL
ANNUAL RAINFALL IN INCHES
80.0
70.0
60.0
50.0
40.0
30.0
20.0
10.0
1860 1870 1880 1890 1900 1910 1920 1930 1940 1950
YEARS A.D.
SOCIAL DISTURBANCES
NUMBER OF INTERNAL DISTURBANCES
450
400
350
300
250
200
150
100
50
0
500 700 900 1100 1300 1500 1700 1900
YEARS A.D.

Generation length affects the manner in which the evolving population responds to particular forms of environmental fluctuation. If many generations pass during the course of one upward or downward turn, the population will tend to track the change with genetic evolution. If many reversals in the environmental conditions occur within single generations and repeatedly through multiple generations, the population is likely to acquire an intermediate or even polymorphic genotype that reflects the statistical properties of the fluctuation. At the same time it may follow each twist and turn with short term cultural adjustments. But in both cases the overall gene-culture fitness depends on the length of the period that the population tracks the environment (Figure 5-5). For the fruit fly *Drosophila pseudoobscura* the onset of a single summer is a secular climatic trend covering several generations, and the frequency of its chromosome inversions changes in perceptible microevolution. For man the same period is a brief seasonal shift having little or no effect on gene or chromosome-type frequencies.

The spatial environments of most species of animals are also heterogeneous. When individuals pass from one habitat to another or from one kind of food item to another at frequent intervals during their lifetimes, that collective portion of the environment is said to be fine-grained. When the organisms remain within one patch or choose one dietary item consistently for relatively long periods, the environment is said to be coarse-grained. The dividing line between the two categories depends on the mobility and longevity of the species. What is fine-grained to a human being is often coarse-grained for an insect or other small animal (Figure 5–6). As Levins (1968) and other population biologists have shown, the

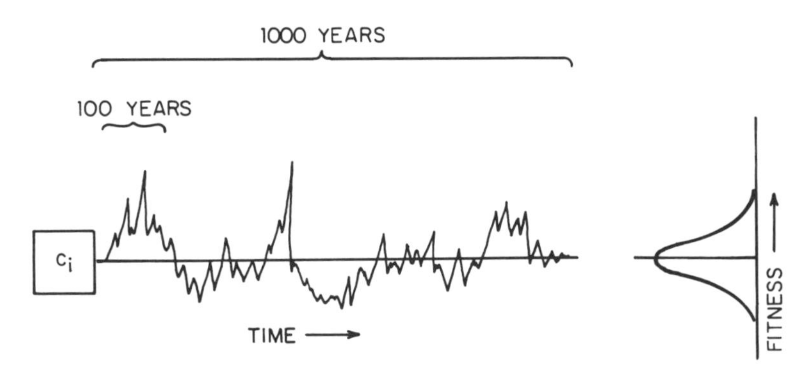

Figure 5–5 The genetic fitness of a culturgen (c_i) in a fluctuating environment. The fitness is not a single value, but deployed in a frequency distribution.

Figure 5–6 Variation in the grain of heterogeneous environments according to the size and mobility of the species. To the human family the tree is part of a fine-grained environment, an entity visited and deserted at frequent intervals. To the colony of weaver ants the tree is a coarse-grained patch, on which it remains for most or all of its life. The ant workers treat leaves as part of a fine-grained division of the tree, but the scale insects they visit (to collect sugary excrement) may reside permanently on one leaf and hence treat it as a coarse-grained patch. (Original drawing by Karen S. Ku.)

degree of fineness of each environment profoundly influences the purely genetic strategies of species adapting to it. The concept of grain is readily extended to include the cultural environment. In a fine-grained environment, for example, the individual moves often among tribes or protocultural groups, encountering different culturgens at relatively frequent intervals.

The molding of gene-culture transmission to heterogeneous environments can be approached using an optimization technique first introduced by Levins (1962, 1968). His fitness set theory provides a convenient digression from our main line of development because it is one component in a set of powerful and refined optimization methods relevant to evolutionary biology (Maynard Smith, 1978; Oster and Wilson, 1978). Despite its simplicity and its ambiguity about the choice of fitness-maximization rule, Levins' technique generates interesting and important predictions about phenotypes best suited to heterogeneous environments. Since it is not yet apparent what sort of optimization rules might hold for the majority of realistic gene-culture models, we have for the most part relied on the more basic equations of gene frequency change elsewhere in the book. On the basis of our observations about the gene-culture landscape, however, we feel that it is appropriate to anticipate the situation somewhat and illustrate the type of gene-culture problem to which fitness-maximization arguments can be extended. In the following development we shall adapt conventional fitness set procedures, the background and limitations of which have been treated elsewhere (Levins, 1962, 1968; Strobeck, 1975; Maynard Smith, 1978). The disclosure of optimization principles that cover many types of gene-culture systems will warrant more detailed optimization models than are justified at present.

A single phenotypic variable x is used to summarize the epigenetic rule design in more concise form. In the two-culturgen case it replaces $u(c_1)$, the probability of selecting c_1 as opposed to c_2; thus $u(c_2) = 1 - x$. In this simplest of cases, knowledge of x provides all of the information required concerning the bias curve.

Now consider an environment that alternates between two states, E_1 and E_2. In E_1 the culturgen c_1 is very adaptive compared to its alternative c_2, while in E_2 it is deleterious. If the gene-culture topography were static and the environment remained in E_1, a total commitment to culturgen c_1 would represent a peak on the landscape toward which the population would be expected to climb. In an environment consisting entirely of E_2, the peak would lie at $x = 0$. Suppose, however, that there is a temporal succession of E_1 and E_2, as for example in the series

$$. . . E_1E_1E_1E_2E_1E_2E_2E_1E_2E_1E_1 . . .$$

The fluctuation can be due either to changes in the environment or to movement of the organisms back and forth between patches in the two states. What is the optimal gene-culture "choice" in such a case?

Referring back to the basic model, we now define the expected fitness of a genotype with phenotype x in any particular generation:

$$W_t(x) = W_t(c_1)x + W_t(c_2)(1 - x), \qquad (5\text{-}15)$$

where $W_t(c_1)$ is the absolute selective value associated with full c_1 usage during the prereproductive period T in generation t, and $W_t(c_2)$ is the value for full c_2 usage. Thus $W_t(x)$ is the mean fitness of an individual in a population composed of pure x-morphs. Rearranging Eq. (5-15), we obtain

$$W_t(x) = [W_t(c_1) - W_t(c_2)]x + W_t(c_2), \qquad (5\text{-}16)$$

which forms a straight line with slope $[W_t(c_1) - W_t(c_2)]$. As depicted in Figure 5–7, the slope is positive when $W_t(c_1) > W_t(c_2)$, which occurs in the E_1 environment, and it is negative if $W_t(c_1) < W_t(c_2)$, which occurs in the E_2 environment. The two conceivable cases are given in the upper and lower pairs of figures respectively: partial culturgen matching occurs when both culturgens confer some amount of genetic fitness, while in complete culturgen matching each of the culturgens has zero fitness in one of the environments.

Examine the right-hand diagrams in Figure 5–7. The $\mathscr{L}$-curve defines the set of admissible loci in fitness space onto which a member of the population can fall. For every x there is a coordinate $[W(x|E_1), W(x|E_2)]$, which is a point on $\mathscr{L}$. The question of interest is, Which point or set of points on $\mathscr{L}$ is best suited to the pattern of environmental change?

In answer it can be said that optimal adaptation to a heterogeneous environment is achieved by one or the other of two general classes of tactics. The first procedure is to invest in a single best-adapted phenotype, which in the case of a gene-culture system is the epigenetic rule parameterized as x. The second procedure is to deploy a mixture of epigenetic rules. To illustrate the latter case we shall suppose that two alternative rules, x_1 and x_2, are present in the population with frequencies ν and $(1 - \nu)$ respectively. The two states can be maintained discretely by one or the other of two devices: either balanced genetic polymorphism en-

PARTIAL CULTURGEN MATCHING

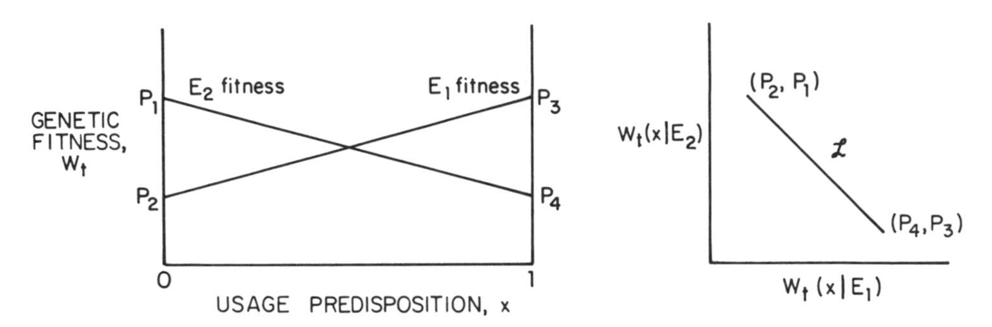

COMPLETE CULTURGEN MATCHING

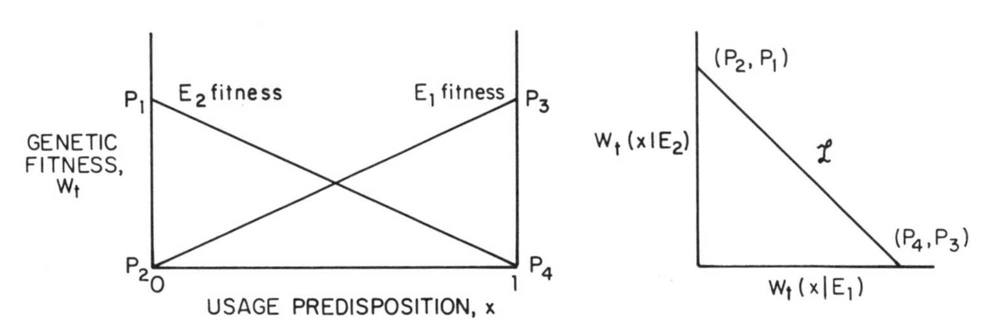

Figure 5–7 The genetic fitness of bias curves, varying by the predisposition to use one culturgen over another, in two environments. The upper pair of diagrams (*partial culturgen matching*) illustrates a case in which bearers of culturgen c_1 are better adapted to environment E_1, but bearers of c_2 have some ability to survive and reproduce also (and the reverse is the case in environment E_2). The lower pair of diagrams (*complete culturgen matching*) illustrates a case in which c_2-bearers have zero fitness in E_1 and c_1-bearers have zero fitness in E_2. The points P_i designate the fitness values obtained when one culturgen is always chosen over another ($x = 0, 1$). $\mathscr{L}$ is the fitness space of the population.

tailing two or more alleles, or else reliance on bifurcation in the developmental pathway, which can be programmed with a single allele or some other monomorphic genetic arrangement (see Figure 5–8). The first case corresponds to ordinary genetic polymorphism (Dobzhansky et al., 1977). The second is found, for instance, in the program of caste differentiation in the social insects (Oster and Wilson, 1978; Brian, 1979) and can be applied by functional analogy to major role differentiations within human societies. Similarly, and regardless of the mechanism underlying the

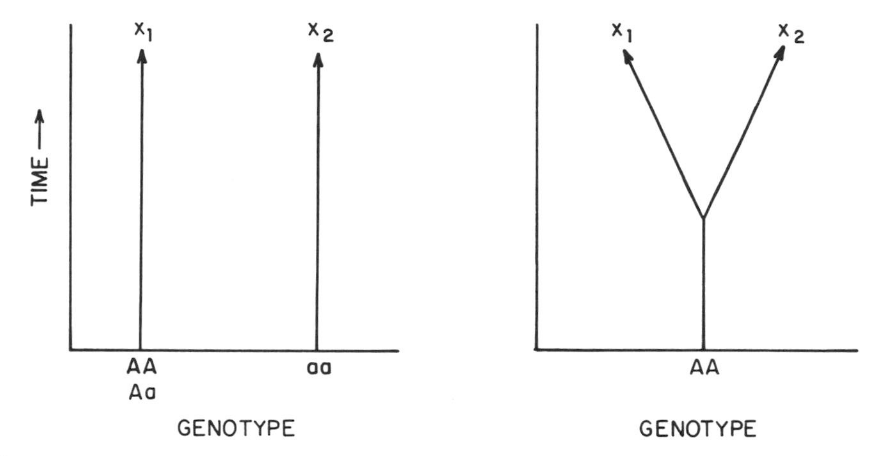

Figure 5–8 Differentiation of usage bias curves (and the combinations of epigenetic rules producing them) within a population. *Left,* the curves originate from different genotypes; *right,* the curves originate from an environmentally triggered bifurcation in development, with the total pattern of developmental response being determined by a single genotype.

polymorphism, the theory can be applied both to cases in which individual-level selection prevails and to those in which group selection is dominant. The special circumstances in which group selection can become important, particularly among highly organized societies and certain types of structured populations, have been discussed by E. O. Wilson (1975), Oster and Wilson (1978), and D. S. Wilson (1980).

The mean fitness in environment E_1 is

$$F_1(\nu) = \nu W(x_1|E_1) + (1 - \nu)W(x_2|E_1), \tag{5-17}$$

while that in environment E_2 is

$$F_2(\nu) = \nu W(x_1|E_2) + (1 - \nu)W(x_2|E_2). \tag{5-18}$$

These equations can be plotted into the right-hand diagrams of Figure 5-7 by forming them into coordinates

$$\begin{aligned} \mathcal{P} &\triangleq [F_1(\nu), F_2(\nu)] \\ &= \nu \cdot \{[W(x_1|E_1), W(x_1|E_2)] - [W(x_2|E_1), W(x_2|E_2)]\} \\ &\quad + [W(x_2|E_1), W(x_2|E_2)], \end{aligned} \tag{5-19}$$

which constitute a vector running for a distance proportional to ν between the points $[W(x_2|E_1), W(x_2|E_2)]$ and $[W(x_1|E_1), W(x_1|E_2)]$ on $\mathscr{L}$. It follows that there are two ways to reach any point on $\mathscr{L}$. The population (that is, the whole breeding population, society, or progeny group) can deploy monomorphs that possess the x-phenotype matching the point. Alternatively, the population can deploy a mixture of morphs such that their x-values straddle the single x-monomorph value just cited. For example, if the optimum monomorph is 0.5, morphs of 0.4 and 0.6 could be deployed equally well. An infinite number of other values can be used to generate a mixture with the same mean intermediate value.

With the bias curve fitness space identified, it is possible to consider at greater length the effects of differing patterns of environmental fluctuation on gene-culture coevolution. An entrée into this problem is provided by the distinction between coarse-grained and fine-grained environments. Suppose that at intervals on the order of a generation or longer, a hunter-gatherer band shifts out of one principal habitat into another. For example, it may move from forest to grassland and back, or stay in one home range as the environment passes through cycles of forest encroachment and deforestation caused in turn by major fluctuations in rainfall. If W is the mean absolute fitness of individuals in the band, the ratio of the number of individuals to the initial number after n generations is W^n. If P is the proportion of generations spent in the first habitat, where the fitness is W_1, and $(1 - P)$ is the proportion spent in the second habitat, where the fitness is W_2, then for the case of purely genetic transmission

$$W^n = W_1^{nP} W_2^{n(1-P)},$$

$$W = W_1^P W_2^{(1-P)}, \tag{5-20}$$

and

$$W_1 = W^{1/P} W_2^{(P-1)/P}.$$

The equivalent expressions in gene-culture transmission are

$$W(x) = [W(x|E_1)]^P [W(x|E_2)]^{(1-P)}$$

and

$$W(x|E_1) = W(x)^{1/P} [W(x|E_2)]^{(P-1)/P}. \tag{5-21}$$

Suppose instead that the portion of the environment in the natural-selection regime of the hunter-gatherer band is fine-grained; for example, two different forms of food items alternate repeatedly during one generation or even during the course of a single day. Through generation after generation there is a more or less constant flow of approximately the same mixture of these food items. The genetic fitness consequently will be based on the summation of the proportions of the two forms of items encountered within generations. In the conventional case of genetic transmission the fitness is the arithmetic mean

$$W = PW_1 + (1 - P)W_2$$

and

$$W_1 = \frac{1}{2}[W - (1 - P)W_2], \qquad (5\text{-}22)$$

and the equivalent expressions in gene-culture transmission are

$$W(x) = P[W(x|E_1)] + (1 - P)[W(x|E_2)]$$

and

$$W(x|E_1) = \frac{1}{P}[W(x) - (1 - P)W(x|E_2)]. \qquad (5\text{-}23)$$

The goal in the optimality problem is maximization of these mean fitness functions. The solution of the problem for particular populations is the intersection of the curves of highest average fitness (Eqs. 5-21 and 5-23) with $\mathscr{L}$, the actual fitness sets of the populations (see Eqs. 5-16 to 5-18 and Figure 5–7). In Figure 5–9 we have performed this operation for the general case of coarse-grained environments. It can be seen that the maximum mean fitness hyberbola is likely to touch the true fitness space at an intermediate position. The population has the option of adopting either a single relatively unselective bias curve or a mixture of selective bias curves whose mean probability of usage equals the probability of usage of the unselective curve.

Figure 5–10 depicts the expected outcome in fine-grained environments. Here the solution is very different from that in coarse-grained environments. The population should choose one or the other of two relatively selective bias curves. Depending on the degree of differentiation of the microenvironments being encountered in a fine-grained fashion, it is possible for the species to evolve the capacity to make only a single response. This extreme result corresponds to the case labeled "complete culturgen matching" at the bottom of Figure 5–7.

Next consider the special but potentially important case in which the alternative culturgens are coadapted. This expression means that assimi-

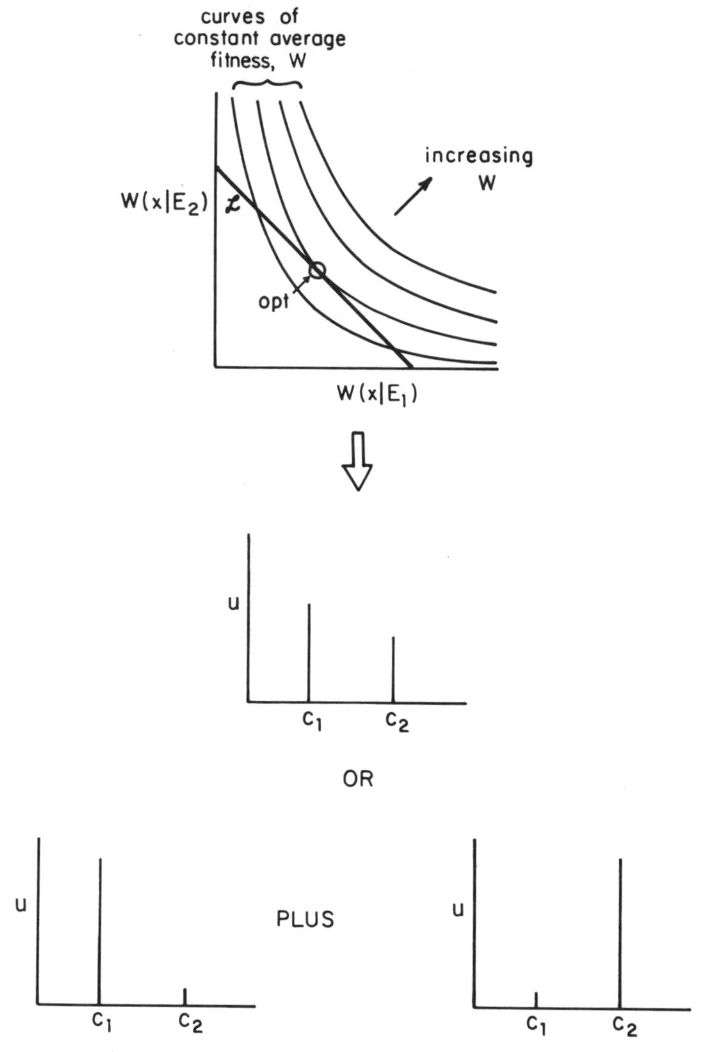

Figure 5–9 Optimization of the genetic usage predisposition in coarse-grained environments. The population should evolve toward a single relatively unselective bias curve or a mix of specific curves that produces a mean equal to the predisposition value of the unselective curve.

lation by the population of one culturgen alone confers less fitness; various members of the society must respond to both culturgens. Hence there is a division of labor even within the realm of a single culturgen category, such as the production of different but related cutting tools to handle two species of prey, where both tools are needed but to

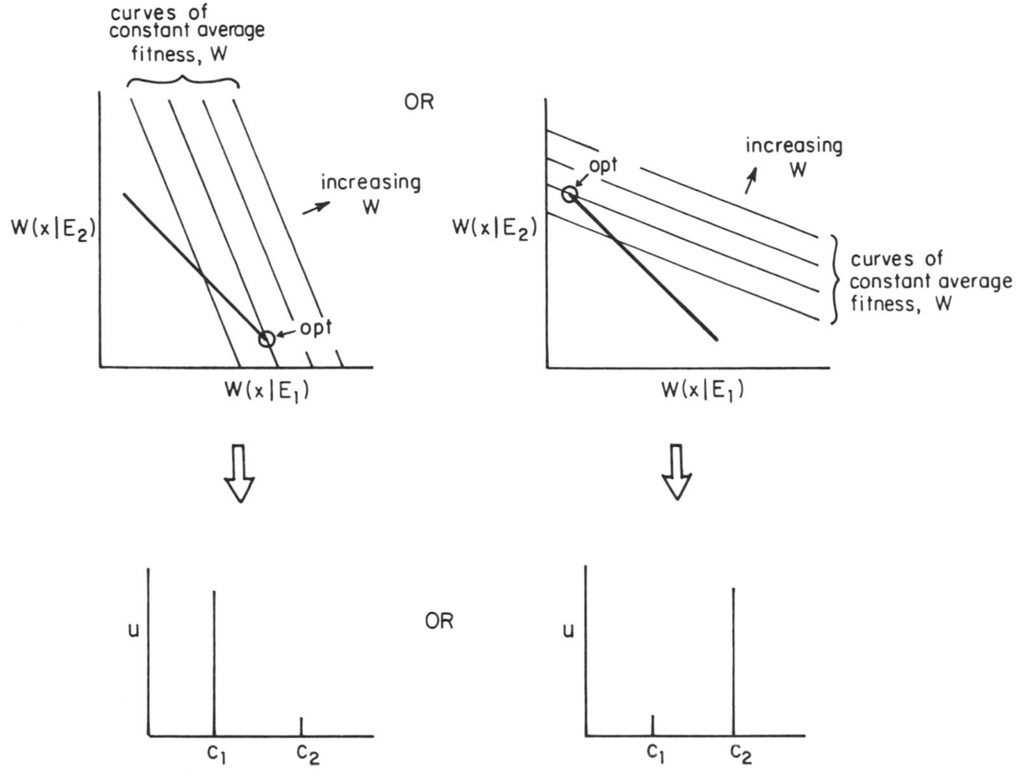

Figure 5–10 Optimization of the genetic usage predisposition in fine-grained environments. The population should evolve toward a single relatively specific bias curve.

different degrees according to the availability of the prey. This is the circumstance depicted in Figure 5–11. Regardless of whether the environment fluctuates in a coarse-grained or fine-grained manner, the optimal solution is the same: a single relatively unselective bias curve. The point $\mathcal{P}$ marking the position of multiple, selective bias curves (Eq. 5-19) lies within $\mathcal{L}$ and is not competitive with the unbiased morph. When the E_1 and E_2 fitness curves are separated far enough on the x-axis, the fitness curve becomes bilobed, instead of single-lobed as depicted in the right-hand panel of Figure 5–11. Small shifts in the slope of the population-average fitness curves (the straight lines in the right-hand panel) can move the optimum point abruptly from one lobe to the next. Thus it is possible under these special conditions for the optimum bias to change markedly and discontinuously with a relatively modest change in the environment.

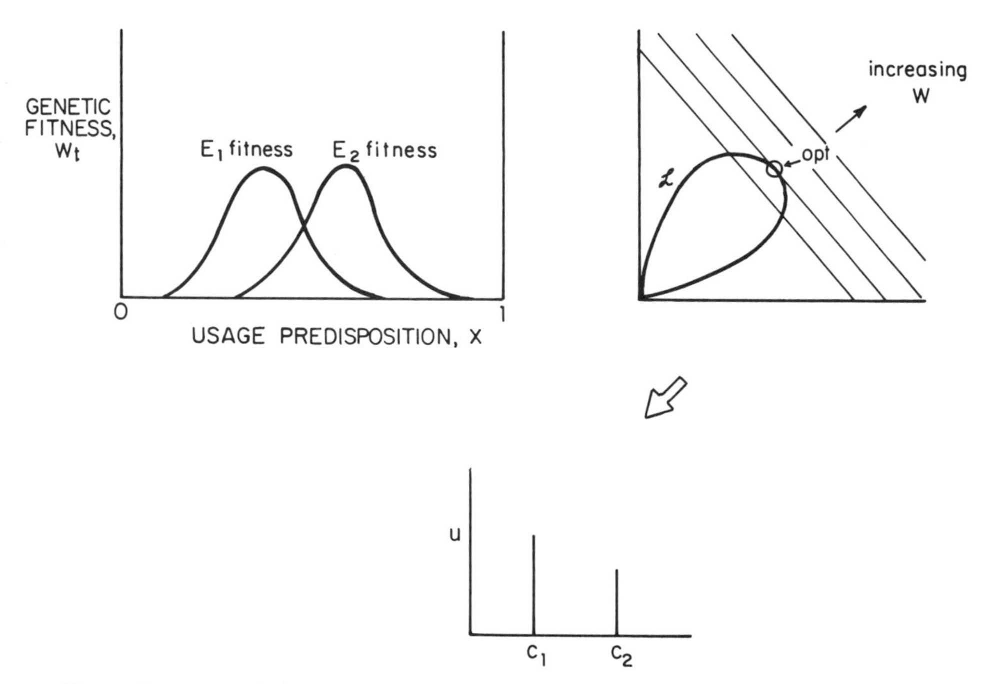

Figure 5–11 Optimization of the genetic usage predisposition in the special case of coadapted culturgens belonging to the same culturgen category.

But as long as the overlap between the E_1 and E_2 curves extends to at least their respective modes, the optimum bias is still intermediate in value.

These various qualitative results are summarized in Table 5–1. For each of the entries the quantitative structure of the usage bias curves is obtained by applying the equations just derived. These curves in turn specify the pattern of culturgen usage in the population, and consequently $P(\xi)$ is fixed as before.

We are unlikely to proceed much further with this mode of optimality analysis until additional details are known concerning the epigenesis of bias rules and environmental fluctuation. Nevertheless, the models do lead to the following fairly general principle of genetic assimilation: *for the class of epigenetic rules considered here, the more uniform the environment encountered by individuals in a life stage through many generations, the more selective will be the learning rules evolved in that life stage*. If members of a particular age group encounter approximately the same stimuli in the same proportions and context generation after generation, then by definition they exist in a fine-grained environment. The species will tend to evolve epigenetic rules that yield specialized

Table 5–1 The optimum bias curves for societies in heterogeneous environments.

Environment	Relationship of culturgens: Not coadapted (division of labor based on culturgens not advantageous)	Relationship of culturgens: Coadapted (division of labor based on culturgens advantageous)
Coarse-grained: changing roughly every generation or less frequently (for instance, the principal environment changes in time, or the group moves to a new habitat)	Either one relatively unselective bias curve or multiple selective bias curves	At low niche variance, one relatively unselective bias curve. At high variance, multiple selective bias curves.
Fine-grained: changing many times within a generation (for example, resources change through daily cycles or seasonally, or the group moves through multiple subdivisions of the environment)	One selective bias curve	At low niche variance, one relatively unselective bias curve. At high variance, one relatively selective bias curve.

responses on the part of the age group to a limited array of stimuli. The group with the most predictable environment is that composed of infants and small children, who also possess the richest array of selective epigenetic rules thus far discovered. It is furthermore true that the array of stimuli and requirements presented by the infant to its mother are nearly invariable, while the mother's attachment to the infant entails some of the few specialized rules thus far discovered in adults. This correspondence appears to support the main thrust of the optimality arguments. Yet it may be the result of biased sampling, since the relatively naive, uncomplicated behavior of the very young makes them the easiest subjects in the search for epigenetic rules.

Adaptation to coarse-grained or heterogeneous environments calls for an intermediate phenotype, which can be achieved by the deployment of either unselective bias curves or a mixture of specialized morphs. In light of this result we can appreciate the significance of the evolutionary step that made possible the secondary epigenetic rules of reentrant decision, choice, and valuation. As Figure 5–1B shows, some individuals are able to retain knowledge of multiple alternatives and choose between them according to context. Epigenetic rules of this kind in effect combine multiple "morphs" into one "phenotype," different morphs being "released" according to circumstance. We can therefore postulate that the invasion of heterogeneous environments tends to result in a selection pressure favoring the more advanced form of cognitive mechanism shown in Figure 5–1B.

The Role of History

The trajectory of gene-culture coevolution is prescribed by the initial gene frequencies and the fitness parameter $\mathbf{W}_{ij} \circ \mathbf{L}_{ij}$. But it is also influenced by the specific history of the evolving population—in particular, whether the features of the environment exercising selection fluctuate moderately or violently, and if violently, whether the catastrophes are frequent or rare, evenly spaced or clumped. Yet the formulation of general laws and the description of specific histories are complementary, not opposed. The laws reveal the mechanism behind the rich surface detail. In conjunction with historical data they can be used to predict the emergence of historical patterns. Of equal importance, the laws reveal what elements of history are necessary in order to explain the patterns.

We shall adapt an approach presented by Haldane and Jayakar (1963) to introduce a limited class of historical patterns to the gene-culture coevolutionary model. Instead of a single fitness value of $\mathbf{W}_{ij} \circ \mathbf{L}_{ij}$, consider the time series in which the value varies from one generation to the next over a period of N generations. In the case of two alleles the data take the following form:

$$\begin{aligned} &\mathbf{W}_{AA,1} \circ \mathbf{L}_{AA}, \mathbf{W}_{AA,2} \circ \mathbf{L}_{AA}, \ldots, \mathbf{W}_{AA,N} \circ \mathbf{L}_{AA} \\ &\mathbf{W}_{Aa,1} \circ \mathbf{L}_{Aa}, \mathbf{W}_{Aa,2} \circ \mathbf{L}_{Aa}, \ldots, \mathbf{W}_{Aa,N} \circ \mathbf{L}_{Aa} \\ &\mathbf{W}_{aa,1} \circ \mathbf{L}_{aa}, \mathbf{W}_{aa,2} \circ \mathbf{L}_{aa}, \ldots, \mathbf{W}_{aa,N} \circ \mathbf{L}_{aa} \end{aligned} \tag{5-24}$$

where 1, 2, . . . , N are subscripts labeling the generation. In the case of A dominant and a recessive, the contemporary terms in the first two lines can be labeled as

$$g_t = \mathbf{W}_{AA,t} \circ \mathbf{L}_{AA} = \mathbf{W}_{Aa,t} \circ \mathbf{L}_{Aa}. \tag{5-25}$$

Returning to the equation of discrete generation dynamics in gene-culture coevolution, where p is the frequency of A and q the frequency of a, we recall that

$$p_{t+1} = \frac{p_t^2 \mathbf{W}_{AA,t} \circ \mathbf{L}_{AA} + p_t q_t \mathbf{W}_{Aa,t} \circ \mathbf{L}_{Aa}}{p_t^2 \mathbf{W}_{AA,t} \circ \mathbf{L}_{AA} + 2p_t q_t \mathbf{W}_{Aa,t} \circ \mathbf{L}_{Aa} + q_t^2 \mathbf{W}_{aa,t} \circ \mathbf{L}_{aa}}. \quad (5\text{-}26)$$

Substituting g_t in this equation and rearranging, we obtain

$$p_{t+1} = \frac{p_t^2 + p_t q_t}{p_t^2 + 2p_t q_t + q_t^2 \dfrac{\mathbf{W}_{aa,t} \circ \mathbf{L}_{aa}}{g_t}}. \quad (5\text{-}27)$$

Now define the function f_t to be

$$f_t \triangleq \mathbf{W}_{aa,t} \circ \mathbf{L}_{aa} / g_t. \quad (5\text{-}28)$$

This is the fitness of aa relative to AA/Aa. It allows us to rewrite Eq. (5-27) in the compact form

$$p_{t+1} = \frac{p_t^2 + p_t q_t}{p_t^2 + 2p_t q_t + q_t^2 \cdot f_t}. \quad (5\text{-}29)$$

Application of Haldane-Jayakar analysis to this modified gene-culture form, by the procedure presented in Appendix 5–1, leads to some conclusions of general interest. For example, it can be shown that the recessive allele a cannot be eliminated during a span of N generations if its arithmetic-average relative fitness $\frac{1}{N}\sum_{t=1}^{N} f_t$ exceeds one, which is by definition the relative fitness of the AA and Aa genotypes over each of the N generations. Hence, for an epigenetic rule prescribed by a homozygous recessive to persist, its average relative gene-culture fitness must exceed that of AA and Aa. It is similarly true that A cannot disappear from the population during the N generations if the *geometric-mean relative fitness* of the competing recessive homozygote aa is less than one, in other words if

$$[f_{t_1} f_{t_2} \cdot \cdot \cdot f_{t_N}]^{1/N} < 1. \quad (5\text{-}30)$$

When these two results are combined, conditions are revealed by which multiple bias curves can be maintained even in the face of what appears at first to be countervailing selection. Suppose that the bearers of culturgens favored by *aa* are more fit most of the time, generation after generation, but more vulnerable to occasional catastrophes. The theory predicts that a polymorphism will result. The allele *a* will be sustained by the higher arithmetic mean fitness of its homozygotes, but kept from spreading by the infrequent catastrophes during the same time period, provided these events depress the geometric mean below unity.

Next consider the consequences of incomplete dominance and heterozygote superiority, in which *Aa* generates a distinct usage bias curve (in turn due to distinctive epigenetic rules) that confers superior fitness. The time series of gene-culture fitness is again given by Eq. (5-24). The series terms can be divided in each generation by $\mathbf{W}_{Aa,t} \circ \mathbf{L}$ to yield

$$\begin{array}{l} f_{AA,1}, f_{AA,2}, \ldots, f_{AA,N} \\ 1 \quad, 1 \quad, \ldots, 1 \\ f_{aa,1}, f_{aa,2}, \ldots, f_{aa,N}. \end{array} \tag{5-31}$$

By a procedure outlined in Appendix 5–2 the following results have been obtained. The *a* allele cannot be lost during the period of *N* generations if the heterozygote-scaled fitness of the homozygote *AA* has a geometric mean of less than unity. Symmetrically, the *A* allele cannot disappear during the same *N* generations if the heterozygote-scaled fitness of the homozygote *aa* has a geometric mean of less than one.

A somewhat similar result has been obtained in the heterozygote-superiority case by Feldman and Cavalli-Sforza (1976) in an interesting diallelic model in which two states occur, "skilled" and "unskilled," the probabilities of which depend on the parent's phenotype and the offspring's genotype. Time series were not employed, so that selection pressures were assumed to remain constant; also, transmission was limited to parents and offspring. Under these special circumstances it was found that a stable genetic polymorphism will result when the heterozygote-skilled parents rear more skilled offspring than homozygote-skilled parents, but only if the skill provides a substantial genetic selective advantage. For example, if a skilled teaching parent raises 80 percent of *Aa* offspring to be skilled as opposed to 70 percent of *AA* and *aa* offspring, the selective advantage of skilled individuals over unskilled ones must exceed one-third in order to achieve a stable genetic polymorphism. This conclusion is in accord with classic population genetics models, although the elevated level of selection required to sustain the polymorphism is an unexpected result of the complex form of transmission specified in this special case.

The time-series models we propose go further by incorporating specific histories. They reveal the interplay of ordinary and catastrophic episodes of selection in the context of diverse modes of enculturation, including the band-wide pattern prevailing in primitive societies (see Williams, 1972a,b). This relationship is likely to have existed during the millions of years of gene-culture coevolution in the preliterate era. Lee (1976), for example, has documented the occurrence of drought years in the home range of the Botswana San people. Eleven years classified as severe (55–69 percent of overall mean annual rainfall) occurred during a forty-six-year period, with one year classified as very severe (less than 55 percent of overall mean rainfall) recorded. Such episodes drastically affect movement of the bands and long term social relationships among them.

The findings from the time-series models also enhance the general conclusion reached earlier from optimality analysis that coarse-grained environments promote polymorphism and flexible response. If that portion of the environment to which the bias curves are directed remains fine-grained long enough, the genotypes prescribing the rules will tend toward homozygosity, and homozygous genotypes that favor strongly biasing rules will prevail. Major environmental change at frequencies of approximately one (or less) per generation "rescue" the population from such specialization, directing it back toward both genetic heterozygosity and nonuniform epigenetic rules. But if the coarse-grained changes are truly rare, so that the geometric fitnesses are insufficiently altered, and the changes are catastrophic as well, the shifts in gene frequency will come too late. A response c_1 strongly favored by an epigenetic rule could be rendered dangerous by a change in the environment, requiring a shift to an alternative response c_2. With the rule prescribing a heavy bias toward the first response in particular, or an intractable fidelity to the category to which people are enculturated, the capacity to change might be inadequate. For example, chronic tribalism, if it is indeed based on firm learning propensities toward xenophobia and aggression (Wilson, 1978), could prove too strong to overcome even when alternative strategies of conflict resolution are available and—in a nuclear age—necessary. Specialized, biased learning rules "fly in the face of common sense" not because of some immanent perversity but because of the particular past trajectories of genetic history.

The Role of Phenotypic Variation

The models presented thus far posit fixed bias curves; for every genotype G_iG_j and culturgen c_k there is a single epigenetic rule value for culturgen usage, $u(c_k|G_iG_j)$. It is more realistic, however, to expect the presence of some purely phenotypic variation in the u values, the circumstance illustrated in Figure 5–12. Several important theoretical questions are posed by this additional condition. Do variable bias curves differ in fitness from fixed bias curves that have the same values of $\bar{u}$? Is the variance itself an adaptive feature, or is it simply noise in the gene-culture transmission? And finally, does the existence of such variation alter the main conclusions drawn earlier concerning the gene-culture adaptive landscape? The answers can be expected to throw light not only on the role of phenotypic variation but also on the distinctions between single-locus and polygenic systems.

Note that in the variable system, as in the fixed system, $u_1 + u_2 = 1$. Let us define $P(u|ij)\,du$ as the probability that a diploid genotype G_iG_j has a value of $u(c_1)$ lying between the values u and $(u + du)$ during the prereproductive period. Integrating over u, we see that for

$$
\begin{array}{lll}
\text{genotype } AA\text{:} & p_t^2 N_t \int_0^1 uP(u|AA)\,du & \text{assimilate } c_1 \\
 & p_t^2 N_t \int_0^1 (1 - u)P(u|AA)\,du & \text{assimilate } c_2 \\
\text{genotype } Aa\text{:} & 2p_t q_t N_t \int_0^1 uP(u|Aa)\,du & \text{assimilate } c_1 \\
 & 2p_t q_t N_t \int_0^1 (1 - u)P(u|Aa)\,du & \text{assimilate } c_2 \\
\text{genotype } aa\text{:} & q_t^2 N_t \int_0^1 uP(u|aa)\,du & \text{assimilate } c_1 \\
 & q_t^2 N_t \int_0^1 (1 - u)P(u|aa)\,du & \text{assimilate } c_2
\end{array}
\qquad (5\text{-}32)
$$

Since by definition $\int_0^1 uP(u|ij)\,du = \langle u \rangle_{ij}$ is the mean value of u under the probability distribution $P(u|ij)$, we can rewrite Eq. (5-32) to state that for

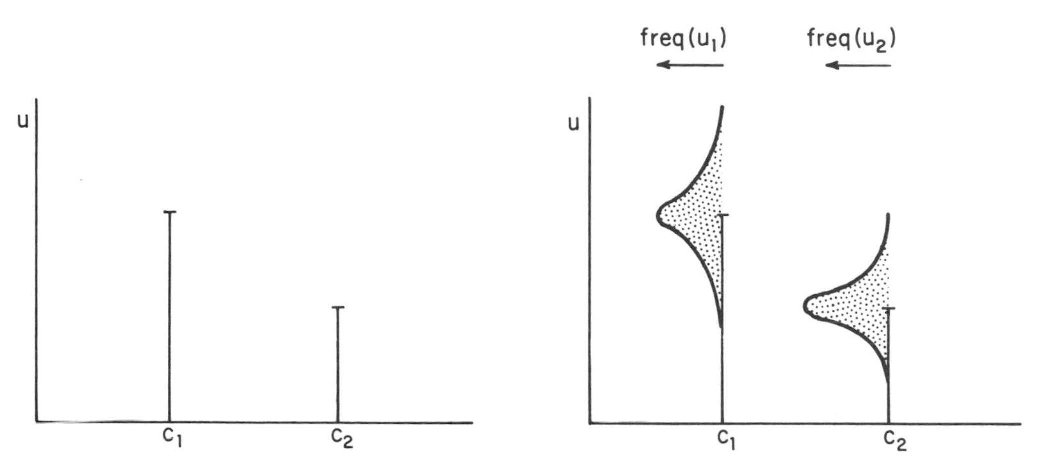

Figure 5–12 The presence and absence of phenotypic variation in the usage bias curves. The bell-shaped curves on the right are the frequency distributions of the usage probabilities u; the mean values of u_1 and u_2 are the same in both diagrams.

$$\begin{aligned} \text{genotype } AA\text{:} \quad & p_t^2 N_t \langle u \rangle_{AA} && \text{assimilate } c_1 \\ & p_t^2 N_t (1 - \langle u \rangle_{AA}) && \text{assimilate } c_2 \end{aligned} \tag{5-33}$$

and so on in parallel expressions for Aa and aa. It follows that the allocation of naive classes is governed solely by $\langle u \rangle_{ij}$, the means of the norms of reaction $P(u|ij)$, and not by any other statistical property of the phenotypic variation. In particular, the variance plays no explicit role.

By tracing the enculturated population through the survival and reproduction schedules prescribed by the genotype and assimilated culturgen, we arrive at the rate of change in frequency of A in the diallelic case:

$$\Delta p = \frac{pq}{2} \frac{d}{dp_t} \ln \langle \mathbf{WL} \rangle \tag{5-34}$$

where $\langle \mathbf{WL} \rangle$ is the hypersurface of the gene-culture landscape, defined as before (Eq. 5-10) except that $\langle u \rangle_{ij}$ replaces $u(c_1|ij)$:

$$\begin{aligned} \langle \mathbf{WL} \rangle &\triangleq p_t^2 \mathbf{W}_{AA} \circ \mathbf{L}_{AA} + 2p_t q_t \mathbf{W}_{Aa} \circ \mathbf{L}_{Aa} + q_t^2 \mathbf{W}_{aa} \circ \mathbf{L}_{aa} \\ \mathbf{L}_{ij} &\triangleq (\langle u \rangle_{ij}, 1 - \langle u \rangle_{ij}). \end{aligned} \tag{5-35}$$

Thus only the mean values, and no other statistical properties of the variation in the learning bias curves, enter into the gene-culture coevolutionary landscape. The general conclusions based on invariant curves are preserved.

This result has further interesting implications. As long as the mean value of u is the same (for example, the probability distribution is symmetric), the variance can drift in a neutral manner from narrow to broad or broad to narrow (Figure 5–13). Even when the variance itself is under genetic control, perhaps as the pleiotropic effect of genes active in other behavioral processes, the conclusion to be drawn is nevertheless that the effects on variance will remain neutral with reference to selection on the epigenetic rule. This permissiveness could theoretically promote a diversity of bias curves within populations, not only between families but within families. In extreme cases it could lead to a quasi-polymorphism of behavioral roles.

If the variance drifts in such a way as to become broader, it will tend to enhance gene-culture coevolution. For it will uncover any new, adaptive, superior values of u that exist, one or more of which can then be incorporated through genetic assimilation as the new central tendency (see Figure 5–14). The genotypes prescribing the mean value of the usage predisposition are altered by natural selection, and they will tend to be accompanied by a novel pattern of variation around the new mean. It is further true that unselective curves, both with $\langle u \rangle$ values close to 0.5 and 0.5, are least restricted in the amount of variance that can be accumulated, whereas comparable but more selective curves, with $\langle u \rangle$ close to 1 and 0, will be most restricted (Figure 5–15). To see why this must be so, consider the extreme case of $\langle u_1 \rangle = 1$, $\langle u_2 \rangle = 0$; here no variance is possible. As a re-

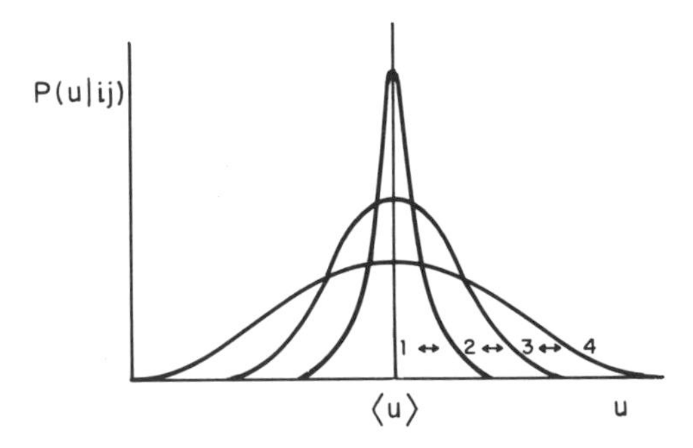

Figure 5–13 Variation in the usage bias curve. The amount is adaptively neutral, as long as the mean remains the same. As a result the variance can drift to either broad or narrow.

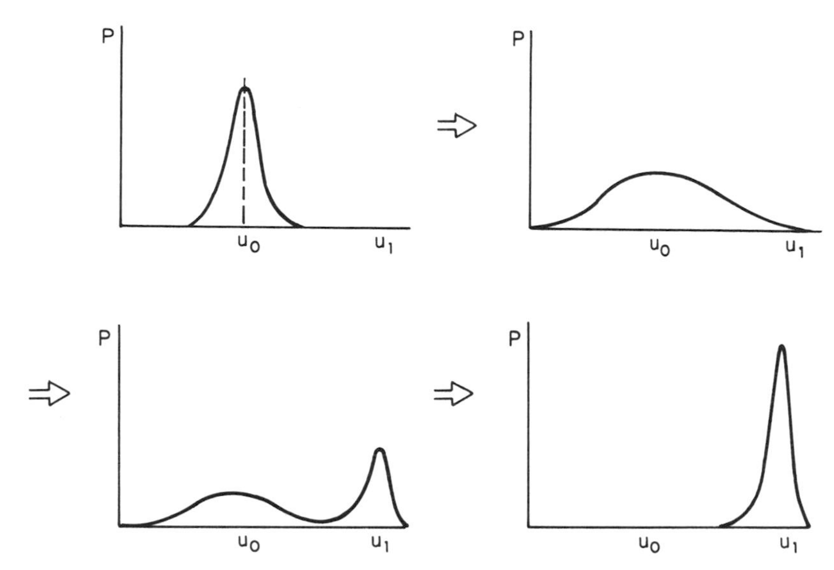

Figure 5–14 Genetic assimilation of a new and superior mean. The genetically based but adaptively neutral broadening of variance in the usage bias curves around a mean u_0 can lead to the "discovery" and capture of a new, adaptively superior mean u_1.

sult, genotypes prescribing unselective bias curves are expected to be more vulnerable to replacement.

It is possible that limits will also be set on the variance of u by the same gene loci determining the mean of u. Alleles on them will be less vulner-

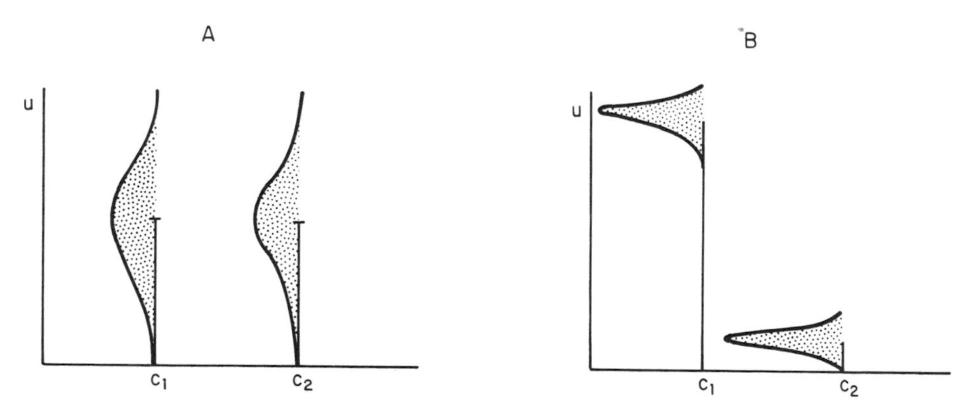

Figure 5–15 Variance in the usage bias curves. The amount is expected to be greatest when epigenesis is relatively unselective (*A*) and to diminish as it becomes increasingly channeled (*B*).

able to replacement and hence persist longer. The same is true for gene loci that prevent the variation from becoming asymmetric and hence reach farther along the u scale in one direction or the other. However, such insulation, being based on second-order selection, is likely to be unstable. Stability is achieved when the mean value of u attained is the most adaptive on the full 0–1 scale. At this point the variance can also become neutral, as we have demonstrated.

Summary

This chapter develops our first concrete model of the coupling between genetic and cultural evolution. In it we demonstrate that an adaptive topography in a simple type of gene-culture coevolution exists and can be described with general dynamical equations. We use the language of conventional population genetics to emphasize the similarities and differences between pure genetic evolution and gene-culture coevolution. By this means we call into service some of the basic theory of population genetics and evolution and thereby clarify the effects on the evolution of epigenetic rules created by environmental heterogeneity and phenotypic variance.

The gene-culture fitness function is then $\mathbf{W}_{ij} \circ \mathbf{L}_{ij}$, where $\mathbf{W}_{ij}$ is the genetic fitness vector of individuals of diploid genotype G_iG_j who possess the culturgens of the series $c_1, c_2, \ldots, c_M$, and $\mathbf{L}_{ij}$ is the epigenetic rule vector. Where the mean gene-culture fitness of a population is designated as $\langle \mathbf{WL} \rangle$, the rate of change of gene frequency in the diallelic, discrete-generation case is

$$\Delta p = \frac{1}{2} \frac{pq}{\langle \mathbf{WL} \rangle} \frac{d}{dp} \langle \mathbf{WL} \rangle .$$

On a static landscape $\langle \mathbf{WL} \rangle$, the population will evolve so as to stabilize gene frequencies (for example, all AA or all aa in the case of two alleles A and a), and its members will saturate on the bias curve prescribed by this final genotypic distribution.

Real environments, however, are heterogeneous in space and time. The true landscape is more accurately described as a viscous seascape, in which peaks subside into depressions and populations must move toward more complex solutions that integrate gene-culture fitness through time. When the environments are coarse-grained, that is, when changes occur on the order of once per generation or less, the average gene-culture fitness is the geometric mean. The maximum potential fitness curve is a hyperbola that touches the real fitness space at a point prescribing a relatively unselective usage bias curve (see Figure 5–9). When the environments are fine-grained, shifting frequently within single generations, the average gene-culture fitness is the arithmetic mean and the maximum potential fitness curve is a straight line that intersects one or the other of the ends of the real fitness space. The result is the genetic prescription of relatively specific usage bias curves (see Figure 5–10 and Table 5–1). These models lead to a general rule: the more uniform the environment encountered by individuals in a particular life stage through many generations, the more specific will be the epigenetic rules evolved for that life stage. This prediction is supported by the fact that the richest arrays of selective epigenetic rules have been detected during early childhood and in other circumstances where encounters are uniform and predictable. However, the correlation is based on a possibly biased sample, and the rule cannot be considered to be established.

Haldane-Jayakar analysis has been applied to the model in order to incorporate a simple form of history. In one case examined, two usage bias curves are prescribed by A- and aa respectively in a diallelic system with complete dominance, where the aa curve is selectively superior generation after generation but is more vulnerable to occasional catastrophes. If the arithmetic gene-culture fitness mean of aa is above one and the geometric mean below one, a not unlikely combination, the two alleles will be maintained in a polymorphic mixture. Thus the idiosyncratic histories of populations affect gene-culture coevolution in ways that cannot be assessed by classic analysis alone. A new interpretation can be placed on the propensity to acquire and retain certain destructive culturgens in spite of a newly acquired rational understanding of their peril.

If the mean value of an inherited usage bias curve is at the adaptive maximum, phenotypic variation in the predisposition has no influence on fitness. Only the mean values, and no other statistical properties in the bias curves, enter into the gene-culture coevolutionary landscape. General conclusions based on the simpler case of invariant bias curves are therefore preserved. In addition, conclusions based on the single-locus model can be extended to many systems of polygenic inheritance generating continuous variation. Because of the adaptive neutrality of phenotypic variance, the variance can drift in such a way as to become broader or narrower, creating greater or lesser amounts of behavioral heterogeneity within populations. The variance will tend to broaden maximally in equiprobable bias curves and minimally in curves with high selectivity. If some degree of selectivity other than the contemporaneous mean value becomes more adaptive, an increase in phenotypic variation will lead to an earlier attainment of this new, more adaptive bias by the population. There will be a tendency to incorporate the bias (and hence shift to a new epigenetic rule) through genetic assimilation, followed by a return to selective neutrality of the variance

Appendix 5–1

Environmental History and the Fate of Dominant/Recessive Alleles

The Haldane-Jayakar method creates an "elastic bounce strategy" in which the gene frequencies rebound when they fall to sufficiently low levels. Having achieved the specified conditions, the frequencies cannot be eliminated. We follow Roughgarden's (1979) development.

The gene frequency laws are, with $\bar{f} = p_t^2 + 2p_tq_t + q_t^2f_t$,

$$p_{t+1} = (1/\bar{f})p_t \tag{5-A1}$$

and

$$q_{t+1} = [(p_t + f_tq_t)/\bar{f}]q_t. \tag{5-A2}$$

Let $z_t \triangleq p_t/q_t$; then

$$z_{t+1} = \frac{p_t/q_t}{f_tq_t + p_t}$$

$$= \frac{z_t(1 + z_t)}{f_t(1 + z_t) + z_t(1 - f_t)}. \tag{5-A3}$$

Hence

$$z_{t+1} = \frac{z_t(1 + z_t)}{z_t + f_t}. \tag{5-A4}$$

We next develop an equation for $\Delta z = z_{t+1} - z_t$:

$$\Delta z = z_{t+1} - z_t = \frac{z_t(z_t + 1)}{z_t + f_t} - z_t = \frac{z_t(1 + z_t) - z_t(z_t + f_t)}{z_t + f_t}$$

$$= \frac{z_t(1 - f_t)}{(z_t + f_t)} = \frac{z_t(1 - f_t) + f_t(1 - f_t) - f_t(1 - f_t)}{z_t + f_t}$$

$$= \frac{(1 - f_t)(z_t + f_t)}{z_t + f_t} - \frac{f_t(1 - f_t)}{z_t + f_t} = (1 - f_t) - \frac{f_t(1 - f_t)}{z_t + f_t}$$

$$= (1 - f_t)\left[1 - \frac{f_t}{z_t + f_t}\right] = (1 - f_t)\frac{z_t}{z_t + f_t}.$$

Therefore

$$\Delta z_t = (1 - f_t)\frac{z_t}{z_t + f_t}. \tag{5-A5}$$

If a cannot be eliminated, then it must increase once it becomes sufficiently rare. But a rare means q_t is small, which implies that $z_t p_t / q_t$ is large.

Consequently we require that z_t decrease when it becomes sufficiently large. For very large z_t, the Δz_t of Eq. (5-A5) simplifies to

$$\Delta z_t \sim 1 - f_t \qquad (a \text{ rare}). \tag{5-A6}$$

Hence $\Delta z < 0$ if the requirement is $f_t > 1$ during that generation. For a persistent net recovery over the N generations, it is required that

$$\sum_{t=1}^{N} \Delta z_t \sim \sum_{t=1}^{N} (1 - f_t) = N - \sum_{t=1}^{N} f_t < 0, \tag{5-A7}$$

yielding

$$\frac{1}{N}\sum_{t=1}^{N} f_t > 1, \tag{5-A8}$$

which is the arithmetic average result.

The strategy for A is slightly different. For A to resist elimination, it must rebound when sufficiently rare. Now A being rare means that $p \ll 1$, and consequently $z_t = p_t/q_t \ll 1$ also. To develop the required formula, A's frequency must be made to *increase* once A is sufficiently rare. We build the quotient

$$\frac{z_{t+1}}{z_t} = \frac{1}{z_t}\frac{z_t(1+z_t)}{z_t+f_t} = \frac{z_t+1}{z_t+f_t} = \frac{1+z_t}{z_t+f_t+z_tf_t-z_tf_t}$$

$$= \frac{1+z_t}{f_t(1+z_t)+z_t(1-f_t)} = \frac{1}{f_t+\dfrac{z_t(1-f_t)}{(1+z_t)}}$$

$$= \frac{1}{f_t\left\{1+\dfrac{[z_t(1-f_t)]}{[f_t(1+z_t)]}\right\}}. \qquad \text{(5-A9)}$$

When A is rare, z_t is near zero and (5-A9) becomes

$$\frac{z_{t+1}}{z_t} \sim \frac{1}{f_t} \qquad (A \text{ rare}). \qquad \text{(5-A10)}$$

Now we cumulate over N generations by multiplying:

$$\frac{z_{N+1}}{z_1} = \frac{z_{N+1}}{z_N}\cdot\frac{z_N}{z_{N-1}}\cdot\frac{z_{N-1}}{z_{N-2}}\cdot\frac{z_{N-2}}{z_{N-3}}\cdot\cdot\cdot\frac{z_2}{z_1}$$

$$\sim \frac{1}{f_N}\cdot\frac{1}{f_{N-1}}\cdot\frac{1}{f_{N-2}}\cdot\cdot\cdot\frac{1}{f_1}.$$

For A to increase over these N generations the necessary condition is that

$$\frac{z_{N+1}}{z_1} > 1. \qquad \text{(5-A11)}$$

Hence $(f_1f_2 \ldots f_N)^{-1} > 1$, which means that

$$f_1f_2 \ldots f_N < 1 \qquad \text{(5-A12)}$$

or equivalently

$$(f_1 f_2 \cdot \cdot \cdot f_N)^{1/N} < 1. \qquad (5\text{-A}13)$$

In words, the geometric mean must be less than one.

Appendix 5–2

Environmental History and the Fate of Alleles with Heterozygote Superiority

We proceed as in Appendix 5–1 to obtain the dynamics

$$z_{t+1} = \frac{f_t^A z_t + 1}{z_t + f_t^a} z_t \qquad (5\text{-A}14)$$

where for brevity $f_t^A \triangleq f_{AA,t}$ and $f_t^a \triangleq f_{aa,t}$.

To explore a allele rebound, we form the quotient z_{t+1}/z_t and rearrange Eq. (5-A14) into

$$\frac{z_{t+1}}{z_t} = f_t^A + \frac{1 - f_t^A f_t^a}{z_t + f_t^a}. \qquad (5\text{-A}15)$$

For a rare, $z_t \to \infty$ and the right-hand side of Eq. (5-A15) simplifies to

$$\frac{z_{t+1}}{z_t} \sim f_t^A \qquad (a \text{ rare}) \qquad (5\text{-A}16)$$

This can also be seen directly from Eq. (5-A14). Thus

$$\frac{z_{N+1}}{z_1} = \frac{z_{N+1}}{z_N} \cdot \frac{z_N}{z_{N-1}} \cdot \cdot \cdot \frac{z_2}{z_1} \sim f_N^A \cdot f_{N-1}^A \cdot \cdot \cdot f_1^A. \qquad (5\text{-A}17)$$

The a allele will rebound if $z_{N+1}/z_1 < 1$, which yields the geometric mean conditions

$$f_1^A f_2^A \cdot \cdot \cdot f_N^A < 1 \qquad (5\text{-A}18)$$

and

$$(f_1^A f_2^A \cdot \cdot \cdot f_N^A)^{1/N} < 1. \qquad (5\text{-A}19)$$

The requirement for the A persistence is similar. For A rare, z_t is $\ll 1$, and taking this limit in Eq. (5-A14) gives

$$\frac{z_{t+1}}{z_t} \sim \frac{1}{f_t^a} \qquad (A \text{ rare}) \tag{5-A20}$$

and thus

$$\frac{z_{N+1}}{z_1} \sim (f_1^a f_2^a \cdot \cdot \cdot f_N^a)^{-1}. \tag{5-A21}$$

For A to rebound during this history, $z_{N+1}/z_1 > 1$ is required. We obtain at once the geometric mean conditions

$$f_1^a f_2^a \cdot \cdot \cdot f_N^a < 1 \tag{5-A22}$$

and

$$(f_1^a f_2^a \cdot \cdot \cdot f_N^a)^{1/N} < 1. \tag{5-A23}$$

CHAPTER SIX

The Coevolutionary Circuit

It has often been said that mind and culture can only be understood holistically. We believe this to be true, but in a sense largely unappreciated by scientists who work exclusively in physiological time as opposed to evolutionary time. The structures of mind and culture are most effectively understood as developmental processes, underwritten by genes whose frequencies are the product of the protracted interaction of social behavior and selection forces working from the environment. Thus to understand culture fully is not just to perceive the rich detail of which it is composed, but to follow through each step in the coevolutionary circuit—from physiological time and cultural evolution, to evolutionary time and genetic evolution, and back again.

As we began to trace the circuit in earlier chapters, we proceeded from the genetic constraints on individual development in the form of epigenetic rules to the way in which the rules are translated upward into societal patterns. We next examined the inverse process, which is the impact of natural selection on learned social behavior as it is transferred from culture back to the epigenetic rules through the genes. In the first approach we dealt with very simple cultures that were based both on usage-independent epigenetic rules and on a highly simplified connection between epigenetic rules, the cognitive processing of information, and the selection forces at work in the system. Thus we established the existence of a gene-culture adaptive landscape and identified the very general features of simple coevolution in heterogeneous environments.

It is time now to attempt a more comprehensive formulation. We wish to incorporate epigenetic rules that use information derived from cultural organization, basing the coevolutionary models more fully on the processes of learning, valuation, and decision. In other words, we shall put the gene-culture translation models into evolutionary time and study the coevolutionary mechanisms that can generate substantial changes in the frequencies of both the culturgens and the associated genes.

The key is a deeper analysis of epigenesis, one that incorporates the cognitive machinery and goes further, to identify the units of inherited cognitive design whose fate under natural selection determines the epigenetic rules. We perform this analysis in the next three sections, beginning with a discussion of developmental neurogenetics and psychogenetics, then moving to an examination of the node-link structures of long term memory as the ultimate biological theater of interplay between genes and culturgens. Then with this new view of gene interaction we are equipped to formalize the required mathematical models and analyze the coevolutionary mechanisms, together with the prime movers that shape the coevolutionary trajectories.

Epigenesis

It has become clear that the future of general microevolutionary studies lies in the refocusing of genetic analysis on development. This is particularly the case for human social behavior, the class of phenotypes most removed from the DNA templates. One can go further and say that the traits most tractable to genetic dissection are not the finished products themselves (kin groups, territoriality, hypergamy), but rather their underlying epigenetic rules. Once these rules have been characterized, they can be traced forward to adult behaviors and then translated into societal patterns, with the aid of the kinds of techniques utilized in earlier chapters. The more selective and inflexible the rule, the more likely it is to yield clear results at this preliminary stage of investigation. Examples of behavioral categories that seem most promising include color naming, incest avoidance, mode of infant carrying, phobias, and facial expressions. Not only can the epigenesis of such behaviors be readily studied at much greater depth than in the past, but mutations affecting these behaviors and the underlying neurobiology may be found and characterized by methods already routinely employed in human genetics.

The amplification of genetic prescription occurs generally in the formation of the nervous system. Much of the brain and sensory system of mammals develops by waves of differentiation of populations of cells. Neighboring zones have distinct origins and differ from one another in cell architecture (M. Jacobson, 1978a,b). In the development of the neocortex of rats and mice, cells originate in waves from the germinal epithelium lining the ventricle, then migrate through upper zones of cells to reach their final destination. Eventually six layers are assembled, in such a way that the cells migrating first form the deepest layer and those migrating last become the outermost layer. The controls that direct this intricate choreography are unknown, but they create a set of positional rules and responses to gradients far less complex than the final patterns themselves. As Bullock and colleagues (1977) state, "It seems most likely . . . that the migrating cells are genetically programmed to move (in unknown ways) toward an appropriate chemical environment, along a chemical gradient or series of gradients, diffusible or built into the cells passed en route. The phenomenon may not be greatly different from that of chemotaxis in leucocytes or bacteria." When the cell primordia approach other types of nervous or muscular tissue, they respond in specific ways by altering the growth and differentiation of their dendrites and more proximate cellular components (Wessells, 1977; Patterson, 1979). In principle a large number of responses can be generated by the interactions of relatively few kinds of elements, each capable of only a limited array of decisions (Hillman, 1979; Szentágothai, 1979).

The relation between genes and neurons in vertebrate brains is more homeomorphic than isomorphic (Changeux and Danchin, 1976; Bullock et al., 1977; M. Jacobson, 1978a; Cowan, 1979). A gene or set of genes does not specify individual neurons. Rather, genes inaugurate programs of growth and migration and the general rules of neuron interaction that lead to brain ontogeny. For example, just two forces are necessary and sufficient to direct the vertebrate neural plate into its keyhole shape: a differential and programmed constriction of the apical surfaces of the plate cells, and movement to the midline by the plate cells overlying the notochord (A. G. Jacobson, 1978). Each neuron is affected by certain rules prescribed by multiple genes, while each gene addressed to the nervous system affects assemblages of cells.

This fundamental relationship is reflected clearly in the nature of the mutants that alter the nervous system. For example, the temporal retinal cells of albino mammals (which are homozygous for a single recessive allele) project their axons across the optic chiasma to the opposite side of the brain, in contrast to the cells of normal individuals, which remain on the same side. In mice of the mutant strain BALB/cJ, the hippocampal mossy fibers form an *intra*pyramidal synaptic field instead of the usual *infra*pyramidal field (M. Jacobson, 1978a). As Figure 6–1 shows, granule and Purkinje cell populations are selectively affected in the reeler, weaver, and staggerer mouse mutants with premature cell death, neuronal

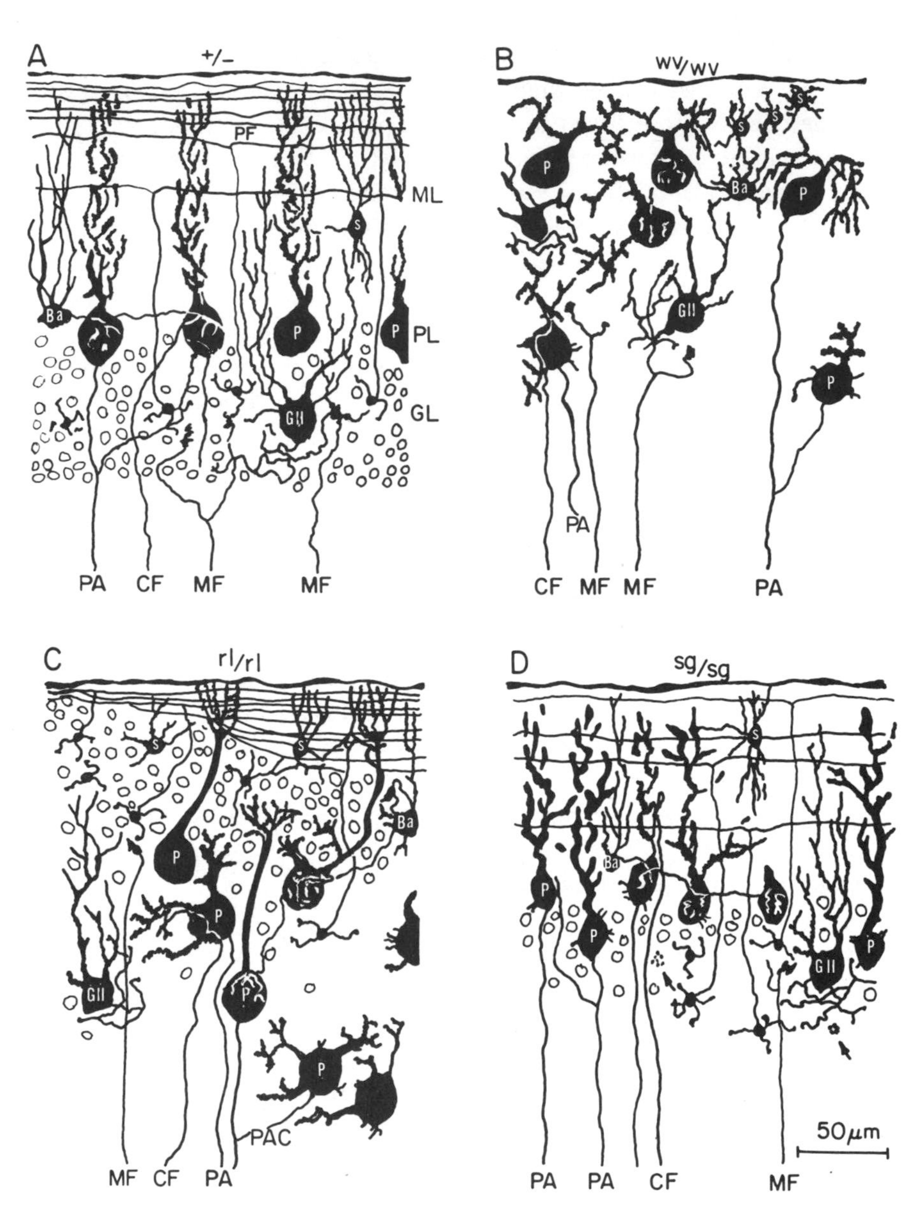

Figure 6–1 Qualitative rearrangements in the mouse cerebellum induced by mutations at a single locus. The resulting behavior patterns are diagnostic for the trait. *A*, normal; *B*, homozygous weaver; *C*, homozygous reeler; *D*, homozygous staggerer. Ba-basket cell; CF-climbing fiber; G-granule cell; GII-Golgi type II cell; MF-mossy fiber; P-Purkinje cell; PA-Purkinje cell axon; PF-parallel fiber; S-stellate cell. (Modified from Rakic, 1979.)

malpositioning, and disturbed synapse formation (Rakic, 1975a,b, 1979; Caviness and Rakic, 1978). In Menkes disease, an inherited cerebellar pathology in man, Purkinje cells develop multiple dendritic trees instead of the normal single arbor (Purpura et al., 1976; Williams et al., 1978). Thus it is easily possible for a single allele to create profound changes in both neural anatomy and behavior. Such gene-controlled changes can be simultaneously widespread and extremely specific. The *sev* mutation in *Drosophila*, for example, disturbs the entire compound eye of the insect, but does so by eliminating precisely the same photoreceptor cell (R_7) from each ommatidium (Hall and Greenspan, 1979).

The developing nervous system acts in many ways like a population of semi-independent organisms. Neurons are produced in excess and there is a "natural selection" among them, with most of the malpositioned or unconnected cells dying out (Changeux and Danchin, 1976; M. Jacobson, 1978a; Cowan, 1979). When a larger amount of target tissue is provided experimentally, the excess cells survive. Thus the weeding process is not due to intrinsic malfunctions of the growing elements but is imposed by properties of their extrasynaptic environment. Variation in the neurons of a particular class, originating in chromosome rearrangements, somatic mutations, or the ordinary statistical noise of development, is followed by a narrowing of the products through cell death. Looked at another way, the selection process constitutes part of the canalization of the brain.

The traditional view that canalization produces a vertebrate brain made up largely of vast but haphazardly connected nerve nets, with fine structure beyond the capacity of the genes, is no longer acceptable. A vertebrate brain grounded on ultrastructural specificity is part of the "quiet neuroscientific revolution" (Szentágothai and Arbib, 1974; Rakic, 1975a,b; Schmitt et al., 1976; Schmitt and Worden, 1979). Neural assemblages in vertebrate cerebral cortex, thalamus, tectum, reticular formation, cerebellum, retina, and olfactory bulb are now seen to consist of interpenetrating modules or arrays of cells, organized in patterns that repeat both in and between organisms (Rakic, 1975a, 1979; Shepherd, 1974; Cowan, 1979; Szentágothai, 1979). Epigenesis achieves a previously unsuspected precision of neuronal form, location, and connectivity, a view supported by anatomical work on isogenic invertebrates. The connectivity patterns of identifiable neurons are remarkably constant both between organisms and between the two sides of the same organism (Levinthal et al., 1975; Goodman, 1978). This neural templating can reach out as far as the second order of dendritic branching (Macagno et al., 1973) and along at least the initial courses of axons (Cowan, 1979). Rat hippocampal cells continue to develop their characteristic form when propagated in tissue culture (Banker and Cowan, 1977). Neural connectivity is equally specific, but includes multiple strategies of cellular addressing. In optic ganglia of flies, connections between ommatidia and lamina cartridges are one-to-one (Strausfeld, 1976; Braitenberg, 1977). In vertebrate brain stem, some stellate cells address other neurons through a characteristic space-filling efflorescence of their dendritic trees (Ramón-Moliner and Nauta, 1966).

The exact rules and genetic mechanisms leading to precision and pattern formation in nervous systems are currently unknown, but the evidence from other developmental models is revealing. These cases underline the fact that embryogenesis of the brain and therefore of the mind itself is an open system from the RNA outward. Furthermore, as we shall see in the next section, there is no fundamental break in this process between tissue formation and behavior. Learning can be defined as epigenesis that takes place largely outside the womb. And much of learning, like morphogenesis, can be construed as a canalizing device, a process that guides members of the species through various sets of stimuli to approximately the same steady state in relation to food, mates, and other aspects of the environment (Bateson, 1976).

The body of *Drosophila* is assembled from a number of parts, each of which arises from a primordial group of cells in the blastula. Each group differentiates as a local field (Benzer, 1973; García-Bellido, 1975). Similarly, the color patterns of the wings of butterflies and moths are determined independently in each wing cell and arise from the deposition of pigment with reference to a definite focus. Nijhout (1978) and other investigators have shown how striking changes in the patterns can arise from relatively minor alterations in the "interpretation landscape," a gradient system whose value at each point in the morphogenetic field is a measure of how positional information is interpreted. Single allelic substitutions and minor cautery on developing wings can cause major changes in the total pattern.

A model system for the precise control of morphogenesis by genes is that of scutellar bristles in *Drosophila*. The recessive sex-linked mutation *scute* (*sc*) reduces the number of bristles on the scutellum from four to one or two. Rendel (1967, 1979) demonstrated that the frequency distribution of flies homozygous for wild type (+ +), heterozygous (+*sc*), and homozygous for *scute* (*scsc*) can be explained by a model in which bristle-making ability differs among genotypes and varies according to a continuous, normal distribution about the mean of each genotype. Bristle-making ability is very similar in + + and +*sc* flies but substantially lower in *scsc* individuals, producing some of the observed dominance relationship of + over *sc*. When Rendel conducted selection experiments in which the number of bristles was varied within *sc*-bearing lines, he found that minor genes alter the number of bristles of *scute* flies. These homozygous individuals proved to be canalized at four bristles when the mean was raised to 3.5. Through the selection on *scute* flies the bristle number of wild-type sibs was eventually raised to over four, and it was further shown that at higher means *sc* + flies had an average of 1.5 fewer bristles than + + flies. By selecting on a further set of modifying genes, Rendel demonstrated that *scute* genotypes can be canalized at two bristles. Richelle and Ghysen (1979) have extended the notion of a bristle-making capacity into a model of the distribution of an actual diffusible, bristle-inducing substance, called a chaetogen. Any cell that receives a concentration of chaetogen above a certain threshold value becomes determined to produce a bristle. The synthesis of the bristle-making substance in turn is an all-or-none response of cells to positional cues within the imaginal disk that contains them. A cluster of cells engaged in making chaetogen will give rise to a single peak of the substance and hence a single cell will be determined as a bristle maker. Even a moderately precise, probabilistic reading of positional information by many cells, conforming to relatively simple rules, can result in the highly accurate positioning of individual bristles. Relatively simple alterations in the suggested parameters of positional reading closely reproduce the bristle patterns of wild-type and mutant strains of *Drosophila* (Ghysen and Richelle, 1979).

Experiments on the genetics of laboratory populations of *Drosophila* and a few other kinds of organisms have shown that when developing organisms are stressed by being placed in extreme environments, the combinations of genes operate so that the final product in canalized traits is still close to the population norm. The epigenetic rules for canalization produce a qualitative response only to specific signals from the environment. The reduction of genetic diversity in the population by selection or inbreeding is accompanied by a reduction of genetic diversity within single organisms—in other words, by the occurrence of homozygosity at a larger number of loci. More extensive homozygosity in turn sometimes causes an increased variation of the affected trait, with larger numbers of individuals deviating from the norm that prevailed prior to the impoverishment. The means by which genetic diversity per se might promote canalization has not been demonstrated, but some authors believe that larger numbers of genes generate a greater number of feedback mechanisms, such as rate-dependent reactions and growth-rate compensation, that are able to control and coordinate development more precisely (for example

Thompson and Thoday, 1979). When these mechanisms are coarsened or eliminated by the simplification of their genetic blueprint, phenotypic variation increases, less fit individuals are produced, natural selection intensifies, and combinations of genes are reassembled that channel individuals back toward the norm. This is probably how specificity in the epigenetic rules of brain structure and social behavior has been shaped, and why the rules are typically under the control of multifactorial inheritance.

Genetic assimilation is related to canalization and developmental homeostasis in the following way. Sometimes the controlling polygenes are lax enough to permit deviations far outside the phenotypic norm. Alternatively, an exceptional change in the environment can evoke an extraordinary developmental response, even in a rigidly canalized system. Because of the superficial resemblance of the resulting forms to mutations, they are called phenocopies. In the original, now classic analysis of the phenomenon, Waddington (1953) obtained *Drosophila* flies lacking crossveins in their wings by the simple procedure of exposing the pupae to heat shock. When the individuals displaying crossveinlessness were then selected for further breeding, the genes most prone to allow the deviant development were automatically favored. In time the deviant phenotype began to appear in many individuals even in ordinary environments. Thus Waddington was able to isolate a strain of *Drosophila* that consistently lost the crossveins even in the absence of heat shock treatments. The trait had been genetically assimilated.

It follows that the multifactorial genetic control of traits is a theater of unusual opportunity for gene-culture coevolution. By cultural experimentation and the continual exploration of new environments, protocultural species are likely to test the potential of the genes that prescribe epigenetic rules of behavior and to produce novel responses that are the equivalent of behavioral phenocopies. When these responses confer selective advantage, the genes possessed by individuals tend to shift into a new frequency distribution, prescribing epigenetic rules that channel development toward the new response. It is possible for relatively few genes to shape the salient features of important brain structures and psychobiological responses because much of development involves the growth and differentiation of cells as groups that conform to holistic patterns. Such a genetic shift can therefore be rapid.

The Ontogeny of Mind

In Chapters 2 and 3 we characterized the key operations of cognitive activity, which are perception, information processing (including short term and long term memory), valuation, and intention. We shall now examine the set of cognitive processes and information storage strategies known as long term memory more closely and suggest how their properties can be acted upon by natural selection to guide genetic evolution.

It is useful to begin with the distinction between episodic and semantic memory first made by Tulving (1972). Episodic memory recalls specific events and thus concerns particular persons, objects, and actions that reenter the conscious mind through a time sequence. Semantic memory recalls meaning in the form of related concepts, which are classes of objects and events, or else their abstracted, symbolic surrogates. Clearly semantic memory originates in episodes and almost inevitably evokes the recall of some of them. But the human mind also has a strong tendency to generalize episodic memory into concepts and higher-level entities, which then constitute the "nodes" or reference points in long term semantic memory. These concepts are not necessarily words. Although many are labeled by words, some are not.

Nodes are almost always linked to other nodes, so that the recall of one summons several others. The links could be of various kinds: operational, in which for example an object and an action are associated; ascriptive, in which particular properties such as color or swiftness are attached to an object or action; denotative, calling up a word or some other symbol; and emotional, evoking feelings that typically are "difficult to put into words." Particular networks of nodes form schemata, the broader ideas, plans of action, and criteria to which the brain constantly refers in generating almost all forms of behavior. They are the meaning structures by which human beings organize new information and make decisions (see the reviews by Lindsay and Norman, 1977, and Wickelgren, 1979a,b; and a consideration of the special properties of visual imagery by Kosslyn, 1980).

In an early, influential study Quillian (1967) proposed a "spreading-activation" theory of long term memory (LTM). His initial purpose was to devise a superior computer-search technique based upon a model of the brain, but subsequently Collins and Quillian (1969), Collins and Loftus (1975), and others transformed the basic technique into a more explicitly psychological theory. In essence, the brain is considered to learn by constructing a growing network of related concepts, which are semantically primed by the particular links created in long term memory. When new episodes and concepts are added, they are processed by a spreading search through the network that attempts to find links with previously established nodes. Thus a novel form of fruit would be quickly classified by its physical properties, its edibility, the circumstances under which it was encountered, and so on, resulting finally not just in its association with other kinds of fruit but also with a repertory of emotional feelings, recollections of previous but similar discoveries, memories of dietary custom, and so forth.

Specific details of various individual network models have been questioned, but the basic conception appears to be consistent with a growing body of experimental evidence. Furthermore, it is our impression that the principal rival hypotheses, embodied in the set-theoretic and feature-comparison models (reviewed by Loftus and Loftus, 1976), are complementary to the conception of a node-link structure rather than exclusionary.

For the purposes of the theory of gene-culture coevolution, node-link structures can be defined as the form in which culturgens reside in long term memory. More precisely, a particular culturgen can be circumscribed, if necessary by the set-theoretic procedure suggested in Appendix 1–1, and sets of nodes and node links mapped onto it. The values associated with a culturgen, or more precisely the memory nodes to which it is linked, can be objectively measured by the semantic differential technique of Osgood and associates (1957) as well as by linguistic analysis. As illustrated in Figure 6–2, subjects are asked to place words or objects according to scales in each of an array of qualities, such as rough-to-smooth and beautiful-to-ugly. The list can be expanded until it encompasses virtually all of the conceivable links of importance in the mind of the person interviewed.

The characterization of long term memory as the activity of semantic networks is consistent with prevailing explanations of several other aspects of behavior and mental process. Physical activities are thought to be guided by motor programs or "sensorimotor schemata" (Lindsay and Norman, 1977). Each performance, such as eating and walking, follows a schema for an organized sequence of motions, during which information received through the sensory system is coordinated step by step with appropriate motor movements. In much of animal behavior the schemata are strictly inborn, having been assembled within the brain during morphogenesis and in the absence of learning. If we stipulate that cognition in the broad sense is information processing by animal brains, then it is reasonable to suppose with Griffin (1976) that even such hardwired networks can constitute a substratum for cognition and be linked to more reflective activities similar to human consciousness. In the case of operant learning by both animals and human beings, sensorimotor schemata are carved out of a wider array of actions performed during the course of exploration and play (Fagen, 1981). Sequences of activities that are rewarded are linked to

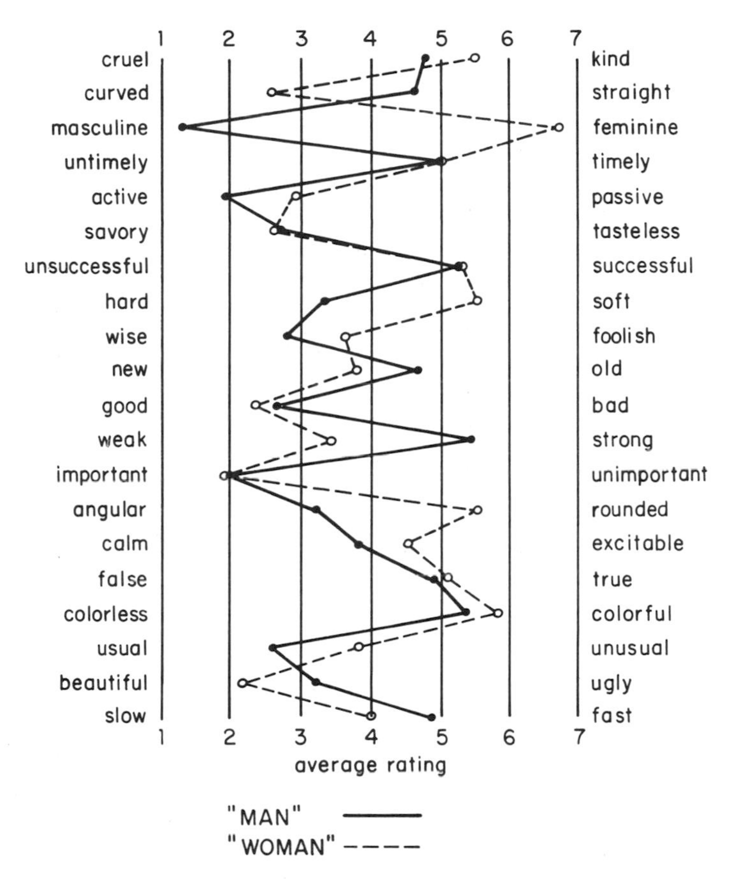

Figure 6–2 The semantic differential technique. Subjects are asked to place words or objects on a subjective scale for each of an array of subjective qualities, permitting the assessment of memory links to particular culturgens. This example compares the concepts of "man" and "woman." (Modified from Lindzey et al., 1975; based on Jenkins et al., 1958.)

goal nodes and repeated at need. In children the sensorimotor stage of development is succeeded by a period of "preoperational thought," in which the sensorimotor schemata can be activated mentally and mental experiments are possible. During subsequent development more advanced feats of abstraction, logic, and mental hypothesis testing are added (Piaget, 1952).

When organisms are predisposed to form certain semantic networks as opposed to others, the result is directed cognition, including prepared learning in the sense used to describe the results of earlier studies of overt behavior (Seligman, 1972a,b; Shettleworth, 1972). In Chapters 2 and 3 we saw that multiple forms of prepared learning occur in human beings and constitute important elements in the epigenetic rules. When an innate bias extends to the way words are linked in the formation of phrases and sentences, the result is deep grammar, an undeniably unique human property (Slobin, 1971; Chomsky, 1972)—but one about which we understand very little.

Still more subtle mental phenomena can be interpreted as the employment of episodic and semantic networks. When an attempt is made to recall an event from the distant past, it is difficult or impossible to remember all of the details. The conscious mind then engages in refabrication (Loftus and Loftus, 1976): it fixes on a few clearly remembered facts and searches through memory for additional plausible nodes to complete a satisfying schema. Once completed, the refabricated schema can become a more enduring feature of long term memory, and the true facts are then difficult to distinguish from the details that were summoned to fill the gaps.

Problem solving entails a related kind of activation spreading through the node-link structures as well as the discovery of new nodes and connections (Newell and Simon, 1972; Simon, 1979). Fresh concepts are attained by mental operations used to solve problems that resist searches through the easy accesses of the mind's problem space. Recent research on problem solving suggests that expertise is based to a large extent upon the memorization of facts, as common observation tells us. However, the expert also possesses a rich store of higher-level schemata that serve as rapid guides to the various parts of the knowledge store. The physical intuition of physicists and engineers, for instance, may consist in large part in the capacity to manipulate rapidly and efficiently chunks composed of many related facts (Larkin et al., 1980). There is good reason to believe that appropriate studies of long term memory and problem solving will eventually illuminate the deep cognitive processes that are loosely labeled talent, judgment, imagination, and creativity. We can expect to refine the diagnosis by Boulding (1978) that "the great difference between biological and social evolution is that, whereas prehuman organisms occupy niches and expand to fill them, the human organism is a niche-expander creating the niches into which it will expand."

Cognition and Genetic Fitness

In this brief examination of brain epigenesis we have traced the ontogeny of cognitive activity from genes forward to long term memory. Virtually every step in this marching order of biological causation is under active investigation by one group of specialists or another, and their relations are becoming increasingly clear. When the translation from epigenetic rules to cultural patterns is added (Chapter 4), it becomes possible to visualize the full extent of the linkage between genes, mind, and culture.

In Figure 6–3 we have sorted the steps of the gene-culture circuit into the four basic levels of biological organization. The first two levels (molecular, cellular) are ascended by anatomical and physiological epigenesis in the manner traditionally perceived within biology. The next level (organismic) is understood by examining the basis of the epigenetic rules of cognition and behavior. The highest level, that of the population and hence of society and culture, can be quantified through the analysis of gene-culture translation. Finally, the patterns of population structure and growth affect gene frequencies through natural selection.

It is in the last step, entailing the process of natural selection, that our theory finds its greatest challenge. The crux is the manner in which selection operates on culture. As we documented in Chapter 1, some culturgens undeniably provide superior genetic fitness over others. But how? In particular, one needs to know the role played by cognition itself in mediating the fitness of culturgens. Until we have some answers, the feedback from culture to genes cannot be reliably investigated with the models of population genetics. We need to characterize the transformation from culturgens to the genetic fitness conferred by sets of culturgens:

$$c_1, c_2, \ldots, c_M \longmapsto W(c_1, c_2, \ldots, c_M).$$

Because culturgens interact with one another in long term memory, this mapping is not trivial.

From recent findings in cognitive psychology we can start to investigate the general form of this relation and go on to provide more realistic elementary genetic models. The key fact is that the brain operates on meaning. When a culturgen is assimilated, it is not merely stored as an isolated element in long term memory, where it awaits recall and use in the manner of a card flipped from a file. It becomes a node linked to other nodes and hence part of a broader meaning structure. An important distinction can be made between "cold" and "hot" links of a node. A completely cold link is purely informational, associating the node with another that corresponds to a physical quality, an action, or definitional relation. A hot link stimulates an emotional feeling. The formation of such links involves circuits within the limbic region, although other connections to the

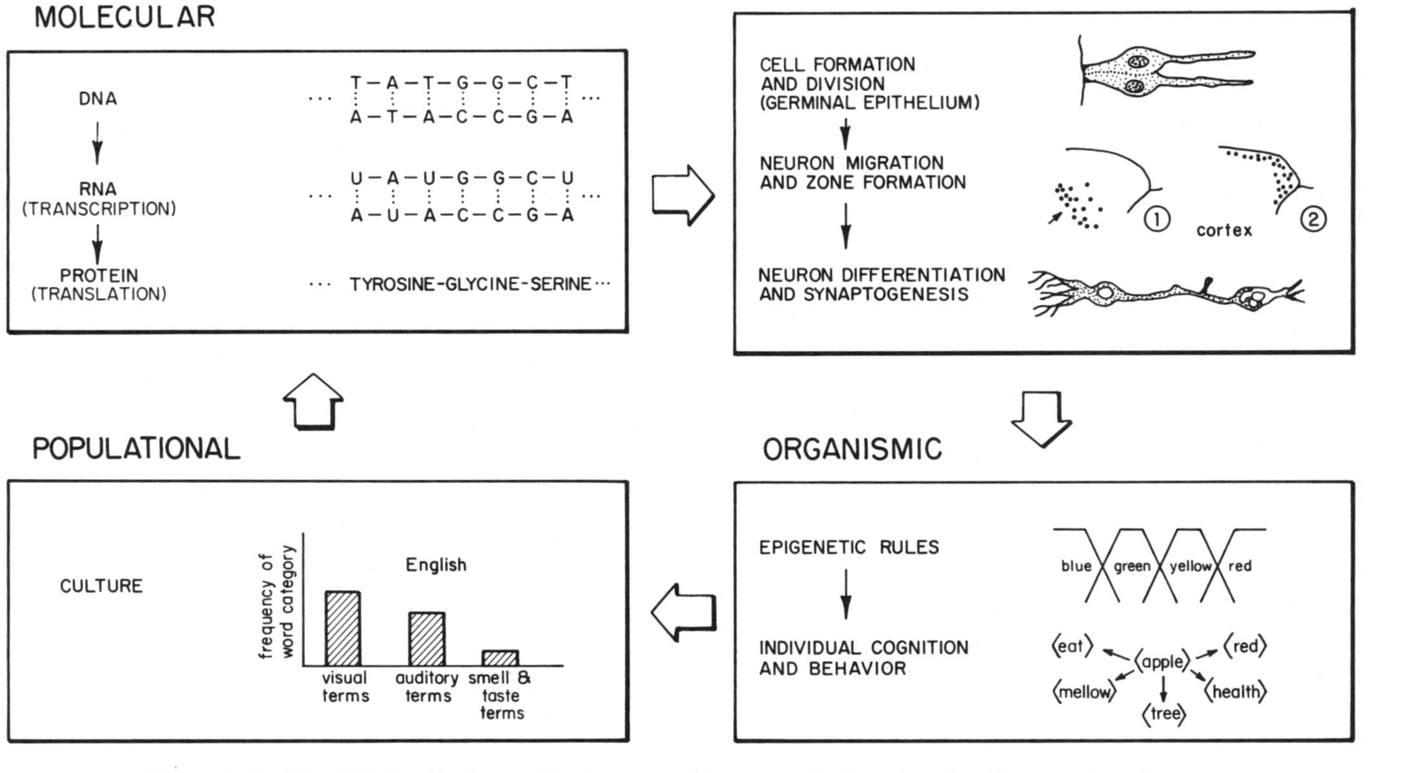

Figure 6–3 The full circuit of causation in gene-culture coevolution, showing the grouping of steps within the four principal levels of biological organization. The molecular, cellular, and organismic steps constitute epigenesis; the transition between the organismic and populational levels consists in gene-culture translation; while population phenomena affect gene frequencies through natural selection.

cortical region exist through the fibers of the brain reward system (Routtenberg, 1978). Most nodes, and therefore most culturgens, possess multiple links of both kinds, and the extent to which they reenter the conscious mind depends on the context in which the concept is evoked. Thus a taxonomist sitting in a museum studies the many biological traits of the tiger (*Panthera tigris*) and associates it with other species of cats into a formal higher taxon, the family Felidae. Little emotion is evoked by this exercise. But the same person walking alone at night in a Thai forest is charged with fear at sensing the nearby presence of a tiger. For the moment at least, his mind makes no effort to continue his classification of the Felidae.

The nodes, therefore, are seen as elements of various chunks, schemata, and plans that are summoned according to circumstance. Some of the schemata are highly abstract, others are sensorimotor and concerned with programs of physical action or the mental conception of such action. Some parts of long term memory are cognitive manifestations of the culturgens and can probably be mapped onto them isomorphically. An artifact is a culturgen that may be used in a precise, invariant manner. A hammer will be struck according to a simple motor plan or sensorimotor schema. But it is also an object with physical properties that allow us to connect it through the recall of concepts to other kinds of hammers, sledges, and related instruments. It can become part of still other, metaphorical concepts—the hammer of God, the futility of John Henry's hammer, harsh construction between the hammer and anvil. Finally, the culturgen can be a pure mentifact, a wholly mental construction based entirely on symbols or imaginary creatures and objects. We can employ the hammer itself as a physical culturgen, as in archaeological analysis, or we can place it within certain behaviors, stories, and metaphors that constitute the other principal categories of culturgens.

The cognitive characterization of culturgens permits a formulation of the relationship between culture and genetic fitness. Fitness is clearly not the product of an artifact lying in a toolshed or a mentifact circulating in the recesses of long term memory. It is a function of explicit behavior, of muscular contraction and the motion of parts of the body. The mind of the human being intervenes to impose an immense new realm of order and process between learning and explicit behavior. It integrates experience while searching for alternative schemata and new ways to achieve its goals. The mind accumulates behavioral options out of experience and ceaseless conscious activity, during which old experiences are relived, new ones imagined, and both evaluated by their affective links. The knowledge structures grow like chemical polymers, adding nodes and links with each new experience, and shedding others as disuse causes them to fade beyond the reach of recall (see Figure 6–4).

Mental activity and outward behavior are based on memory, the episodic and semantic networks of nodes and links. The networks are built up in response to experience in forms organized by the epigenetic rules. This canalization results in part from sensory screening, which causes perception to be limited to narrow segments of the immense arrays of

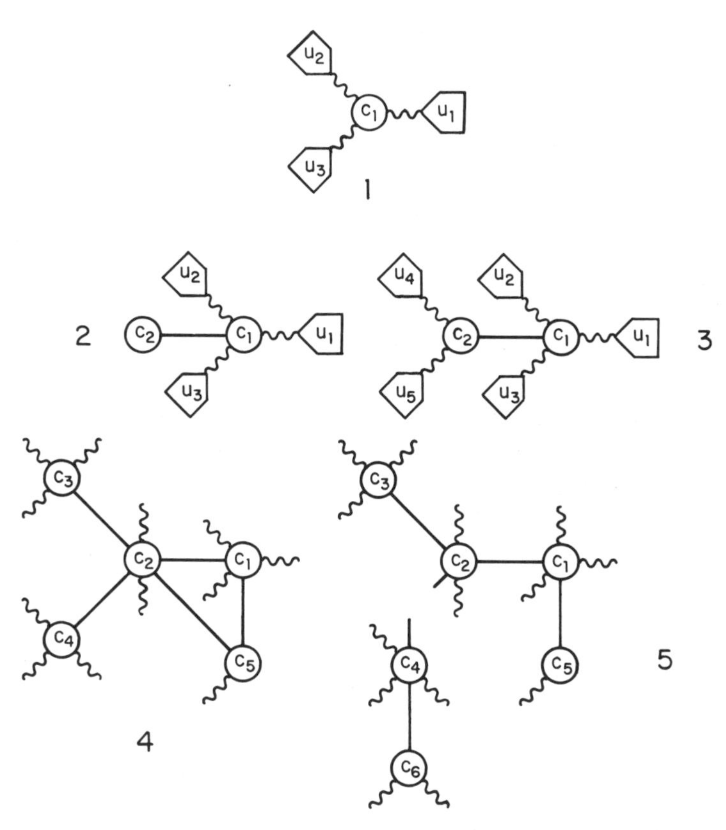

Figure 6–4 The development of the mind, envisioned as a polymer-like growth of semantic networks in long term memory. Many of the nodes, designated here as c_i, form the core stimulus associations of the culturgens. Others, designated as u_i, comprise the emotional associations or hot semantic nodes linked to the culturgens and are used in valuation. Node-link structures can be built up by gradual accretion of nodes, as suggested in sequence 1–4. They can also be altered by the breaking of memory links and the formation of new links by the separated components, as shown in diagram 5.

physical stimuli impinging on the body. It results in part from a tendency for certain nodes in long term memory to form links differentially to other nodes associated with the limbic and brain reward system, so that they are more likely to become positively or negatively reinforcing. Thus bonding results from the virtually automatic positive linking of mothers and infants during their initial contacts, while snakes, heights, and other typical objects of phobias are almost equally likely to form negative links. Finally, the canalization results in part from the peculiar constraints that lead to algebraic rules of risk estimation and other forms of decision making, as described in Chapter 3. Canalization in the generation of the semantic networks leads to a substantial convergence of the forms of mental activity among the members of societies and even among people belonging to different cultures. Idiosyncracies in concept formation and valuation obviously distinguish one human being from another. But the epigenetic rules are sufficiently tight to produce a broad overlap in the mental activity and behavior of all individuals and hence a convergence powerful enough to be labeled human nature (see Wilson, 1978).

Culture can be heuristically defined as the cognitive and behavioral outcomes of the totality of shared culturgens defined in this new sense. Reification and symbolization are seen as devices for creating and codifying culturgens for more efficient processing, storage, and recall. Language is the means whereby the culturgens are labeled and swiftly juxtaposed to assemble and communicate vastly more complex knowledge structures, such as narratives, instruction, and art. Language further serves to transmit *meaning* from one person to another, from one long term memory into another, quickly and efficiently. Under the influence of the epigenetic rules, the culturgens shared in such a manner will tend to possess similar core meaning and to evoke similar behavior.

The components of long term memory can be regarded as forming a hierarchy. Nodes and sets of closely linked nodes constitute chunks or schemata, and sets of schemata can be delimited from other parts of long term memory. All of the culturally induced nodes and links stored in the memory of an individual constitute his received portion of the culture. For purposes of analysis it is useful to denote as *knowledge structures* any set of nodes and links in long term memory, from a single node to chunks to the entire contents of long term memory susceptible to cultural mediation. Segments of knowledge structures can be mapped onto culturgens and in many instances are isomorphic with them. Given the rules that can be inferred about the formation of knowledge structures, we are in a position to begin with their simplest forms, much as the geneticist begins with elementary diallelic inheritance, and gradually expand their size and complexity.

The full sequence connecting culture to genetic fitness is suggested in Figure 6–5. An individual is enculturated by exposure to the culturgens of other members of the society. This information is incorporated into long term memory under the influence of the epigenetic rules, and the resulting knowledge structures become the culturgens of the learning individual.

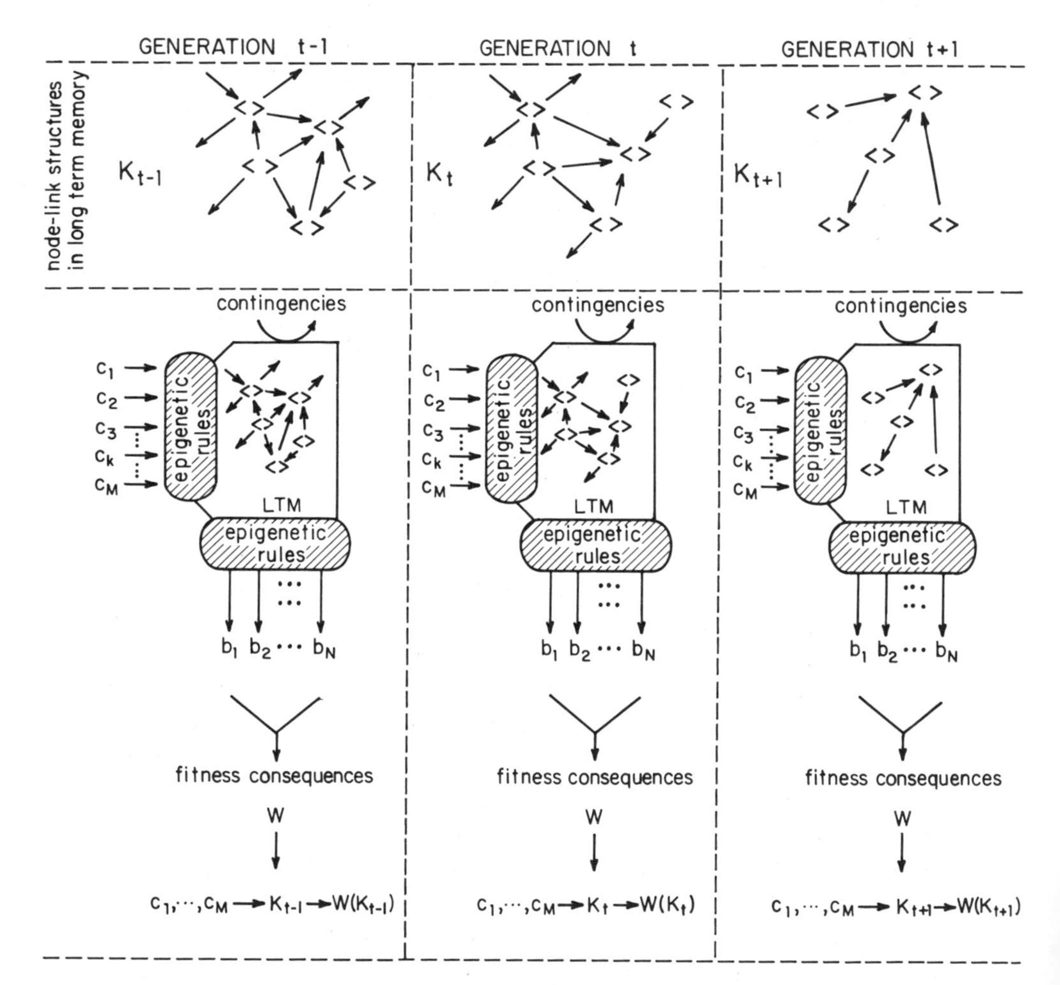

Figure 6–5 The complete sequence linking culture and genetic fitness. Culturgens ($c_1, \ldots, c_M$) are encoded in long term memory in the form of polymer-like knowledge structures (K) and later translated into behavioral acts; both processes are guided by the epigenetic rules. The behavioral acts ($b_1, \ldots, b_N$) determine genetic fitness.

Other epigenetic rules influence his recall, valuation, and decision making. The resulting behavioral acts then determine his genetic fitness. We can therefore speak of a fitness function $W(K)$ of particular organisms bearing specific knowledge structures, segments of which in turn can be traced back through socialization to the culturgens and in many instances regarded as identical with them.

In the determination of genetic fitness, a diversity of selection pressures test the behavioral acts emanating from the mind. Selection, by imposing differential mortality and fecundity, tests not only the behavioral acts but also the knowledge structures of long term memory, the culturgens composed of the knowledge structures, and the epigenetic rules that guided the formation of all of these elements.

The knowledge structures themselves can theoretically be inherited as hardwired processes within the brain. Such purely genetic transmission occurs in many lower animals. But in *Homo sapiens,* this is the case in at most a tiny minority of behaviors, and these are predominantly reflexive and autonomic in nature. Almost all of human behavior is based on knowledge structures that are learned to some degree. What is inherited are the storage and processing capacities of long term memory and the epigenetic rules. These are the physiological properties of the mind most directly subject to genetic evolution, and they are properly made the principal focus of the natural selection models.

Within gene-culture theory such models can be given mathematical form. Thus notions about mechanism, otherwise vague, can be made precise, and we are able to articulate with maximum clarity the chains of inference that connect our postulates with assertions about real systems. Furthermore, we can propose minimal standards that realistic models of gene-culture coevolution must fulfill. They include first of all the perception of a complete circuit, one that loops upward from genes through epigenetic rules and cognition to macroculture, then back through the fitness effects of behavior to the genes. Second, they incorporate postulates about cognitive and behavioral mechanisms. The existence of the epigenetic rules means that the mind is an active entity shaping its own growth, not a passive receptacle of impressions. Culturgens do not enter long term memory like kernels of grain filling an urn. Knowledge structures are composed of *interacting* nodes, of which many or most correspond to culturgens. A third condition of realism is the recognition that genes produce the epigenetic rules that under environmental boundary conditions direct the assembly of knowledge structures and ultimately of the whole mind. As a result the prescriptors of cognition and behavior cannot be assessed as "x percent genes" and "y percent culture." The true causation is fully epigenetic and not decomposable in any naive additive sense.

In addition, individuals perceive group properties, and there is consequently a feedback from macroculture to cognition. Individuals are seen to reify the institutions of their society, while the products of the reification learning rule color the algebra of thought. The phenomena to be predicted by coevolutionary models are more than the patterns of change in gene frequency; they are also cultural history as expressed in the ethnographic distributions.

Gene-culture theory therefore goes far beyond conventional notions about the evolution of a "capacity for culture." It specifies the key step required to understand either euculture in man or protoculture in other animals: the evolutionary study of epigenetic rules. It provides the body of data required to begin the program and assembles the necessary mathematical techniques from a variety of disciplines. With these tools we can start to reconstruct the trajectories in organic evolution that lead eventually to culture.

At the present time almost nothing is known about human evolution from this vantage point. Progress has been impeded by the lack of both an adequate data base and evolutionary postulates powerful enough to encompass the human mind. As a consequence there is an urgent need for precise models that treat part or all of the gene-culture circuit. The evolutionary models discussed in the previous chapter and the similarly incomplete models treated by other authors (to be reviewed presently) only begin to fill this gap. It is our belief that further progress in the subject requires models designed according to the criteria set forth above.

In this chapter we develop and explore one such family of models. We focus on individual fitness rather than group fitness, because it is both more tractable and evidently the level that dominated during at least the first steps leading to euculture in the primate line. The model, which reveals the workings of a novel and complex selection mechanism, has proved mathematically challenging in spite of our simplifications. It nevertheless achieves a complete tracing of the circuit of coevolution and reveals some of its distinctive phenomena.

Previous Studies on Coevolution

Before proceeding to the mathematical formulation, we may find it useful to consider previous studies on the coevolution of genes and culture.

Many biologists and social scientists have written on this subject, especially during the past ten years. Their contributions are summarized and very briefly evaluated in Table 6–1. The listing is not exhaustive, but we believe it to be representative of the diverse forms of research conducted and to include the principal substantive advances to the present time.

Four of the research efforts have approached issues relevant to our own conception of gene-culture coevolution. S. T. Emlen (1976, 1980) argued that even when behavioral patterns are purely cultural in transmission, they are likely to be adaptive in a biological sense and hence difficult to distinguish from patterns that are transmitted by partial or pure genetic means. The same principles of population biology can be applied to both; the theory of "ecological sociobiology," which interprets social patterns as adaptations to the environment, can in this view be pursued without immediate reference to the role of genetic constraints in the development of behavior.

Durham (1976, 1978, 1979) developed a similar argument: pure culture can be, and in general is expected to be, Darwinian. "Simply stated, my hypothesis is that the selective retention in biological *and* cultural evolution generally favors those attributes which increase, or at least do not decrease, the ability of individual human beings to survive and reproduce in their natural and social environments. This perspective has the advantage of explaining both how human biology and culture can be adaptive in the same sense . . . and how they may interact in the evolution of human attributes" (1979:54). For Durham coevolution means the joint and potentially independent contributions of genes and culture to behavior, rather than coevolution in the sense employed by biologists of two systems mutually altering each other (see for example Janzen, 1980). The mechanisms through which biology and culture might be coupled were therefore not pursued in his analysis.

In a set of ingenious models Boyd and Richerson not only examined the outcomes of coevolution in the strict sense but visualized genes and culture as two systems engaged in a kind of game played for control of the behavioral phenotype (Boyd and Richerson, 1976; Richerson and Boyd, 1978). They recognized correctly that phenotypes maximizing cultural fitness (that is, the most rapid transmission of the culturgens) might not be the same as those maximizing genetic fitness (the most rapid transmission of the genes). There is consequently a "struggle" between genes and culture for the preferred steady-state behavior of the population. In this process the genes generally hold the upper hand, because the "capacity for culture" is under genetic control and hence adjusted to optimize

Table 6–1 A synopsis of previous work on the relation between genetic and cultural evolution.

Reference	Problem addressed	Methods	Major result(s)	Comments
Dahlberg (1947)	Genetic predisposition toward class membership	Single-locus model of population genetics with individual-level selection. Cognition, social interaction, gene-culture coevolution not explicitly considered	In a steady-state society a gene promoting success will be concentrated in upper socio-economic classes within a relatively few generations	An interesting if greatly oversimplified model that has not been explored further, although a similar idea was developed intuitively with reference to IQ by Herrnstein (1971)
Campbell (1965)	Directive forces of cultural evolution, with special reference to social organization	Microevolutionary argument by analogy to organic evolution	Processes of variation and selection in social systems parallel those in genetic evolution	Anticipates later adaptation of natural selection theory to human culture
J. M. Emlen (1967)	Genetic contributions to human social behavior	Macroevolutionary argument; posits role of natural selection but does not contain an explicit model. Cognition and mode of culturgen transfer not considered	Natural selection affects which behaviors are positively and which negatively reinforcing, and hence determines those forms of social behavior most likely to be transmitted by culture	Foreshadows later findings on epigenetic rules and directed learning
Alexander (1971, 1979a,b)	Evolution of the capacity for social learning and culture; adaptiveness of social patterns; "reconcilia-	General arguments from sociobiology and evolutionary theory, with emphasis on kin selection and nepotistic be-	Support for the general hypothesis that "human individuals, like other organisms, have evolved to interpret	Foreshadows but does not express an explicit theory of gene-culture coevolution

	tion'' of Darwinism and social theory	havior.	their best interests (not necessarily consciously) in terms of reproductive maximization''	
Wilson (1975, 1978)	Genetic contributions to human social behavior	Micro- and macroevolutionary arguments from models of individual and group selection. Cognition and mode of culturgen transfer not explicitly considered	Directed learning evolves as a genetic product and channels cultural evolution. Ethical values derived from genetically evolved controls in limbic system	Foreshadows but does not develop current gene-culture theory
S. T. Emlen (1976, 1980)	Similarities between genetic and cultural adaptation in the shaping of adaptive behavior	General arguments and documentation from behavioral ecology and sociobiology of both animals and human beings. Specific models of gene-culture coevolution not pursued	In response to particular environments, genetic and cultural evolution will yield social patterns that are similar if not identical. Therefore ecological sociobiology can be studied independently of the mode of transmission	See text
Durham (1976, 1978, 1979)	Coevolution of human biology and culture	Micro- and macroevolutionary arguments. Gene frequencies, cognition, and social interaction not explicitly considered	Cultural traits are retained according to the same criterion as natural selection. Slack can be introduced into time-energy budgets by cultural adaptations	See text

(*continued*)

Table 6–1 (continued)

Reference	Problem addressed	Methods	Major result(s)	Comments
Feldman and Cavalli-Sforza (1976, 1977, 1979)	Gene-culture coevolution; the general phenomenon of "complex transmission" of both genes and purely phenotypic variants	Single-locus, single-culturgen model, variance analysis; includes teaching of offspring by parents. Cognition, epigenetic rules not considered	Conditions deduced for the maintenance of gene frequency polymorphism through heterozygote advantage in complex transmission. Dependence of the polymorphisms on the parameters of the learning process. Existence of oscillatory and multiple stable gene frequency equilibria	See text (Chapters 5 and 6)
Boyd and Richerson (1976); Richerson and Boyd (1978)	Gene-culture coevolution. Evolution of the capacity for culture and strength of genetic constraints on culture	Static optimization theory. Coevolution visualized as a game between genes and culture. Cognition, epigenetic rules, and social interaction not explicitly considered	Enumeration of circumstances in which phenotypes of a population evolve to either the genetic optimum, the cultural optimum, or an intermediate state, under the influence of a variety of fitness functions	See text

Rice et al. (1978); Cloninger et al. (1979a,b)	The genetic and cultural factors that influence transmission of a trait from parent to offspring; techniques for separation of the effects of the two classes of factors	Multifactorial genetics modified to include cultural transmission. Detailed consideration of intrafamilial pathways of cross-generation information transfer. Cognition, epigenetic rules, and social patterns not explicitly considered	Formulation of mathematical techniques to measure correlations between parents and offspring, between siblings, and between kin-group members and neonates with reference to cultural transmission	Nonevolutionary; natural selection not considered
Karlin (1979a-d; 1980a,b)	Dynamics of phenotypes controlled by both genes and culturgens	Models of multifactorial genetics in discrete generations, including consideration of both individual and sexual selection. Correlation equations, usually linear, between genotypes and between phenotypes. Cross-generation communication, usually between parents and offspring. Cognition, epigenetic rules, and social patterns not explicitly considered	Extension of models of multifactorial inheritance to simultaneous consideration of many genotypic, phenotypic, and cultural characteristics. Idea of the "vector phenotype." The models emphasize phenotype-phenotype correlations	

(continued)

Table 6–1 (continued)

Reference	Problem addressed	Methods	Major result(s)	Comments
Thomas (1971); Jochim (1976); Winterhalder (1977); Binford (1978); Hames (1979); Reidhead (1979, 1980); Smith (1979, 1981); Keene (1981)	Individual behavior and culturgen usage as adaptations to the environment	Optimal foraging theory; linear programming; optimality and satisficing concepts from behavioral ecology. Models tested with ethnographic data. Emphasis on maximization of net energetic profit. Gene-culture coevolution implied in some cases but not explicitly formulated	The human mind assigns greater utility to culturgens and behaviors that have high reproductive value (or its equivalent in energetic yield) in particular environments	See discussion of behavioral ecology (Chapters 3 and 5)
Pulliam and Dunford (1979)	Evolution of learning strategies with special reference to human behavior	General arguments from sociobiological and learning theory. Nonsocial foraging problems evaluated numerically.	Natural selection designs learning behavior in such a way that adaptive acts are positively reinforcing and nonadaptive acts are negatively reinforcing	

Bonner (1980)	Evolution of culture in animals	Comparative analysis of data from zoological literature	This first synthesis in an evolutionary perspective shows that animal protocultures are surprisingly widespread and diverse	
Fagen (1981)	Evolution of play and capacity for innovation	Single-locus, discrete-generation models of population genetics, in which the alleles prescribe phenotypes that are high or low in innovative capacity. Culturgens spread by imitation	The first systematic theory of the genetic basis of innovation. Among the results: social learning puts a drag on the spread of *innovator* alleles because noninnovators can profit from the culturgens introduced by the more costly *innovator* genotype. Critical innovation rates are required to sustain the *innovator* allele; episodes of innovation and spread of new culturgens throw allele frequencies into oscillation	Play and innovation are important components of gene-culture coevolution and are likely to assume a larger role in more advanced models with expanding cultural arrays

genetic fitness. When the phenotypes can be produced by genes alone, culturally transmitted behaviors will be allowed only when they result in a genetic optimum. The Boyd-Richerson models pertain to steady-state resolutions of the gene-culture "game" and are almost wholly free of content from genetics, neurobiology, and psychology. As a consequence the models do not provide means for measuring and relating phenomena and cannot be used in the literal analysis of human social behavior. Nevertheless, game-theoretic and optimization techniques are potentially fruitful in the study of human behavioral evolution. This is especially so if particular uses of gene-culture transmission can be identified and characterized as evolutionarily stable strategies, which are adaptations adopted by a critical proportion of the evolving population and thereafter unopposed by any parallel adaptations immediately available to the population (Maynard Smith, 1974, 1976; Dawkins, 1980).

An explicit coevolutionary model has been studied by Feldman and Cavalli-Sforza (1976) in connection with their more general analysis of "complex transmission," or the compound inheritance of genetic and purely phenotypic variation (see also Feldman and Cavalli-Sforza, 1977, 1979). Their model stipulates a single culturgen—"skilled"—which can be attained from the unskilled state. The capacity to acquire the skill is dependent on the genotype of the offspring and the skilled or unskilled state of the parents. A balanced polymorphism can exist in the case of two alleles when the heterozygotes learn the skill more readily than do the homozygotes, but only if the skill provides a moderately strong selective advantage. We have discussed this result in Chapter 5, in connection with the general problem of adaptation to heterogeneous environments. The Feldman and Cavalli-Sforza model is another step in the right direction. However, because this model is fundamentally ad hoc (in other words, proposed and utilized independently of postulates about underlying mechanism), to date it has been only weakly heuristic. Furthermore, it lacks explicit treatment of many features that we believe are important for understanding gene-culture coevolution, such as cognitive information processing, choice among culturgens, epigenetic rules, and sensitivity of individuals to trends manifest in cultural patterns.

Modeling the Coevolutionary Circuit

Early in our own study we realized that coevolution cannot be effectively grasped unless the entire circuit of causation is traced as a virtually complete process, from the genome of individual organisms to the cultural pattern of the society as a whole and back around again. The reason is that the cultural society is a heterarchy instead of a pure hierarchy; the epigenetically guided actions of the individual members create the cultural patterns, but the patterns influence the actions and, ultimately, the frequencies of the underlying genes themselves. Thus in order to model the process effectively, it is necessary to identify and to interlock the full array of phenomena that lead in small steps completely around the gene-culture circuit: the gene frequencies, epigenetic rules, socialization, cognition, cultural pattern, and sensitivity to usage pattern by the remainder of the society. We found this task to be difficult but not intractable.

In the family of models to follow, the interaction of genes and culture is characterized from existing information about cognition and sensitivity to usage pattern in human societies. The interaction is extended into evolutionary time in order to envisage changes not only in cultural pattern but also in the gene ensembles and epigenetic rules. A life cycle is incorporated in which the offspring produced in a large, randomly mating population are socialized by both their peers and the older generation. Alternative culturgens are learned and evaluated through exploration, play, and observation. Later they are employed during a prereproductive period in the gathering of a resource, which can be broadly defined according to circumstances to mean either food and other limiting resources or territorial ownership and other modes of control by which resources can be gathered in an unimpeded manner. Individuals choose between two culturgens under the influence of epigenetic rules. Genetic variation is permitted in the degree of bias in the epigenetic rules, the amount of sensitivity to peer and parental usage, and the function by which resources are converted to genetic fitness.

A number of phenomena, some of them previously undiscovered, are revealed by our investigation of the model. The tabula rasa state, in which no innate bias exists in culturgen choice, is shown to be unstable, easily replaced by any of a wide range of biasing epigenetic rules. The time scales for this replacement are calculated. Sensitivity to usage pattern increases the rate of evolution in epigenetic biasing and hence the genetic assimilation of culturgen use. This catalytic effect might have contributed to the rapid evolutionary increase in human brain size associated with the onset of gene-culture coevolution. Culture slows the rate of genetic evolution, but coevolution still occurs rapidly enough for the genes to track many culturgen changes. On the basis of the results of the model we propose a "thousand-year rule" for this type of system: the alleles of epigenetic rules favoring the features of more successful culturgens can largely replace competing alleles within as few as fifty generations, or on the order of a thousand years in human history.

Finally, coevolution is revealed to be based on a novel form of frequency-dependent selection. Under certain conditions, in particular when genetic fitness declines after a certain amount of resource has been harvested, selection leads either to stable genetic polymorphisms or to chaotic fluctuations in gene frequency. In the latter regime, coevolutionary history would be both unpredictable and arbitrarily complex in a well-defined sense. The genes would not move toward an optimal genotype. We suggest that this phenomenon can enhance genetic diversity, division of labor, and individuality among highly eucultural organisms.

The Model Life Cycle

The core of the model is the life cycle introduced in the evaluation of the gene-culture adaptive landscape (Chapter 5) but now elaborated to introduce aspects of individual cognition and response to the social environment. The details are summarized in Figure 6–6. For analytic convenience we have modified the human life cycle in two ways. First, the population is made large and randomly mating, in order to exploit the deterministic equations of population genetics. Second, the generations are quasi-discrete, meaning that adults persist long enough to enculturate juveniles but die before the next breeding phase takes place. These features are not profound deviations from the human condition, and they permit a simpler and more abstracted description of the life cycle that can be utilized in initial studies of populations of both human beings and protocultural species.

The ethnographic curve is adjusted during each generation according to the genotypes of the individuals and the culture of their parents. The curve, which constitutes the cultural pattern, is calculated for the two-culturgen case, that is, for the classic anthropological case of binary choice and binary classification. The culturgen choice by each individual and the usage pattern of the rest of the society determines the effectiveness of the individual in exploiting the environment. The effectiveness in turn determines the individual's genetic fitness. Variation in genetic fitness among the members of the society is translated into an alteration of the gene frequencies in succeeding generations. It is not necessary at this stage of the analysis to make a distinction between individual and inclu-

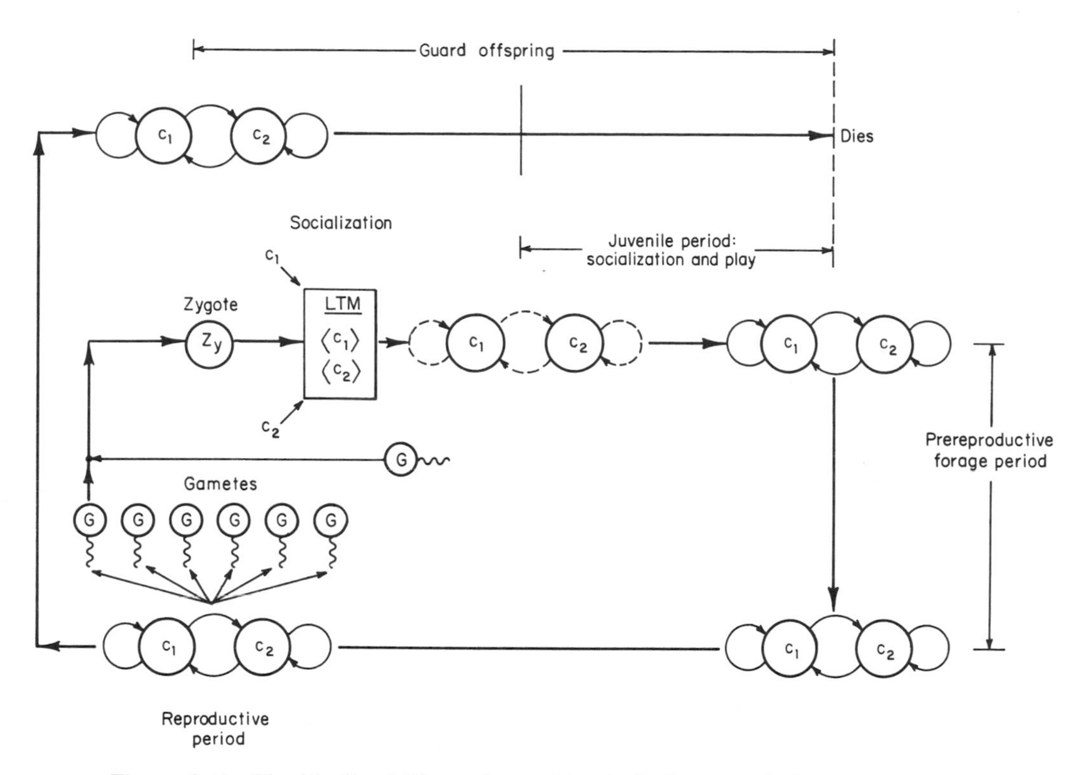

Figure 6–6 The idealized life cycle used to study the coevolutionary process.

sive fitness, so that special effects of group and kin selection can be ignored.

The offspring, then, are produced during a discrete reproductive period toward the end of the life cycle. They proceed to assimilate both of two culturgens, c_1 and c_2, into their long term memory. The exact form of their knowledge structures is determined to a large extent by the structures held by the adults and the manner in which the adults enculturate the offspring. The juveniles also record impressions of how many adults use each of the two culturgens. During exploration and play they assess the culturgens still further by means of lessons, games, practice, and conversation. The learning can be direct, in which case the juvenile employs the culturgen with or without tutelage, or it can be purely observational in form, consisting in unrewarded learning that occurs when one individual watches the activities of others (see reviews in Alcock, 1969; Hinde, 1970; and Bateson, 1976). During exploration and play the juveniles also observe one another, adding to their perception of the uses and value of the culturgens. In later life this information will be consulted repeatedly to help make decisions about the serious uses of the culturgens.

Thus the result of socialization, of imitation, and of exploration and play is a knowledge structure that has two capacities. First, it contains the schemata corresponding to the culturgens themselves; and second, it contains information that partitions the environment into situations calling for culturgen c_1 or for culturgen c_2. For simplicity we assume that this classification exhausts the set of possible events.

In order to evaluate the competing culturgens, the juvenile mentally connects their use with the various contingencies of the events, such as prey to be hunted, shelters to be built or enemies to be confronted. The event classification, by which the individual matches culturgens with environmental contingencies, is composed of mutually interpenetrating sets, as visualized in Figure 6–7A. Because the sets overlap, the decisions made by individuals are probabilistic rather than deterministic (Figure 6–7B). It is conceivable that an organism with unlimited cognitive resources could achieve a flawless separation of such events (Figure 6–7C). However, all living organisms, including man, have a finite capacity and limited time in which to make decisions.

In Chapter 3 we reviewed the evidence that the human brain has responded evolutionarily to this framework of constraints by the use of the principles of fuzzy logic, which is represented by the convention of overlapping sets of event classification. Fuzzy logic translates directly into the existence of finite culturgen transition probabilities, a basic condition used in our earlier models of gene-culture coevolution. To take a concrete example, imagine a Polynesian fisherman selecting between two kinds of hooks. His position off the reef and the water conditions there suggest the species of fish most likely to be available. His judgment is based on memories of the gradients of marine environments in which he and others of his village have fished before. From one extreme set of conditions to another, the usefulness of one of the kinds of hook rises while that of another declines. For many parts of the fishing grounds the choice can never be absolutely clear. The fisherman consequently has a certain probability of selecting the first kind of hook and another probability of selecting the second kind.

The development of the knowledge structures and their later valuation are constrained by the epigenetic rules. Other epigenetic rules shape the way in which the patterns of culturgen use among parents and peers are incorporated into the final decision probabilities (see Chapters 2 and 3). Genetic changes in the society as a whole consequently influence the strategy by which the final probabilities are calculated.

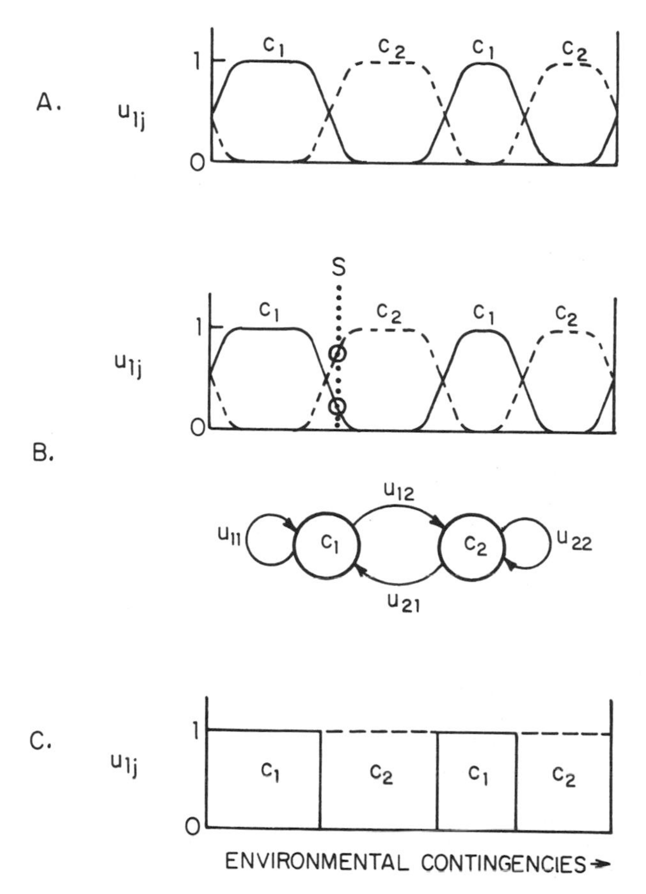

Figure 6–7 Cognitive matching in the solution of problems. During socialization and play the learning individual classifies different environmental contingencies into overlapping sets of culturgen usage. The curves in this figure give the likelihood u_{1j} that an individual previously using culturgen c_1 will classify any specific environmental contingency S as being appropriately met with c_j, where $j = 1, 2$. Panel *A* shows the culturgen cells. Panel *B* illustrates the exposure of the individual to a particular environmental contingency S and the role of the resulting transition probabilities in determining the probabilities of choice between the two competing culturgens. The vertical dotted line cuts cell c_1 at u_{11} and cell c_2 at u_{12}. The totally disjunct categories of a "perfect" (and highly unlikely) decision system are shown in panel *C*.

The adults of the previous generation die at the end of the enculturation period and their offspring enter the prereproductive period. During this time the group members use the culturgens to gather a resource that subsequently will be turned into gametes. These simple expressions are meant to serve as a shorthand for larger and possibly more complicated classes of phenomena. To gather a resource means either literally to accumulate food in a direct manner or else to secure a richer subsequent harvest through the establishment of territory, attainment of rank in a hierarchy, formation of economic and political alliances, or other devices. To turn the resource into gametes means to contribute to the zygotes that form the next generation; the number thus deployed depends on the success of the individual during the period of environmental exploitation. The choice of culturgens determines both the strategy of exploitation and the reproductive success.

Each foraging bout begins with a new decision concerning the culturgens, such as the hook to be employed, the kind of foraging team to be organized, the means of protection against inclement weather, and so on through many diverse categories. On the basis of the decision the individual uses the chosen culturgen of a given category to gather resources at a fixed rate throughout the episode. The beginning moments of new episodes are exponentially distributed through time, a condition stipulated and explained in Chapter 4 (see Figure 4–8). The foraging environment remains constant, so that the manner in which it is classified by individuals (Figure 6–7B) remains always the same. Under this special condition the decision probabilities are constant and the Markov approximation to gene-culture translation is accurate.

The accumulation of resource for the purposes of reproduction stops at the end of the prereproductive phase. In the next phase the resource is converted to gametes according to a fixed fertility rule and emitted into a random mating system. The fertility rule tallies the benefits resulting from the resource harvested, after the benefits have been discounted by the costs incurred from the construction, maintenance, and use of the cognitive processors responsible for socialization, reification, and decision making. The variation in fertility due to alternative culturgen usage is the sole cause of variation in absolute fitness among the genotypes in the population. It is straightforward to extend the model to include mortality effects in the neonatal, juvenile, and reproductive periods, but for clarity we do not include these additional parameters here.

The features of the relevant epigenetic rules, and hence the transition probabilities $u_{ij}(\xi)$, are controlled by a single major gene with two alleles, A and a (Figure 6–8). The corresponding processors can differ in accuracy of operation, in complexity, in cost of construction and maintenance, and in cost of use per computation. We expect that a higher degree of accuracy will generally require more complex and expensive processors, resulting in trade-offs among feasible system designs during the evolution of the brain.

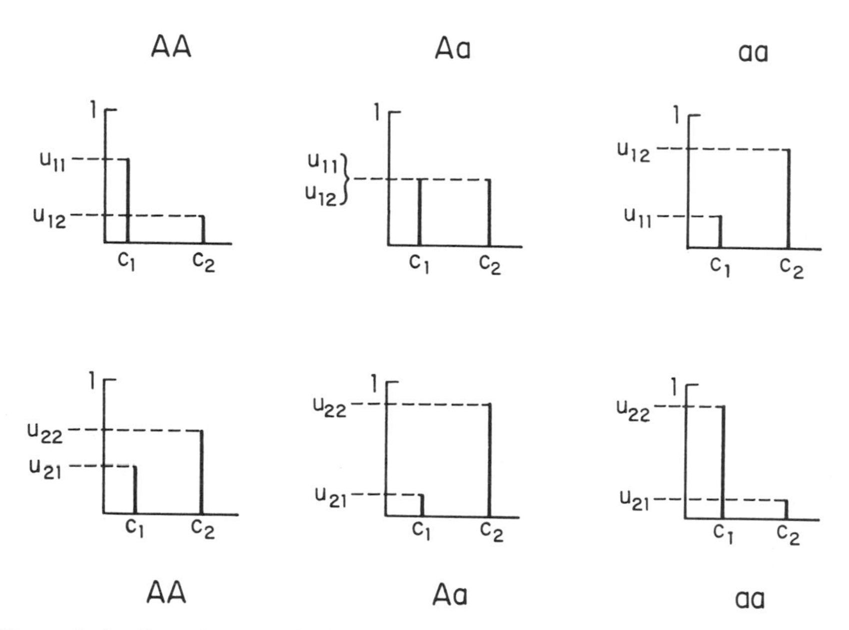

Figure 6–8 Genetic control of the epigenetic rules. In the elementary coevolutionary model being developed, the transition probabilities u_{ij} and the transition rates v_{ij} are prescribed by a single major gene with two alleles, *A* and *a*. (The values of u_{ij} indicated in the diagram are illustrative.)

While the genes prescribe the epigenetic rules, which in turn shape the decision behavior, the culturgens determine the reward. In the model this relation has been made fully restrictive: an individual of genotype *AA* using c_1 does exactly as well as one of genotypes *Aa* and *aa* using the same culturgen. Although it is possible to include pleiotropic effects, we have limited the action of the gene to the epigenetic rules. In this simplified version the model exposes the key role of culture as the interface between phenotype and environment. It also emphasizes the basic nature of genetic changes that affect the neurobiological subsystems within the individual.

As the juvenile period ends and the adult prereproductive period begins, the ethnographic curve of adult behavior forms. We have chosen the case in which the prereproductive period is sufficiently long for the ethnographic curve to be at steady state during most of its span. This condition requires the prereproductive period to contain many decision points at which culturgens can be switched, as is the case in human life history. Each generation is therefore characterized not only by a particular gene frequency distribution but also by a steady-state ethnographic curve, which specifies the frequency distribution of specific patterns of culturgen usage.

In Chapter 4 we showed that the ethnographic curve is grounded in the epigenetic rules. The transition rates v_{ij} in the individual choice of culturgens provide the translation from genes to culture. In the present model the ethnographic curve and the gene frequency curve combine in a coevolutionary system. Specific gene frequencies produce specific distributions of individual epigenetic rules. The subsequent decisions about culturgen use not only generate the ethnographic curve but also fix the reproductive success of the group members. As the gene frequencies change, so do the distribution of epigenetic rules and the ethnographic curve. But the epigenetic rules feed on information derived in part from the ethnographic curves of both the current and preceding generations. Variation in reproductive success induces a change in gene frequencies across generations, which in turn induces a shift in the ethnographic curve.

The model will permit exact specification of this coevolution. Equations of motion can be written for the gene frequencies p_t of the population and for the variables ν_t corresponding to the expected frequencies of culturgen use by the society as a whole. The two sets of equations form a coupled system, where

$$p_t = \text{function of genes, culturgens} \tag{6-1a}$$

$$\nu_t = \text{function of genes, culturgens} \tag{6-1b}$$

and they will be seen to be driven by a novel and remarkably effective selection mechanism.

Multiple Culturgens and Innovation

The cognitive definition of culturgen provides the means for enhanced realism in the two-culturgen model. We have just shown that it is possible to build a substantial model on the simplest, most rigid conception of two culturgens, in which the two forms must be judged distinct by virtually any criterion employed and no innovation occurs to add further culturgens. Although this is a useful approximation to some real-world circumstances, the model is not actually so restrictive. Gene-culture theory encompasses problems involving many culturgens (recall for example Appendix 4–3, and see our discussion of culturgen biogeography in Chapter

7). In this section we note how the present model applies to the many-culturgen case.

In the simplest extension to the case of three or more culturgens, the array can be clustered into two sets, composed of either one very distinctive form versus all the rest, or else two subpopulations of culturgens that differ most from each other (see Appendix 3–1). Indeed, the members of the two subpopulations might be closely similar or identical in terms of one or more shared, perceivable properties. The properties make up a *core* or *feature cluster* by means of which culturgens can be sorted and analyzed. If this is the case, then c_1 and c_2 become symbols for two different assemblages of core features that can be preferred, and the epigenetic rules determine the frequencies with which the c_1-heuristic and c_2-heuristic will be utilized. The data available to us (see Chapter 3) indicate that many of the valuation heuristics at work in human beings function in this manner.

Suppose that an epigenetic rule increases the use of a heuristic in which a specific core of features are valuated positively. Then if these features correlate with reproductive success, the epigenetic rule is likely to be at an evolutionary advantage with respect to an epigenetic rule under which more arbitrary or reproduction-curtailing culturgens are favored. Many culturgens can be involved, as long as each one can be valuated according to heuristics c_1 and c_2. The intensity of the evolutionary competition will decrease as the overlap between the two sets of core features increases, and it will increase as the superior rule moves closer to the "optimal core."

Innovation can be handled in a similar way. Since the number of specific culturgens is arbitrary in this extension of the model, the culturgen populations can expand or contract. In the dynamics of culturgen change any invariant or slowly changing properties of the culturgens act as identifiers that can attract epigenetic rules. The net reward gathered by an individual under such circumstances varies with the available culturgens. The simplest case, which applies in the sections to follow, stipulates that culturgen turnover, if present, produces a negligible change in expected reward. The model can be extended to innovation-dependent rewards by introducing phenomenological equations that make the reward structures for the c_1- and c_2-heuristics into functions of time.

We do not suggest these refinements in the ethnology of binary classification in order to foreclose a better analysis of the innovation process in future, second-generation models of gene-culture coevolution. Such an advance is clearly needed, and a stepping-stone is already provided in the theoretical treatment by Fagen (1981) of play and innovation. Rather, we wish to emphasize the power and flexibility of two-"culturgen" models, whose utility parallels the two-allele case in theoretical population genetics. It is equally notable that the classification of culturgens is an important form of investigation in itself, since it can shed light on the innate processes of cognition during the formation of knowledge structures.

The Gene-Culture Translation

In deriving the ethnographic curves from the epigenetic rules, we continue to use the notation introduced in Chapter 4. The u_{km} are the transition probabilities between culturgens k and m, and the v_{km} are the transition rates, or probabilities per unit time. The superscript ij denotes the genotype G_iG_j. Given a mean time τ_k^{ij} between decisions for a G_iG_j individual using culturgen c_k, we have for $k \neq m$ the relation $u_{km}^{ij} = \tau_k^{ij} v_{km}^{ij}$. The u_{km}^{ij} and v_{km}^{ij} can be influenced by the cultural patterns of both the parental and the offspring generation. It will be convenient to denote this dependence in several equivalent ways: first, through the variable n_1, the total number of individuals using culturgen c_1; second, through the frequency $\nu_1 = n_1/N_t$, where N_t is the total number of individuals in the current generation; third, through the order parameter $\xi = 1 - 2n_1/N_t = 1 - 2\nu_1$ (recall Chapter 4); and finally, through the variable n_1^{ij}, which is the number of c_1- users in the genotype G_iG_j subgroup of the culture. Hereafter we will write ν_1 as ν. Thus for the purposes of the model

$$u_{km}^{ij}(n_1) = u_{km}^{ij}(\nu) = u_{km}^{ij}(\xi) = u_{km}^{ij}(n_1^{AA}, n_1^{Aa}, n_1^{aa}) \qquad (6\text{-}2)$$

and similarly for the v_{km}^{ij}. We shall use this notation interchangeably as demanded by context.

Since the cultures of two generations impact the epigenetic rules, the relation $u_{km}^{ij} = u_{km}^{ij}(\xi_t, \xi_{t-1})$ holds by the end of the juvenile play period. Throughout the prereproductive and reproductive periods ξ_{t-1} plays a parametric role. For the time being we retain it implicitly, writing $u_{km}^{ij}(\xi_t, \xi_{t-1})$ as $u_{km}^{ij}(\xi_t)$ or $u_{km}^{ij}(\xi)$. Similar statements hold for the other conventions shown in Eq. (6-2).

The process of gene-culture amplification in this model is more complicated than in the culture group analyzed in Chapter 4, because with three genotypes present the system is no longer homogeneous. The population is made up of organisms that differ in their rules of decision, yet still interact with one another. The natural ethnographic curve for such a population is

$$\mathscr{G}(n_1^{AA}, n_1^{Aa}, n_1^{aa}, t) = \text{the probability that, at time } t, \; n_1^{AA} \text{ genotypes } AA \text{ are using culturgen } c_1, \text{ and so on.} \tag{6-3}$$

The $\mathscr{G}(t)$ is a probability function on the three-dimensional state space of coordinates $(n_1^{AA}, n_1^{Aa}, n_1^{aa})$, where $0 \leq n_1^{AA} \leq N^{AA}$, $0 \leq n_1^{Aa} \leq N^{Aa}$, and $0 \leq n_1^{aa} \leq N^{aa}$ (see Figure 6–9A). In generation t

$$N^{AA} = p_t^2 N_t, \qquad N^{Aa} = 2p_t q_t N_t, \qquad N^{aa} = q_t^2 N_t. \tag{6-4}$$

The time scale of observation is sufficiently fine that within any small interval dt at most one individual can switch culturgens. Then the only allowed transitions are those between state vectors $(n_1^{AA}, n_1^{Aa}, n_1^{aa})$ and $(n_1^{AA\prime}, n_1^{Aa\prime}, n_1^{aa\prime})$ such that

$$n_1' = \sum_{(ij)} (n_1^{ij})' = \sum_{(ij)} n_1^{ij} \pm 1 = n_1 \pm 1, \tag{6-5}$$

in other words, between nearest neighbors (Figure 6–9B). Using the arguments developed in Chapter 4, one readily finds that $\mathscr{G}(t)$ obeys the equation of motion

$$\frac{d}{dt}\mathscr{G}(n_1^{AA}, n_1^{Aa}, n_1^{aa}, t) =$$

$$- [n_1^{AA} v_{12}^{AA}(n_1^{AA}, n_1^{Aa}, n_1^{aa}) + n_1^{Aa} v_{12}^{Aa}(n_1^{AA}, n_1^{Aa}, n_1^{aa}) + n_1^{aa} v_{12}^{aa}(n_1^{AA}, n_1^{Aa}, n_1^{aa})$$

$$+ n_2^{AA} v_{21}^{AA}(n_1^{AA}, n_1^{Aa}, n_1^{aa}) + n_2^{Aa} v_{21}^{Aa}(n_1^{AA}, n_1^{Aa}, n_1^{aa})$$

$$+ n_2^{aa} v_{21}^{aa}(n_1^{AA}, n_1^{Aa}, n_1^{aa})] \, \mathscr{G}(n_1^{AA}, n_1^{Aa}, n_1^{aa}, t)$$

$$+ (n_2^{AA} + 1) v_{21}^{AA}(n_1^{AA} - 1, n_1^{Aa}, n_1^{aa}) \, \mathscr{G}(n_1^{AA} - 1, n_1^{Aa}, n_1^{aa}, t)$$

$$+ (n_2^{Aa} + 1) v_{21}^{Aa}(n_1^{AA}, n_1^{Aa} - 1, n_1^{aa}) \, \mathscr{G}(n_1^{AA}, n_1^{Aa} - 1, n_1^{aa}, t)$$

$$+ (n_2^{aa} + 1) v_{21}^{aa}(n_1^{AA}, n_1^{Aa}, n_1^{aa} - 1) \, \mathscr{G}(n_1^{AA}, n_1^{Aa}, n_1^{aa} - 1, t)$$

$$+ (n_1^{AA} + 1) \, v_{12}^{AA}(n_1^{AA} + 1, n_1^{Aa}, n_1^{aa}) \, \mathscr{G}(n_1^{AA} + 1, n_1^{Aa}, n_1^{aa}, t)$$

$$+ (n_1^{Aa} + 1) v_{12}^{Aa}(n_1^{AA}, n_1^{Aa} + 1, n_1^{aa}) \, \mathscr{G}(n_1^{AA}, n_1^{Aa} + 1, n_1^{aa}, t)$$

$$+ (n_1^{aa} + 1) v_{12}^{aa}(n_1^{AA}, n_1^{Aa}, n_1^{aa} + 1) \, \mathscr{G}(n_1^{AA}, n_1^{Aa}, n_1^{aa} + 1, t) \tag{6-6}$$

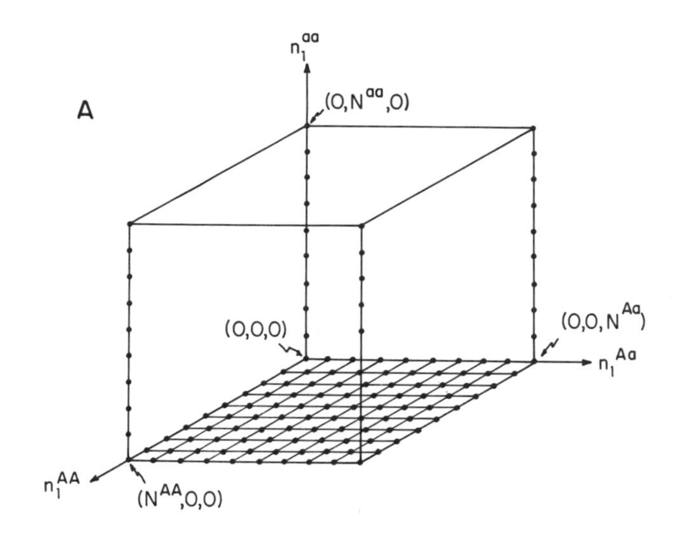

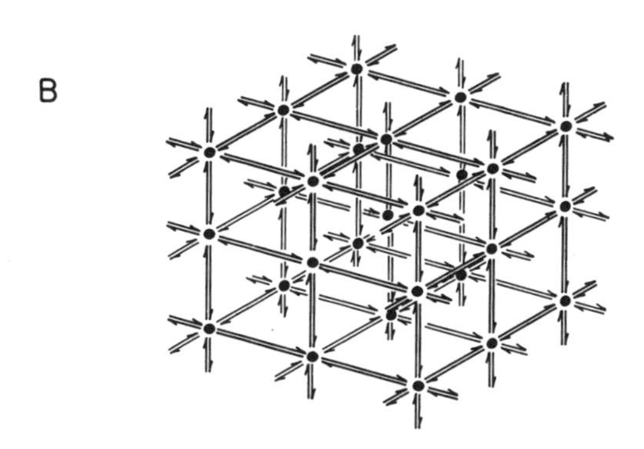

Figure 6–9 The ethnographic state space of a population composed of three genotypes (AA, Aa, aa) that affect the epigenetic rules of choice between culturgens. Each dimension is the number n_1^{AA}, n_1^{Aa}, or n_1^{aa} individuals of a particular genotype possessing culturgen c_1. The number can range from 0 to N^{ij}, the total number of individuals G_iG_j in the population. A, representation of a complete state space; B, the possible transitions within a portion of the state space. The population can at most shift from its original position to an adjacent position during any small time interval dt.

for $0 < n_1^{AA} < N^{AA}$, $0 < n_1^{Aa} < N^{Aa}$, $0 < n_1^{aa} < N^{aa}$, with similar equations for the boundary states.

By assumption, the duration T of the prereproductive interval is sufficiently long that the ethnographic curve is at steady state for all but a negligible fraction of T. From Figure 6–9 it is clear that every state $(n_1^{AA}, n_1^{Aa}, n_1^{aa})$ is connected to every other state $(n_1^{AA'}, n_1^{Aa'}, n_1^{aa'})$ by at least one path of allowed transitions joined chain-like, end to end. Since in addition $v^{ij}_{km}(\xi)$ and $v^{ij}_{mk}(\xi)$, $k \neq m$, are both nonzero, we conclude that a steady state $\mathscr{G}(n_1^{AA}, n_1^{Aa}, n_1^{aa})$ of Eq. (6-6) exists, is unique, and constitutes the single attractor for every initial ethnographic distribution of the system (Schnakenberg, 1976).

The equation of motion (6-6) can in principle be solved exactly for the steady-state ethnographic distribution $\mathscr{G}(n_1^{AA}, n_1^{Aa}, n_1^{aa})$. The necessary techniques are discussed in standard sources such as Schnakenberg (1976) and Haken (1977). However, the resulting formulation is not concise and generally requires the aid of the computer. In this initial study we need an approach that gives concise, readily grasped, analytic solutions of Eq. (6-6) even if some accuracy is lost. The key properties of $\mathscr{G}(n_1^{AA}, n_1^{Aa}, n_1^{aa})$ are thereby opened to immediate study. Based on the existence proof for the steady state, an approximation holds for culture groups which are not too small and which are not close to transition thresholds that shift the ethnographic curve from unimodality to multimodality. Under these circumstances $\mathscr{G}(n_1^{AA}, n_1^{Aa}, n_1^{aa})$ will be unimodal and sharply peaked, and throughout the prereproductive period the usage pattern variables ξ, n_1, and ν deviate only slightly from their modal values. Furthermore, the distribution mean will lie very close to the mode and we have the relations

$$\nu \sim \bar{\nu}, \qquad \xi \sim \bar{\xi}, \qquad n_1 \sim n_1, \tag{6-7}$$

and so on, where the overbar denotes mean value. The steady-state solution of (6-6) can then be approximated by replacing ξ with $\bar{\xi}$ in the assimilation functions. The culture is treated as a group of organisms rearranged from interacting with one another to interacting with an independent, *mean field of cultural order*. The mean-field technique is a powerful approximation for related systems (see for instance Reif, 1965:430–435). We use it self-consistently; we shall show that the value of $\bar{\xi}$, $\bar{\nu}$, or $\bar{n}_1$ is not added to the theory as an extra unknown but can be calculated directly. Furthermore, the methodology is exact in the special case, exemplified by incest, in which the group members are independent decision makers and the epigenetic rules consequently are independent of current values of ξ.

The basic distribution is then

$$\mathscr{I}^{ij}_k(t|\bar{\nu}) = \text{the probability that an organism of genotype } G_iG_j \text{ is using culturgen } c_k \text{ at time } t \text{ given that the mean cultural order is } \bar{\nu}. \tag{6-8}$$

The equations for the $\mathscr{I}_k^{ij}(t|\bar{\nu})$ are

$$\frac{d}{dt}\mathscr{I}_1^{ij}(t|\bar{\nu}) = -\,v_{12}^{ij}(\bar{\nu})\,\mathscr{I}_1^{ij}(t|\bar{\nu}) + v_{21}^{ij}(\bar{\nu})\,\mathscr{I}_2^{ij}(t|\bar{\nu})$$

$$\frac{d}{dt}\mathscr{I}_2^{ij}(t|\bar{\nu}) = v_{12}^{ij}(\bar{\nu})\,\mathscr{I}_1^{ij}(t|\bar{\nu}) - v_{21}^{ij}(\bar{\nu})\,\mathscr{I}_2^{ij}(t|\bar{\nu}) \tag{6-9}$$

which have the steady-state solution

$$\mathscr{I}_1^{ij}(\bar{\nu}) = v_{21}^{ij}(\bar{\nu})/v^{ij}(\bar{\nu})$$

$$\mathscr{I}_2^{ij}(\bar{\nu}) = v_{12}^{ij}(\bar{\nu})/v^{ij}(\bar{\nu}) \tag{6-10}$$

where

$$v^{ij}(\bar{\nu}) \triangleq v_{12}^{ij}(\bar{\nu}) + v_{21}^{ij}(\bar{\nu}). \tag{6-11}$$

The approach to this steady state is exponentially fast on time scales compared to the rate at which decisions are made:

$$\mathscr{I}^{ij}(t) = \mathscr{I}^{ij}(t_0)\exp\left[\mathscr{V}^{ij}(t - t_0)\right] \tag{6-12}$$

where $\mathscr{I}^{ij}(t)$ is the vector $[\mathscr{I}_1^{ij}(t|\bar{\nu}), \mathscr{I}_2^{ij}(t|\bar{\nu})]$ and

$$\exp\left[\mathscr{V}^{ij}(t - t_0)\right] = \frac{1}{v^{ij}(\bar{\nu})}\begin{bmatrix} v_{21}^{ij}(\bar{\nu}) & v_{12}^{ij}(\bar{\nu}) \\ v_{21}^{ij}(\bar{\nu}) & v_{12}^{ij}(\bar{\nu}) \end{bmatrix} + \frac{\exp\left[-v^{ij}(\bar{\nu})(t - t_0)\right]}{v^{ij}(\bar{\nu})}\begin{bmatrix} v_{12}^{ij}(\bar{\nu}) & -v_{12}^{ij}(\bar{\nu}) \\ -v_{21}^{ij}(\bar{\nu}) & v_{21}^{ij}(\bar{\nu}) \end{bmatrix}. \tag{6-13}$$

Thus after many decision points entailing a culturgen switch, the second term will be negligible compared to the first term and $\mathscr{I}^{ij}(t)$ will approach the steady state (6-10).

Since the group members are decoupled from one another (although still sensitive to the mean usage field), we can write

$$\mathscr{G}(n_1^{AA}, n_1^{Aa}, n_1^{aa}) \sim \mathscr{G}(n_1^{AA})\,\mathscr{G}(n_1^{Aa})\,\mathscr{G}(n_1^{aa}), \tag{6-14}$$

a product of three independent, single-variable ethnographic distributions. In turn, the $\mathcal{G}(n_1^{ij})$ can be expressed in terms of the $\mathcal{I}_k^{ij}(\bar{\nu})$. Since $\mathcal{I}_k^{ij}(\bar{\nu})$ is the probability of observing a c_k-user when the individual chosen is of genotype G_iG_j, and $\mathcal{G}(n_1^{ij})$ is the probability that the genotype G_iG_j subgroup contains exactly n_1^{ij} such c_1-users, $\mathcal{G}(n_1^{ij})$ must be the binomial distribution

$$\mathcal{G}(n_1^{ij}) = \binom{N^{ij}}{n_1^{ij}} [\mathcal{I}_1^{ij}(\bar{\nu})]^{n_1^{ij}}[1 - \mathcal{I}_1^{ij}(\bar{\nu})]^{(N^{ij}-n_1^{ij})}. \tag{6-15}$$

For each $N^{ij} \gtrsim 25$ the binomials are well approximated by Gaussians and

$$\mathcal{G}(n_1^{AA}, n_1^{Aa}, n_1^{aa}) \sim \prod_{(ij)} [2\pi N^{ij}\, \mathcal{I}_1^{ij}(1 - \mathcal{I}_1^{ij})]^{-1/2}$$

$$\cdot \exp\, [(n_1^{ij} - N^{ij}\, \mathcal{I}_1^{ij})^2/2N^{ij}\, \mathcal{I}_1^{ij}(1 - \mathcal{I}_1^{ij})]. \tag{6-16}$$

Self-consistency requires that

$$\bar{n}_1 = \sum_{n_1^{AA}, n_1^{Aa}, n_1^{aa}} n_1\, \mathcal{G}(n_1^{AA}, n_1^{Aa}, n_1^{aa}|\bar{\nu}). \tag{6-17}$$

But $n_1 = n_1^{AA} + n_1^{Aa} + n_1^{aa}$, so that

$$\bar{n}_1 = \sum_{n_1^{AA}, n_1^{Aa}, n_1^{aa}} (n_1^{AA} + n_1^{Aa} + n_1^{aa})\, \mathcal{G}(n_1^{AA}, n_1^{Aa}, n_1^{aa}|\bar{\nu})$$

$$= \sum_{n_1^{AA}} n_1^{AA}\, \mathcal{G}(n_1^{AA}|\bar{\nu}) + \sum_{n_1^{Aa}} n_1^{Aa}\, \mathcal{G}(n_1^{Aa}|\bar{\nu})$$

$$+ \sum_{n_1^{aa}} n_1^{aa}\, \mathcal{G}\, (n_1^{aa}|\bar{\nu}). \tag{6-18}$$

In Eqs. (6-17) and (6-18) the conditional dependence of $\mathcal{G}$ on the mean frequency $\bar{\nu}$ is indicated explicitly, rather than implicitly as in the preceding equations. By (6-15) each of the three sums in (6-18) is the mean of a binomial distribution with probability parameter $\mathcal{I}_1^{ij}(\bar{\nu})$ over N^{ij} individuals. Thus

$$\sum_{n_1^{ij}} n_1^{ij}\, \mathcal{G}(n_1^{ij}) = N^{ij}\mathcal{I}_1^{ij}(\bar{\nu}) \tag{6-19}$$

and we obtain the implicit equation for $\bar{n}_1$:

$$\bar{n}_1 = N^{AA} v_{21}^{AA}(\bar{\nu})/v^{AA}(\bar{\nu}) + N^{Aa} v_{21}^{Aa}(\bar{\nu})/v^{Aa}(\bar{\nu}) + N^{aa} v_{21}^{aa}(\bar{\nu})/v^{aa}(\bar{\nu}). \tag{6-20}$$

Expressing this relation in terms of the mean usage frequency $\bar{\nu} = \bar{n}_1/N_t$ we find that in generation t

$$\bar{\nu} = \frac{p_t^2 v_{21}^{AA}(\bar{\nu})}{v_{12}^{AA}(\bar{\nu}) + v_{21}^{AA}(\bar{\nu})} + \frac{2p_t q_t v_{21}^{Aa}(\bar{\nu})}{v_{12}^{Aa}(\bar{\nu}) + v_{21}^{Aa}(\bar{\nu})} + \frac{q_t^2 v_{21}^{aa}(\bar{\nu})}{v_{12}^{aa}(\bar{\nu}) + v_{21}^{aa}(\bar{\nu})}. \tag{6-21}$$

Equation (6-21) is a *coevolutionary equation* for the culture pattern of c_1. We note three significant aspects of its structure. First, it is explicitly dependent on the gene frequencies. Second, Eq. (6-21) is an *implicit equation* for $\bar{\nu}$ and in general will be nonlinear. It will not ordinarily yield analytic solutions for $\bar{\nu}$. Third, (6-21) incorporates a dependence on the $\bar{\nu}$ of the previous generation. Thus the history of both genetic and cultural evolution are important in the model.

Equation (6-21) completes our formulation of the cultural dynamics in the mean-field approximation. The first step in determining the dynamics of the gene frequencies is a calculation of the resources gathered during the prereproductive period.

The Reward Structure

The reproductive success of the members of the society is a function of the quantity of resource they harvest during the prereproductive period T. To revert to the case of the Polynesian fishermen, use of the more appropriate of two hooks at various selected sites around and beyond the reef will yield more fish, and ultimately a higher reproductive potential, than use of the less appropriate hook. In parallel manner, one particular code of land ownership, or one form of the sexual division of labor, or one kind of exchange system among relatives will yield a higher resource during the prereproductive period than its alternative.

An individual of any genotype G_iG_j using culturgen c_k obtains the resource at the rate of J_k units per unit time. However, this amount is not the net harvesting yield, because the cognitive processing required to evaluate usage at each decision point requires brain tissue, time, and energy. The computation apparatus must first of all be built from neurons; its maintenance requirements in energy (or resource) units may be designated the *bearing costs* L^{ij}. Whenever the apparatus is employed at a decision point to carry the individual from usage state $k = 1, 2$ to usage state $m = 1, 2$, a cost in energy (or resource) units is exacted that may be termed the *transition cost* C^{ij}.

For prereproductive periods that contain many decision points, the net expected resource gathered is given by the asymptotic expression (Howard, 1971a,b)

$$R_k^{ij}(T) = g^{ij}T + {}^{\circ}R_k^{ij}, \qquad k = 1, 2 \tag{6-22a}$$

where g^{ij} is the expected gain or net resource harvested per unit time by an individual or genotype G_iG_j. Following the general theory of semi-Markov reward processes, we find that

$$g^{ij} = \sum_{k=1}^{2} \mathscr{P}_k^{ij}[(J_k - L^{ij}) + (\tau^{ij})^{-1} \sum_{m=1}^{2} u_{km}^{ij}C^{ij}] \tag{6-22b}$$

for our system. The quantities ${}^{\circ}R_k^{ij}$ are constants that contain the effects of initial conditions, in particular the first culturgen used during the prereproductive period. We shall evaluate ${}^{\circ}R_k^{ij}$ shortly. For large values of T assumed in the model $g^{ij}T \gg {}^{\circ}R_k^{ij}$ and it is accurate to simplify Eq. (6-22) to the k-independent expression

$$R^{ij}(T) = g^{ij}T. \tag{6-23}$$

We lose little generality by taking $\tau_1^{ij} = \tau_2^{ij} = \tau^{ij}$, which is the equivalent of saying that the mean time between successive decisions is the same for both c_1-users and c_2-users. We can therefore estimate the total transition cost during the prereproductive period as $C^{ij}T/\tau$, where τ is viewed as independent of the genotype. We also visualize the bearing costs as real costs in resource units and hence use the net fluxes

$$J_1 - L^{ij}, \qquad J_2 - L^{ij} \tag{6-24a}$$

rather than the gross fluxes

$$J_1, \qquad J_2 \tag{6-24b}$$

to obtain the reward equation

$$R^{ij}(T) = \langle J_{\text{net}} \rangle^{ij} T + C^{ij} T/\tau$$

$$= (J_1 - L^{ij}) \mathscr{I}_1^{ij} T + (J_2 - L^{ij}) \mathscr{I}_2^{ij} T + C^{ij} T/\tau$$

$$\triangleq R^{ij}_{\text{gross}} - L^{ij} T + C^{ij} T/\tau. \tag{6-25}$$

The angular brackets denote an average over the frequencies $\mathscr{I}_1^{ij}$ and $\mathscr{I}_2^{ij}$.

For short prereproductive periods the asymptotic formula (6-22) is not accurate and one must use the exact solution

$$\begin{bmatrix} R_1^{ij}(T) \\ R_2^{ij}(T) \end{bmatrix} = \frac{T}{v^{ij}(\bar{\nu})} \begin{bmatrix} q_1^{ij} v_{21}^{ij}(\bar{\nu}) + q_2^{ij} v_{12}^{ij}(\bar{\nu}) \\ q_1^{ij} v_{21}^{ij}(\bar{\nu}) + q_2^{ij} v_{12}^{ij}(\bar{\nu}) \end{bmatrix}$$

$$+ \frac{1}{[v^{ij}(\bar{\nu})]^2} \begin{bmatrix} (q_1^{ij} - q_2^{ij}) v_{12}^{ij}(\bar{\nu}) \\ (q_2^{ij} - q_1^{ij}) v_{21}^{ij}(\bar{\nu}) \end{bmatrix}$$

$$+ \frac{\exp[-v^{ij}(\bar{\nu})T]}{[v^{ij}(\bar{\nu})]^2} \begin{bmatrix} (q_2^{ij} - q_1^{ij}) v_{12}^{ij}(\bar{\nu}) \\ (q_1^{ij} - q_2^{ij}) v_{21}^{ij}(\bar{\nu}) \end{bmatrix} \tag{6-26}$$

where the q_k^{ij} are the rates

$$(J_k - L^{ij}) + (\tau^{ij})^{-1} \sum_{m=1,2} u_{km}^{ij} C^{ij}. \tag{6-27}$$

Equation (6-26) follows from the balance equations

$$\frac{dR_k^{ij}(t)}{dt} = q_k^{ij} + \sum_{m=1}^{N} a_{km}^{ij} R_m^{ij}(t) \tag{6-28}$$

$$a_{km}^{ij} = \begin{cases} v_{km}^{ij}, & k \neq m \\ -\sum_{m' \neq k} v_{km'}^{ij}, & k = m \end{cases} \tag{6-29}$$

obtained for the reward process from equations (6-9) (chapter 8 of Howard, 1960).

The second term on the right side of (6-26) is composed of the constants $^{\circ}R_k^{ij}$. The first term is the linear growth term, and the third term is an exponentially decaying term corresponding to the approach of the ethnographic curve to the steady state. We note that its time scale for decay is proportional to $[v^{ij}(\bar{\nu})]^{-1}$ and conclude that the asymptotic formula (6-22) is suitable for the long prereproductive period of the model life cycle.

The Fertility Map

Let F^{ij} be the fertility map that takes the net reward $R^{ij}(T)$ into the gametes produced during the reproductive period:

$$F^{ij}:R^{ij}(T) \longmapsto F^{ij}[R^{ij}(T)] \qquad (6\text{-}30)$$

such that the number of gametes emitted by each genotype G_iG_j is $2F^{ij}[R^{ij}(T)]$. Gamete union is at random, and the survival of the gametes is independent of the gene carried. Also, the number of offspring produced is linearly proportional to the number of gametes emitted. To express the last relation in more realistic terms, we can say that the number of gametes emitted is directly proportional to the number of sexual unions that produce children. There is increasing evidence from human studies that the "correct" choice of culturgens, leading to social and economic success in the opinion of the people employing those culturgens, results in more such mating and hence higher reproductive rates in at least the economically more primitive societies (see for example Alexander et al., 1979; Chagnon, 1979; and Irons, 1979).

Let p_{t+1} be the frequency of allele A in the zygotes subsequently formed. Then

$$p_{t+1} = (F^{AA}p_t^2 + F^{Aa}p_tq_t)F^{-1} \qquad (6\text{-}31)$$

and

$$N_{t+1} = N_tF \qquad (6\text{-}32)$$

where

$$F \triangleq F^{AA}p_t^2 + 2F^{Aa}p_tq_t + F^{aa}q_t^2. \qquad (6\text{-}33)$$

Although Eqs. (6-31) through (6-33) have the general form familiar from conventional population genetics, we note that the F^{ij} are not arbitrary constants. Instead, they are functionals of the reward structure $R^{ij}(T)$.

Through these quantities the F^{ij} depend on the epigenetic rules. The epigenetic rules in turn are sensitive to the patterns of culturgen usage, and this behavior links the reproductive success of the group members to their culture. Inspection of Eq. (6-21) reveals that additional features characterize this intricate selection mechanism. The mean usage pattern $\bar{\nu}$ depends on the gene frequencies. But the reward $R^{ij}(T)$ is composed in part of the term $\langle J_{\text{net}} \rangle^{ij}$ given by the sum $(J_1 - L^{ij})\, \mathscr{I}_1(\bar{\nu}) + (J_2 - L^{ij}) \cdot \mathscr{I}_2(\bar{\nu})$. Thus the reward structures and fertilities are also dependent on gene frequency, and frequency-dependent selection is operating. We arrive at an important new feature of the coevolutionary process: insofar as gene frequency trajectories can stabilize at or around interior points $0 < p_t < 1$ in frequency-dependent selection, the possibility exists that one of the outcomes of reification learning rules or other processes that allow individuals to track macrocultural patterns is the maintenance of genetic diversity underlying cognitive systems. Later we shall note ways in which stabilization can occur.

This very interesting form of frequency dependence does not appear to be an artifact of the mean-field approximation. The exact epigenetic rules depend on the macrocultural variable n_1:

$$v_{km}^{ij} = v_{km}^{ij}(n_1). \qquad (6\text{-}34)$$

where

$$n_1 = n_1^{AA} + n_1^{Aa} + n_1^{aa}. \qquad (6\text{-}35)$$

Although the quantities n_1^{AA}, n_1^{Aa}, and n_1^{aa} seem rather irreducible, let us note that they can be written in the completely equivalent form

$$n_1^{ij} = \nu^{ij} N^{ij} \qquad (6\text{-}36)$$

where ν^{ij} is clearly the frequency variable n_1^{ij}/N^{ij}. But then

$$n_1^{AA} = \nu^{AA} p_t^2 N_t \qquad (6\text{-}37)$$

and so on, which reveals the explicit dependence of the v_{km}^{ij} on the gene frequencies. In words, this result reminds us that in the population, n_1^{AA} is a function of two key variables: the propensity of an *individual AA* to use c_1, and the total *number* of AA's present. The latter is an explicit function of the gene frequencies, so frequency-dependent selection is unavoidable given (6-34).

The General Coevolutionary Equations

Given Eqs. (6-21) and (6-31), we are ready to write down the coevolutionary equations that summarize the dynamics of gene frequencies and culturgen frequencies in this model. There are two stages in the life cycle at which it is natural to formulate the coevolutionary equations. The first is the start of a new generation, $t + 1$. At that moment the genes being tracked lie in the zygotes, which have yet to be enculturated, while the culturgens are the exclusive possession of the postreproductive parents in generation t. Let ν_{t+1}^P denote the c_1-usage frequency that characterizes the parental generation. Then from (6-21) and (6-31) the coevolutionary equations are

$$\begin{aligned}
\nu_{t+1}^P &= p_t^2 v_{21}^{AA}(\nu_{t+1}^P, \nu_t^P)/[v_{12}^{AA}(\nu_{t+1}^P, \nu_t^P) + v_{21}^{AA}(\nu_{t+1}^P, \nu_t^P)] \\
&\quad + 2p_t q_t v_{21}^{Aa}(\nu_{t+1}^P, \nu_t^P)/[v_{12}^{Aa}(\nu_{t+1}^P, \nu_t^P) + v_{21}^{Aa}(\nu_{t+1}^P, \nu_t^P)] \\
&\quad + q_t^2 v_{21}^{aa}(\nu_{t+1}^P, \nu_t^P)/[v_{12}^{aa}(\nu_{t+1}^P, \nu_t^P) + v_{21}^{aa}(\nu_{t+1}^P, \nu_t^P)] \\
p_{t+1} &= F^{-1}[p_t^2 F^{AA}(\nu_{t+1}^P, \nu_t^P) + p_t q_t F^{Aa}(\nu_{t+1}^P, \nu_t^P)].
\end{aligned} \tag{6-38}$$

The second natural stage is the prereproductive period. The parents die as it starts, so both genes and culturgens are confined to a single generation. The mean usage frequency is $\bar{\nu}_{t+1}$, determined by Eq. (6-21). Since there is no mortality, the gene frequencies that characterize the prereproductive adults are the same as the zygotic gene frequencies of (6-31). Then with the stipulation that p refers to the prereproductive stage, we have

$$\begin{aligned}
\bar{\nu}_{t+1} &= p_{t+1}^2 v_{21}^{AA}(\bar{\nu}_{t+1}, \bar{\nu}_t)/[v_{12}^{AA}(\bar{\nu}_{t+1}, \bar{\nu}_t) + v_{21}^{AA}(\bar{\nu}_{t+1}, \bar{\nu}_t)] \\
&\quad + 2p_{t+1} q_{t+1} v_{21}^{Aa}(\bar{\nu}_{t+1}, \bar{\nu}_t)/[v_{12}^{Aa}(\bar{\nu}_{t+1}, \bar{\nu}_t) + v_{21}^{Aa}(\bar{\nu}_{t+1}, \bar{\nu}_t)] \\
&\quad + q_{t+1}^2 v_{21}^{aa}(\bar{\nu}_{t+1}, \bar{\nu}_t)/[v_{12}^{aa}(\bar{\nu}_{t+1}, \bar{\nu}_t) + v_{21}^{aa}(\bar{\nu}_{t+1}, \bar{\nu}_t)] \\
p_{t+1} &= F^{-1}[p_t^2 F^{AA}(\bar{\nu}_t, \bar{\nu}_{t-1}) + p_t q_t F^{Aa}(\bar{\nu}_t, \bar{\nu}_{t-1})]
\end{aligned} \tag{6-39}$$

and $\bar{\nu}_t$ is determined by Eq. (6-21). System (6-38) is both convenient and somewhat more concise than (6-39), and results to be discussed below refer to the zygotic stage.

Specification of the Parameters

In order to make concrete application of the general coevolutionary equations, it is necessary to specify their parameters. From the myriad possibilities we have selected examples that appear to be qualitatively realistic on the basis of existing information from developmental and social psychology. And from the various alternatives that appear realistic we have chosen those that are among the simplest and most tractable.

Epigenetic rules. Exponential assimilation functions were used, because of evidence that they occur in some categories of behavioral development (see Chapter 4). The assimilation function was modified further in a way that accommodates some amount of history and the propagation of tradition. The formula tightly captures the potential trade-off during socialization between the tendency to watch the age-peer group and the tendency to watch the older generation:

$$\begin{aligned} u_{21}^{AA} &= u_{21,0}^{AA} \exp\{-\alpha_{21}^{AA}[\beta_{21}^{AA}\xi_t + (1 - \beta_{21}^{AA})\xi_t^P]\} \\ u_{12}^{AA} &= u_{12,0}^{AA} \exp\{+\alpha_{12}^{AA}[\beta_{12}^{AA}\xi_t + (1 - \beta_{12}^{AA})\xi_t^P]\} \end{aligned} \qquad (6\text{-}40)$$

where $0 \leq \beta_{km}^{AA} \leq 1$, and similarly for the genotypes *Aa* and *aa*. The usage pattern of the peer group is ξ_t, with ξ_t^P the pattern of the parental generation. The parameter α_{km}^{ij} specifies the sensitivity to the culturgen usage patterns ξ_t and ξ_t^P; when it is zero, the individual ignores the patterns.

The mean-field approximation as we have used it here restricts the assimilation functions to conditions such that the ethnographic curve is unimodal and sharp. In order to guarantee these properties we have limited the values of the α_{km}^{ij} to $0 \leq \alpha_{km}^{ij} \leq 0.2$. An examination of the ethnographic curves calculated in Chapter 4 indicates that in this parameter range the criteria of unimodality and sharpness are readily fulfilled. Such small values of the α_{km}^{ij} mean that the organisms are weakly coupled, a circumstance under which the method, which approximates the culture as a group of mutually independent individuals interacting with a static social institution, is relevant. Our highly conservative strategy therefore precludes exploration of those regions in parameter space where α_{km}^{ij} is large and the organisms are strongly interacting. For some choices of the assimilation function exact solutions to the full master equation (6-6) are feasible, given detailed balance methods similar to those employed in Chapter 4. However, the subject lacks a general theory of reward for interacting systems of this type; in order to estimate the $R^{ij}(T)$ using existing methods, then, our small-α, weak-interaction restriction appears to be a useful step to make at this time.

As a result the epigenetic rules (6-39) cover more ground than is apparent at first sight. Small values of the α^{ij}_{km} or a stipulation of weak interaction imply that *any* $u^{ij}_{km}(\xi)$ or $v^{ij}_{km}(\xi)$ is only weakly dependent on ξ. Thus each of them can be approximated by a first-order Taylor expansion, which is a linear, monotonically increasing or decreasing function of the usage parameter. Equations (6-40) are subject to these conditions and thus are representative of all assimilation functions when the α^{ij}_{km} are small. We have retained the forms (6-40) rather than linear functions because they model in a concise way the keeping of residual, nonlinear terms.

The fact that the assimilation functions can be linearized also leads to an approximate, analytical solution of the implicit equation (6-21) for $\bar{\nu}$. Let us write

$$E^{ij}(\bar{\nu}) \triangleq v^{ij}_{21}/[v^{ij}_{12} + v^{ij}_{21}] \tag{6-41}$$

and expand around $\bar{\nu} = 0$:

$$E^{ij}(\bar{\nu}) \sim E^{ij}(0) + \frac{dE^{ij}}{d\bar{\nu}}(0)\bar{\nu} + \text{negligible higher-order terms.} \tag{6-42}$$

Define $e^{ij} \triangleq E^{ij}(0)$ and $e^{ij}_1 \triangleq dE^{ij}(0)/d\bar{\nu}$. Then

$$\bar{\nu} = p_t^2(e^{AA} + e_1^{AA}\bar{\nu}) + 2p_t q_t(e^{Aa} + e_1^{Aa}\bar{\nu})$$
$$+ q_t^2(e^{aa} + e_1^{aa}\bar{\nu}).$$

Therefore

$$\bar{\nu} - \bar{\nu}(e_1^{AA}p_t^2 + e_1^{Aa}2p_tq_t + e_1^{aa}q_t^2) = p_t^2e^{AA} + 2p_tq_te^{Aa} + q_t^2e^{aa}$$

and

$$\bar{\nu} = \frac{\langle e\rangle^G}{1 - \langle e_1\rangle^G} \tag{6-43}$$

where $\langle\cdot\rangle^G$ denotes the gene frequency average.

We now have an exact formula for the mean c_1 culturgen usage $\bar{\nu}$ in weak member-to-member interaction. Trial work has shown that in the case of $\alpha^{ij}_{km} \lesssim 0.2$, Eq. (6-43) is accurate to within several percentage points of the solutions to the exact equation (6-21), calculated either numerically using Newton-Raphson methods or graphically from intercept plots. Equation (6-43) is therefore useful in simulations that track the gene-culture trajectories over hundreds or thousands of generations and require fast, efficient algorithms for $\bar{\nu}$.

The parameter β^{ij}_{km} in Eq. (6-40) specifies the individual's focus of attention. When $\beta^{ij}_{km} = 1.0$, attention is directed entirely at peers; when $\beta^{ij}_{km} = 0$, it is directed entirely at parents, storing in long term memory the impact of the usage pattern maintained by the parents during the play period. For $\beta^{ij}_{km} < 1.0$, there is a transmission between generations not only of culturgens but also of information pertaining to patterns of culturgen usage. For simplicity we have taken all β^{ij}_{km} as equal with reference to both the genotypes G_iG_j and the transitions km. Similarly, we have taken $\alpha^{ij}_{km} = \alpha$.

Our simulations have shown that for $0 \leq \alpha \leq 0.2$ the variation of β over [0, 1] has little qualitative effect on the overall course and rate of evolution. Therefore, unless explicitly indicated, the graphs that follow pertain to the case $\beta = 1$.

Phenotype cost. It is necessary to provide a measure of the cost of cognition. Let us define a cost-free tabula rasa state as follows:

$u^{ij}_{12,0} = u^{ij}_{21,0} = 0.5$	Complete indifference to c_1 and c_2; random selection, no further processing, and no cost
$\alpha^{ij}_{12} = \alpha^{ij}_{21} = 0$	No ability to perceive or utilize usage patterns, and no cost
$\beta^{ij}_{12} = \beta^{ij}_{21} = 1$	No attention to parental usage, and no cost

Then any genotype can be described by a vector of six components $(u^{ij}_{12,0}, u^{ij}_{21,0}, \alpha^{ij}_{12}, \alpha^{ij}_{21}, \beta^{ij}_{12}, \beta^{ij}_{21})$, and the tabula rasa state is (0.5,0.5,0, 0,1,1). We take the bearing cost L^{ij} and transition cost C^{ij} on the part of the cognitive processors as monotone rising functions of their distances d from the tabula rasa vector. Specifically,

$$L^{ij} = e^{\gamma^{ij} d} - 1$$

and

$$C^{ij} = e^{\delta^{ij} d} - 1. \tag{6-44}$$

We shall discuss some of the implications of these rising cost functions in a subsequent treatment of the causes of the rarity of the eucultural state in evolution (Chapter 7). By trial and error it was found that the values $\gamma^{ij} = 1$ and $\delta^{ij} = 0.1$ give realistic behavior over the range of the epigenetic rules we wish to explore. They result in bearing and transition costs per unit time on the order of 1 to 10 percent of the reward rate. These loads approximate the size of the drain that would be put on a total energy budget by an organism with a ratio of brain weight to body weight in the primate range. For every generation $\tau = 0.1$ and $T = 10$ time units.

Resource yield. Culturgen c_1 was selected as the more efficient of c_1 and c_2. The figures that follow, unless otherwise noted, show the evolutionary behavior of the population when $J_1 = 1$ and $J_2 = 0.2$, so that the efficiency of c_1 is five times that of c_2.

Fertility function. After some experimentation we settled on a basic relation that gives a decreasing fertility return-to-scale as more resources are harvested:

$$F^{ij} = F^{ij}_{\max}[1 - \exp(-b^{ij}R^{ij})]. \tag{6-45}$$

This function models the intuitive notion that because of the existence of other biological constraints, fertility must eventually level off with higher resource yields R^{ij}. In particular, we expect the processing of the resource to be slowed by increasing difficulties in transport, storage, and processing (see also Oster and Wilson, 1978). Since only the epigenetic rules are considered to be distinguished by the genotypes, $F^{ij}_{\max}$ was set equal for all genotypes. Also, b^{ij} was given the value of 0.1, which places the society members short of the asymptotic, "saturated" portion of the fertility curve.

Conclusions from the Model

We have shown that it is possible to capture all the key steps in the full coevolutionary circuit in the form of a model that yields explicit gene frequencies and culture patterns. Even prior to specification of the parameters, this formulation demonstrates that natural selection in gene-culture coevolution is *frequency dependent*. Equations (6-21) and (6-38) express the key relations of the dependence: the rate of culturgen change is a function of the frequencies of the genes that underwrite the epigenetic rules, and the rate of change in these frequencies is in turn a function of the proportions of the culturgens. The form of selection is unique. It also produces some remarkable phenomena, which can be explored with the exactly specified form of the model. In the sections to follow we shall interpret the most important of these effects.

Pure tabula rasa is an unlikely state. In all cases examined, directed cognition due to genetically biased epigenetic rules replaced undirected cognition, in other words the pure tabula rasa state, when the two were set in competition (see Figures 6–10 through 6–13). This result confirms the separately derived demonstration in Chapter 1 that even with relatively modest culturgen innovation rates a tabula rasa species will almost always depart eventually from its strategy of pure cultural transmission, and revert to some form of gene-culture transmission.

Sensitivity to usage pattern increases the rate of genetic assimilation. The rise in the slopes of the rate curves of Figure 6–10C means that when individuals are more sensitive to usage by other members of the society, the evolutionary replacement of inferior epigenetic rules is accelerated. The same effect is manifested in the enhanced replacement rates of Figures 6–11C and 6–12C, which incorporate usage sensitivity ($\alpha = 0.2$). These rates are higher than those in the curves of Figures 6–11B and 6–12B, generated under identical conditions except for the absence of usage sensitivity ($\alpha = 0$).

The relation can be expressed in a different way by saying that genetic assimilation is generally hastened by sensitivity to usage pattern. Suppose that two culturgens are adopted by a species for the first time. One culturgen provides higher fitness than the other, but at first there is no clear preference between them because no biasing rule exists in the cognition of the species members. In time new genotypes appear by mutation or immigration and direct cognition in favor of the more efficient culturgen; the constituent alleles now compete with the older, tabula rasa alleles. In species where individuals are already sensitive to the usage patterns of other members of their society, the rate of replacement is increased. In other words, genetic assimilation of the favored culturgen proceeds more quickly.

The catalytic relationship between usage sensitivity and genetic assimilation may in fact be reciprocal. If favorable culturgens spread more rapidly among genotypes with greater usage sensitivity, those genotypes themselves will be favored by second-order selection, and sensitivity will tend to increase. In Chapter 1 we characterized the autocatalytic quality of gene-culture coevolution in the human species, which led to an extraordinarily rapid evolutionary increase in brain size. It is possible that the

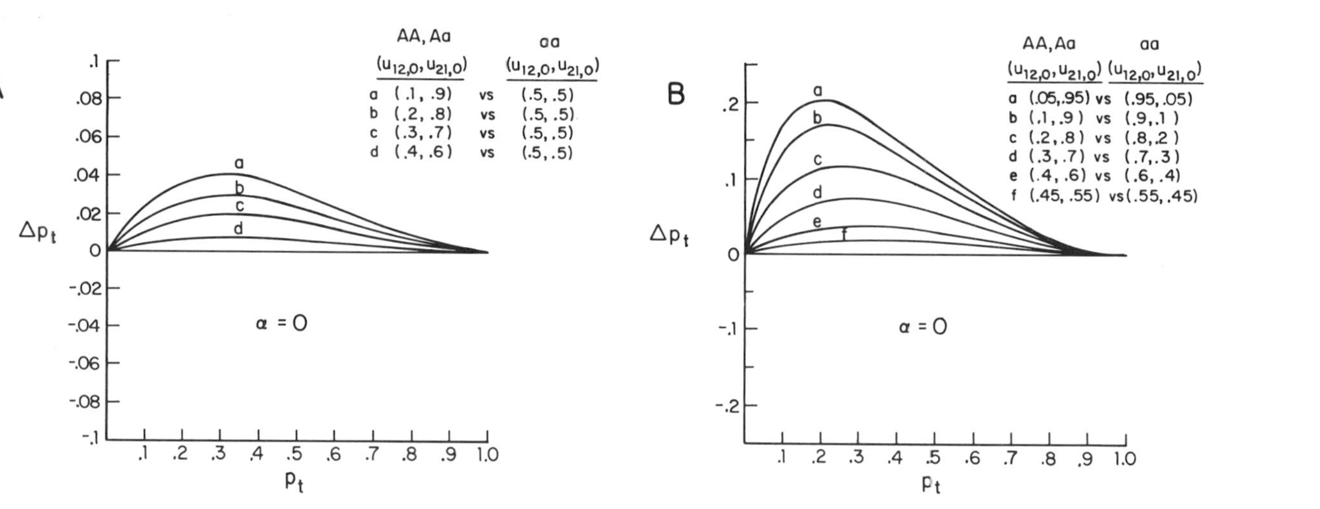

Figure 6–10 Evolutionary competitions between organisms bearing different epigenetic rules. The curves show the change in one generation, Δp_t, expected if the initial gene frequency is p_t. Thus $p_{t+1} = p_t + \Delta p_t$ and the entire manifold of possible trajectories can be read off from these diagrams. The allele coding for the c_1-biased epigenetic rule is dominant. Curves *a* to *d* in panel *A* show competitions between tabula rasa organisms and organisms with various degrees of directed cognition. Curves *a* to *f* in panel *B* show competitions between organisms biased toward c_1 and organisms biased toward c_2. This case models the situation in which a new, more efficient culturgen (c_1) suddenly appears in a population adapted to culturgen c_2. Panel *C* shows the effects of nonzero values of α at the representative point $p_t = 0.1$. Thus Δ is Δp_t when $p_t = 0.1$. The evolutionary competitions are those treated in panel *B*. The influence of $0 < \alpha < 0.2$ is slight.

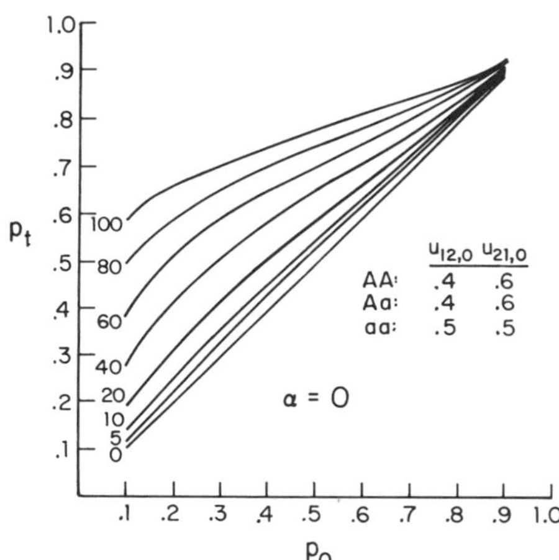

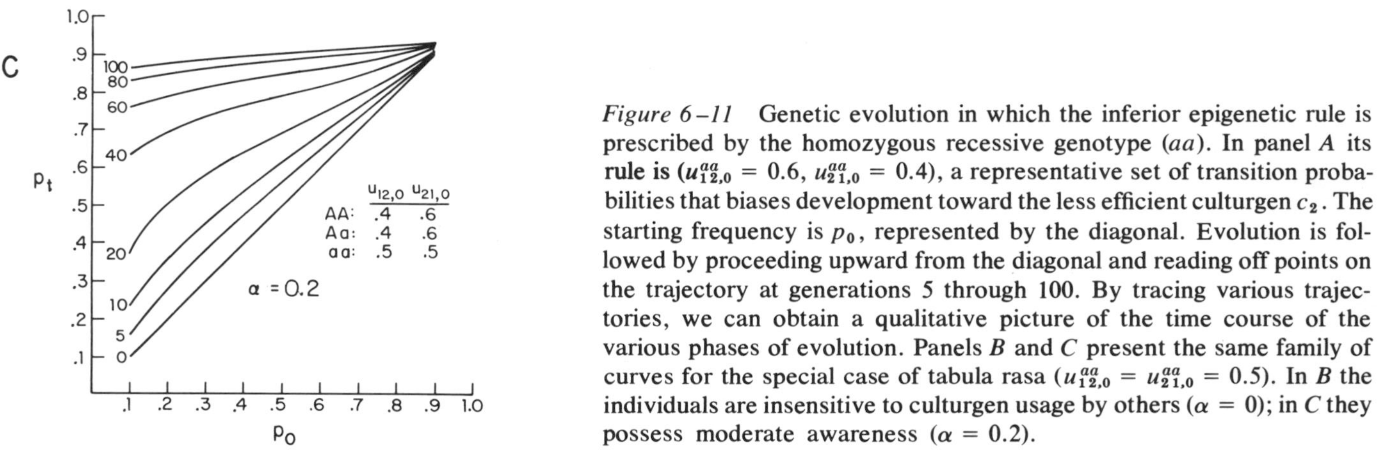

Figure 6–11 Genetic evolution in which the inferior epigenetic rule is prescribed by the homozygous recessive genotype (*aa*). In panel *A* its rule is ($u_{12,0}^{aa} = 0.6$, $u_{21,0}^{aa} = 0.4$), a representative set of transition probabilities that biases development toward the less efficient culturgen c_2. The starting frequency is p_0, represented by the diagonal. Evolution is followed by proceeding upward from the diagonal and reading off points on the trajectory at generations 5 through 100. By tracing various trajectories, we can obtain a qualitative picture of the time course of the various phases of evolution. Panels *B* and *C* present the same family of curves for the special case of tabula rasa ($u_{12,0}^{aa} = u_{21,0}^{aa} = 0.5$). In *B* the individuals are insensitive to culturgen usage by others ($\alpha = 0$); in *C* they possess moderate awareness ($\alpha = 0.2$).

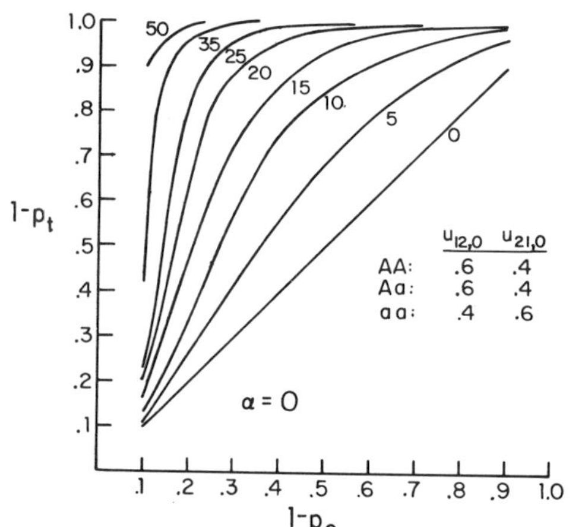

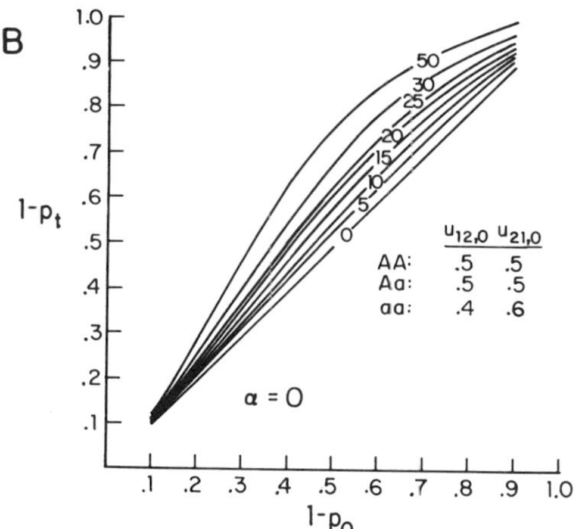

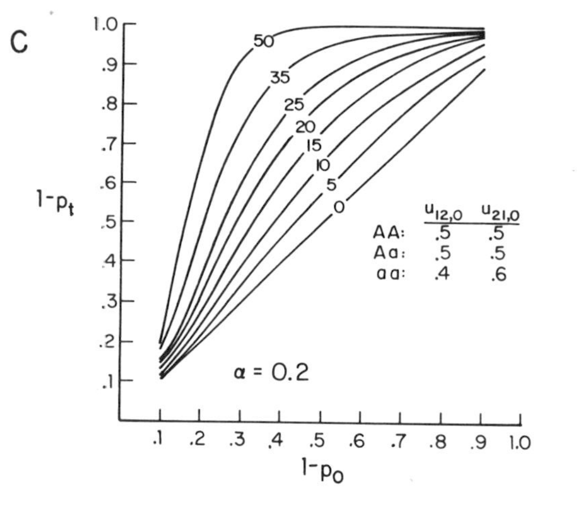

Figure 6–12 Genetic evolution in which the inferior epigenetic rule is determined by a dominant allele (*A*). The conventions are as in Figure 6–11.

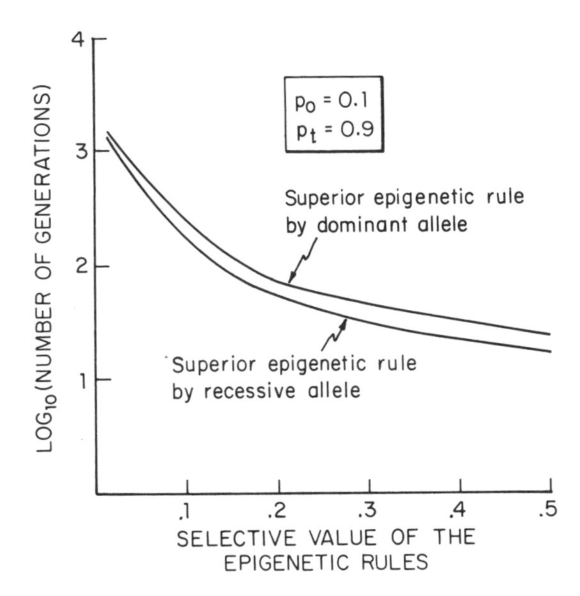

Figure 6–13 Evolutionary times required for partial allele replacement (from $p_0 = 0.1$ to $p_t = 0.9$ for the favored allele) under different selection pressures and where the epigenetic rules are prescribed by dominant and recessive alleles respectively. The selective value is defined as $1 - F^{\text{sup}}/F^{\text{inf}}$, where F^{sup} is the gametic yield of the superior epigenetic rule and F^{inf} the yield of the inferior epigenetic rule.

mechanisms of usage sensitivity played a key role in this unique episode. Human beings are acutely sensitive to macrocultural patterns around them by virtue of the processes of reification and symbolization, which permit a more rapid processing, storage, and evaluation of information, including perceptions concerning social behavior. Thus the autocatalysis might have proceeded through the mutual reinforcement of assessment of the social environment, which was enhanced by reification, and genetic assimilation of an ever expanding array of culturgens, which broadened and sharpened the skills of tool use and social manipulation.

Culture slows the rate of genetic evolution. At the same time that usage sensitivity and reification accelerate the overall progress of coevolution, cultural transmission itself tends to slow genetic evolution *within* the coevolutionary process. Compare, for example, the rates of genetic change in the A (tabula rasa) versus B (biased genotypes) panels of Figures 6–10 through 6–12. In each case the rate is slower where tabula rasa genotypes are present, for individuals with tabula rasa genotypes hold on to a larger share of the favored culturgens. Those who adopt c_1 as opposed to c_2

(one-half do so in our model) succeed at least as well as individuals with other genotypes who adopt c_1. When genotypes are not tabula rasa and bias learning away from c_1, they are replaced more quickly by genotypes that bias learning in the direction of c_1.

However, the resistance of tabula rasa genotypes is only the extreme case of the general principle that cultural transmission slows the rate of replacement of disfavored genotypes. *Any* propensity to acquire the more successful culturgen, in other words any value of u_{21} above zero, will slow the rate of replacement of the prescribing genes below what would be the case if $u_{21} = 0$. When $u_{21} = 0$ and $u_{12} = 1$, we have the extreme case of pure genetic transmission of culturgens, comparable to the instinctive behavior of lower animals. The rate of replacement of the genotype prescribing u_{12} is the maximum possible, all other circumstances being equal, and the trajectories of gene frequency change can be described by the equations of conventional population genetics. It is also true that anything less than strict determinism by the favored genotypes will slow their increase, because the propensity of even a small percentage of individuals to adopt the less successful culturgen will reduce the margin of selective advantage.

Changes in gene frequency during the coevolutionary process can nevertheless be rapid. Although gene-culture transmission slows the rate of change in gene frequency below that possible in pure genetic transmission, the rate can still be much higher than intuitively expected. Under some circumstances genetic evolution can proceed as rapidly as cultural evolution.

An inspection of the curves of Figures 6–10 through 6–13 illustrates this principle very well. Even when the innate bias is mild in comparison with the biases demonstrated in human developmental studies (see Table 4–1), the rate of change in gene frequency can be high enough to achieve partial replacement of one allele by another within as few as ten generations, or about two hundred to three hundred years. Thus in some societies genetic evolution can occur during periods of time when there is relatively little culturgen innovation and change.

We suggest the existence of a rough fifty-generation rule for populations able to exploit highly efficient new culturgens. For mankind in particular this amounts to a *thousand-year rule:* during a period of this length substantial genetic evolution can occur in the epigenetic rules of cultural transmission, resulting in such effects as the genetic assimilation of culturgen preference and the assimilation of bias toward specific decision heuristics. The estimate is by order of magnitude. Only under extreme conditions can an allele be totally or nearly substituted in as few as one hundred or two hundred years; but substitutions can occur under a wide range of conditions in a thousand years.

The thousand-year rule is an intriguing result. During more than 99 percent of the four-million-year history of mankind, people lived in hunter-gatherer bands in which cultural evolution proceeded relatively slowly. Rapid culturgen innovation began in a few populations only about thirty thousand years ago, and even then the turnover of principal culturgen types sometimes required millennia. Similar cultural conservatism, extending to centuries or even to millennia, has persisted to the present time in a few hunter-gatherer and economically primitive agricultural and herding societies. Furthermore, we must remember that any invariant properties of rapidly innovated culturgens provide stable cores of features that can be recognized during cognition and used to choose particular responses. The opportunities have existed, through time spans tens and even hundreds of times sufficiently long, for the genes to track culture and bias the epigenetic rules to favor the most successful forms of culturally transmitted behavior.

This is not to say that every nuance of culturally transmitted behavior in prehistory was hardened genetically in the form of very specific epigenetic rules. The rate of coevolution depended also upon the amount of genetic variance underlying the epigenetic rules. Selectivity in turn was affected by the amount of heterogeneity in the environment, which promoted either mixed rules or more general single rules. But the results of our model do suggest that time has been more than adequate for substantial coevolution and the establishment of some degree of epigenetic bias in virtually every category of cultural behavior.

In order to provide a clearer picture of the extent of this genetic assimilation, a more detailed theory of genetic tracking is needed than the one provided in Chapter 5. The following principles are likely to be included. Within the thousand years usually required, tracking will be closest and the epigenetic rules most selective when only a relatively few well-defined culturgens or culturgen cores can possibly exist, as in the cases of the limited, clear-cut choices of incest avoidance, personal bonding, recognition of kin, and territorial definition and defense. The same trend will be enhanced when the environment is most nearly uniform in space and time, at least with reference to the contingencies that are met by the culturgens of interest. In contrast, genetic tracking will be least advanced when culturgens are numerous, not clearly distinguishable, or possess very similar or identical selection values. Such culturgens are likely to turn over at intervals much shorter than a thousand years, and hence details about them will be less subject to genetic assimilation. Some culturgens within a major behavioral category are likely to be too evanescent for tracking at the same time that others are basic and their feature cores stable enough to be tracked, creating composites of epigenetic rules with greatly varying selectivity. Thus it may be significant that particular features of women's dress have fluctuated rapidly in most Western societies (see Chapter 4) —but not the wearing of clothing itself, or the main body-conforming features of the dress, or the uses of fashion in the communication of tribal membership and tribal status.

As the opportunities for genetic tracking are improved, cognition is likely to evolve so as to sharpen culturgen definition by enhanced recognition of core features. This improvement can facilitate a higher degree of programming in the formation of knowledge structures and hence reification. Such *self-reinforcement* of epigenetic rules might create prototype knowledge structures that serve as the basis of the archetypes recognized in Jungian psychoanalytic theory. If there is any substance in this speculation, the archetypes may be the more complex knowledge structures that tend to develop in a wide range of environments and make repeated appearances in dreams and myths.

Gene-culture coevolution can promote genetic diversity. As the summary equations (6-21) and (6-38) demonstrate, selection within the coevolving system is frequency dependent. For the values of α and the fertility rule used thus far, the effects of frequency dependence are unimpressive. But qualitatively new phenomena leading to sustained polymorphisms appear when one begins to incorporate the effects of culturgen-culturgen linkage in cognition, division of labor, and economic exchange. This condition widens the array of conceivable circumstances under which intermediate gene frequencies can be stabilized and hence sustains a larger amount of genetic diversity within populations.

Consider the fertility rule depicted in Figure 6–14. Genetic fitness does not increase monotonically without limit as more and more resources are harvested. Instead, there is a limit R_{max} beyond which fitness begins to decrease. If this suppression is relatively weak, the allele biasing individuals toward the more successful culturgen will proceed on to fixation. If it is sufficiently strong, the competing alleles will stabilize at an intermediate frequency. And if it becomes even stronger, the gene frequencies enter a chaotic regime in which they fluctuate widely. These three outcomes are illustrated in Figures 6–15 through 6–18.

Fitness suppression is a common phenomenon in human societies. In

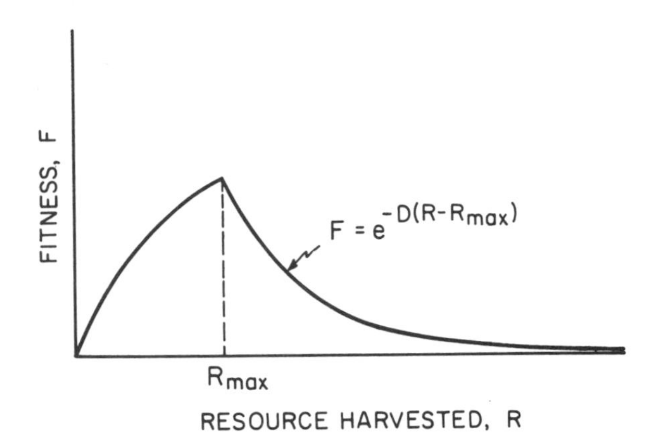

Figure 6–14 The suppression model of gene-culture coevolution. Beyond a certain amount of resource harvested (R_{max}), genetic fitness F begins to decline because of suppressing influences from the society or environment.

the !Kung San hunter-gatherer bands of the Kalahari, excessive attempts to raise personal status or to accumulate excessive amounts of material goods are met with ridicule and hostility. The result is the maintenance of nearly egalitarian societies (Lee, 1969; John Pfeiffer, personal communication). However, when !Kung bands settle near larger communities of other tribes, they become more openly self-serving and possessive (Patricia Draper, personal communication). At least some other hunter-gatherer peoples, such as the Tiwi of Australia, appear to tolerate higher levels of economic or social success.

In economically more complex societies, specialization and division of labor introduce another kind of suppression effect. Excessive production of goods and services leads to intensifying competition, unstable markets, and ultimately a reduction in absolute benefits to the specialized producers. Rising costs in transport, storage, and processing can also play an inhibitory role. It is apparent that regardless of the cost functions at lower levels of production, a threshold level R_{max} must exist beyond which the absolute return in benefits begins to drop. The result will be not just a spreading of economic and social roles, as expected from elementary economic theory, but a diversification of the genetic basis that underwrites the capacity to assume each role separately.

This result does not imply the partitioning of human-like societies into genetic castes. Rules of exogamy and the opportunities in most societies for some amount of socioeconomic mobility and occupational change militate against such an extreme phenomenon (Wilson, 1975). Indeed, even the caste system of India, which is the strongest and most elaborate on

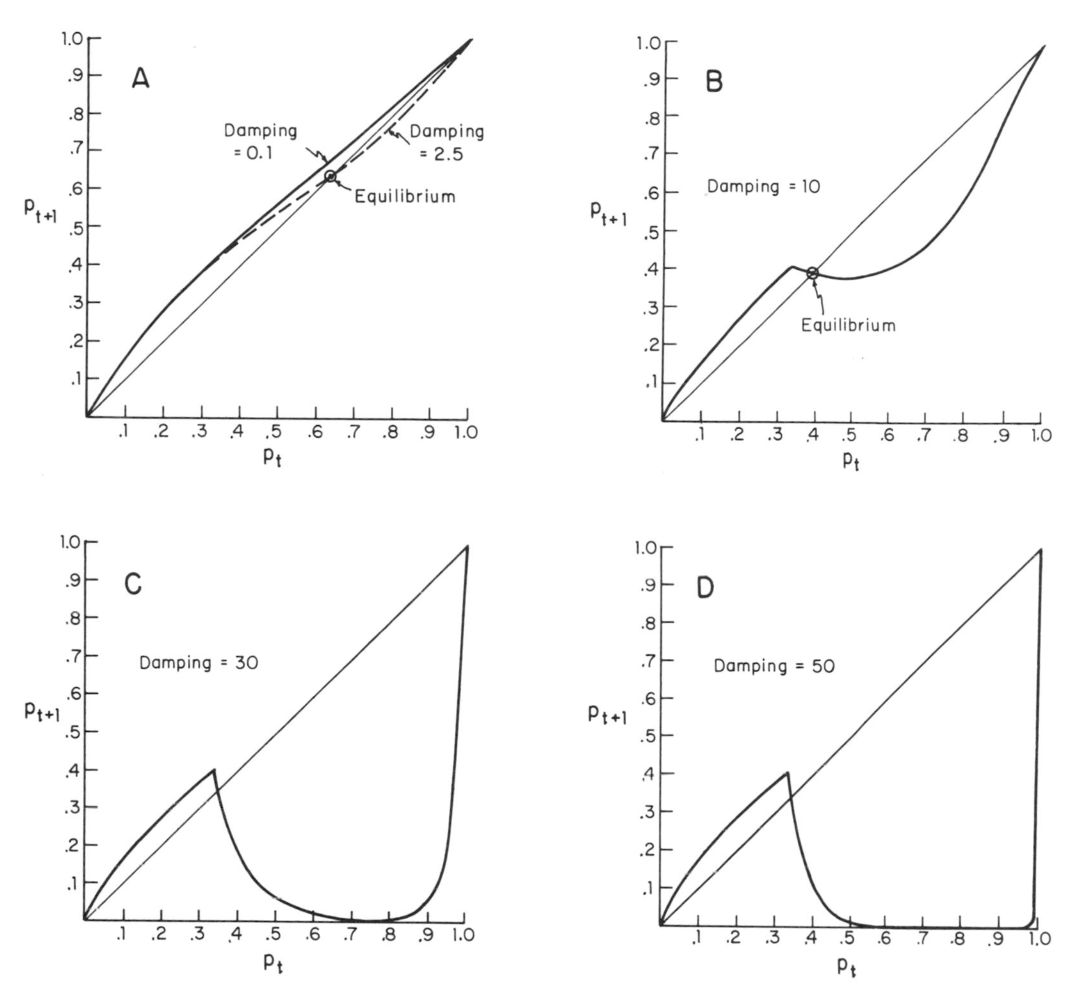

Figure 6–15 Gene frequency changes in the suppression model of gene-culture coevolution. The diagonals (*thin lines*) represent the condition in which no genetic change occurs from one generation to the next. The other curves (*thick lines*) represent the amount of change due to different degrees of suppression: above the diagonal, gene frequencies increase; below it, they decrease. *A* and *B*: at low suppression (damping constant $D = 0.1$) alleles biasing individuals toward successful culturgens proceed to fixation, but at a somewhat higher level (damping constant $D = 2.5$ and 10) they proceed to a stable equilibrium with the alternative allele. At still higher levels (*C* and *D*) the gene frequencies enter chaotic regimes (see Figures 6–16 to 6–18). In the present figure and those that follow, the evolutionary competition takes place between a dominant allele *A* and a recessive allele *a*. When allele *A* is present the innate biases take the values $u^{AA}_{12,0} = u^{Aa}_{12,0} = 0.2$ and $u^{AA}_{21,0} = u^{Aa}_{21,0} = 0.8$. In the homozygous state *aa*, however, these biases change to $u^{aa}_{12,0} = 0.8$ and $u^{aa}_{21,0} = 0.2$. In all genotypes the reification parameter is $\alpha = 0.2$ and the parameter of attention structure is $\beta = 1.0$.

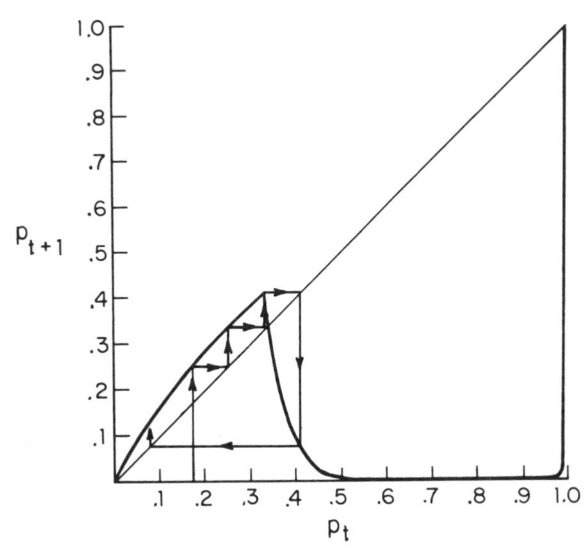

Figure 6–16 A Li-Yorke trajectory in the gene frequency changes under a high level of suppression. A cycle of period 3, and hence chaos, exists in the model. See Li and Yorke (1975) and May (1976). Under such conditions the structure of the trajectories of both gene and culturgen frequencies is very complicated. For every integer $i = 1, 2, 3, \ldots$, there is an initial frequency p_0 of A such that the gene frequency is periodic in time, with period i. Furthermore, there is an uncountable number of initial gene frequencies for which the evolutionary trajectories wander erratically. This wandering behavior can be quantified statistically (see Figure 6-18).

earth and has been in existence over two thousand years, is maintained largely if not entirely by cultural conventions; members of different castes differ from one another only slightly in blood type and other measurable anatomical and physiological traits. What this suppression-diversification hypothesis does suggest is the existence of a higher level of genetic diversity concerned with the epigenetic rules of many of the principal social and economic roles in some gene-culture systems. It also implies a high degree of individuality among group members in the genetic basis of behavior. Whether or not human beings have crossed such a threshold in their gene-culture coevolution remains to be seen.

The Significance of Coevolutionary Models

Even in its restricted form, the coevolutionary model we have employed has yielded several principles of general interest. If the thousand-year rule

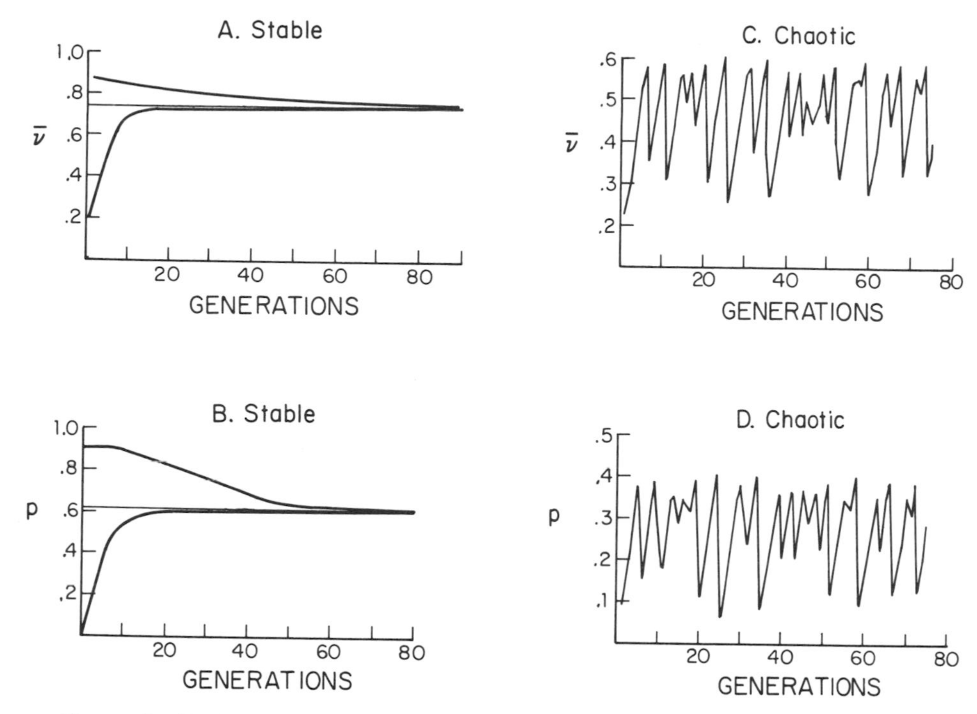

Figure 6–17 Changes in gene frequency and culturgen usage frequency in the suppression model of gene-culture coevolution. *A* and *B*: at lower levels of damping ($D = 2.5$ is illustrated), the trajectories approach steady-state polymorphisms. *C* and *D*: at higher levels of damping ($D = 50$ is illustrated), the trajectories occupy a chaotic regime.

and the suppression-diversification effect in particular apply to human beings, they indicate an evolutionary process radically different from that generally accepted in biology and the social sciences. The conventional view is that significant genetic evolution requires thousands of years and largely came to a halt in human populations thirty thousand years ago or more, after which cultural evolution took over as virtually the sole agent of change. But the coevolutionary model shows that substantial genetic evolution of cognitive traits can occur within only one thousand years and is very likely to have proceeded right into modern times. The conventional view also sees genetic variation in cognitive ability and perceptual and motor skills as noise, the result of random fluctuation around the species norm. Such variance has been enhanced by gene flow between populations and the cumulative effects of various pathological mutations. But the coevolutionary model reveals that part of the variation may stem

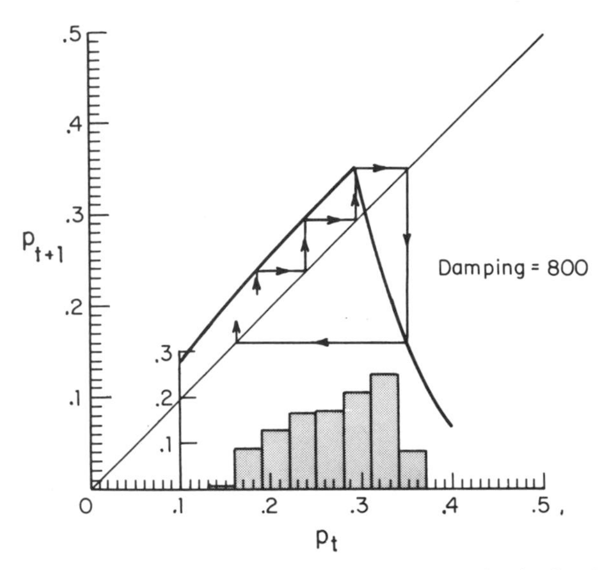

Figure 6–18 The statistical distribution of gene frequencies in the chaotic regime of the suppression model. Although the individual gene frequency trajectories are erratic, they show well-behaved statistical properties. Here, the histogram shaded gray gives the frequency with which a single evolutionary trajectory visited each of ten intervals 0.3 unit wide on the interval [0.1, 0.4]. Frequencies were tracked for 5,000 generations in the model, using Eq. (6-43) to provide a rapid evaluation of the cultural usage parameter $\bar{\nu}$. This histogram samples the stable distribution (Li and Yorke, 1975) approached by all initial gene frequencies under these conditions. Similar behavior obtains for $\bar{\nu}$. A Li-Yorke trajectory is superimposed upon the iteration curve $p_{t+1}(p_t)$.

from a peculiar diversification effect that in turn results from diminishing returns-to-scale in the conversion of behavior into genetic fitness. If this effect exists, the genetic individuality of human beings is part of an adaptation that has resulted in a more efficient functioning of the society as a whole.

These and other principles can be explored and tested with the aid of families of explicit coevolutionary models. The particular version we have employed was chosen to demonstrate that such studies are capable of tracing the full coevolutionary circuit. Once sufficient empirical information is accrued on the properties of each of the steps in the circuits, such as the assimilation function and fertility map, realistic simulations of gene-culture coevolution will contribute to a richer and more solidly based theory of human biology.

Summary

In this chapter we incorporate the full coevolutionary circuit into the formal theory in order to envisage simultaneous evolutionary changes in genes, mind, and culture. The account begins with the principles of epigenesis in the nervous system. Among the most important of these is the generalization that mutations inaugurate changes not in single neurons but in the field gradients and properties of cell form that affect whole populations of neurons. Only a relatively few such developmental rules are needed to reach precision in neuronal form, location, and connectivity. Although these rules are under the control of polygenes, the introduction of single new alleles can induce major modifications in specific features of brain structure and behavior.

The exact rules of pattern formation in the human brain are still unknown, but some of their general features can be inferred from the results of studies of epigenesis in other organisms. In some cases the complex form and patterning of cells appear to depend upon the diffusion of inducing substances. The patterns are canalized by the interactions of genes. Genetic assimilation is most rapid when this canalization is less restrictive, or else when extreme environmental events occur frequently enough to generate many phenotypes markedly different from the population norm. It follows that the polygenic control of traits is a theater of unusual opportunity for gene-culture coevolution. By cultural experimentation and the continual exploration of new environments, species are more likely to test the potential of the genes that prescribe epigenetic rules of cognitive development and to produce novel results. When the behaviors eventually emitted also confer selective advantage, the genes tend to shift into different frequency distributions that prescribe epigenetic rules biased toward the new responses.

Neuronal and cognitive development are tightly linked processes that continue unbroken from the early embryo to full maturity. Learning can be regarded as cognitive epigenesis within and beyond the womb. Recent studies on long term memory have cast considerable light on the assembly of the cognitive structures that form the schemata by which the mind recalls information, evaluates new contingencies, solves problems, and directs motor activity. We utilize this new conception to relate culturgens to mental activity and by this means to evaluate their genetic fitness as the outcome of learning and reasoning (see the summary in Figure 6–5).

We then construct a model that includes the main steps around the entire coevolutionary circuit. A life cycle is stipulated in which the offspring are socialized by both their age peers and the parental generation. The young learn and evaluate all the culturgens of the society by means of exploration and observation. At some future date they use this stored information to exploit the environment (see Figure 6–6). In particular, individuals learn and later choose between two culturgens under the influence of epigenetic rules that are prescribed by two alleles. Variation is permitted in the degree of bias in the epigenetic rules, the amount of sensitivity to peer and parental usage, and the function by which resources garnered during exploitation of the environment are converted into genetic fitness.

The time scales of genetic change in this type of coevolution, along with a diversity of interesting effects (some of them previously unsuspected), have been revealed by investigation of the model. The tabula rasa state, in which no innate bias exists in culturgen choice, is shown to be unstable, easily replaced by any one of a virtually unlimited range of biasing epigenetic rules. Sensitivity to usage pattern increases the rate of evolution in epigenetic biasing and hence the genetic assimilation of culturgen use. This catalytic effect might have contributed to the rapid evolutionary increase in human brain size associated with the onset of gene-culture coevolution. Culture slows the rate of genetic evolution, but coevolution still occurs quickly enough for the genes to track many forms of cultural evolution. We have summarized our inferences by a thousand-year rule: the alleles of epigenetic rules favoring more successful culturgens can largely replace competing alleles within as few as fifty generations, or on the order of one thousand years in human history. It is thus possible that substantial genetic evolution occurred during historical times and continues even today.

Finally, gene-culture coevolution appears to be based on frequency-dependent selection. Under certain conditions, and in particular when genetic fitness declines after a certain amount of resource has been harvested, selection leads to such intermediate evolutionary states as stable genetic polymorphisms or chaotic fluctuations in gene frequency. This phenomenon may have enhanced genetic diversity, division of labor, and individuality among human beings.

CHAPTER SEVEN

The Biogeography of the Mind

Culture can be analyzed in a novel way with the aid of the theory and methods of biogeography. We can think of the human mind as an island, into which culturgens immigrate like species of organisms, and where they occasionally evolve into new forms ("innovation") or become extinct. Although the analogy is unavoidably crude, it leads to unexpected insights concerning the size and diversity of cultures.

Consider that the larger the island, meaning the larger and more powerful the capacity of long term memory, the lower the extinction rate of node-link structures and the larger the eventual equilibrium number of active culturgens. The more distant the island, meaning the greater the isolation of the individual organism from surrounding cultures, the lower the immigration rate of culturgens into long term memory and the smaller the equilibrium number of culturgens. Although a steady-state cultural diversity can under some circumstances become extremely large, that level will nevertheless be below the pool of culturgens available to the society. In the case of literate societies, the level will settle far below the pool. The mind is limited in the number of culturgens it is willing to entertain. Furthermore, there is competition among the culturgens recognized as distinct alternatives within each culturgen category, and those most highly valued extinguish their competitors. Often particular representatives of various culturgens link together into strong holistic ensembles that prevent colonization of long term memory by new culturgens and hence stabilize not only the size of the individual's share of the culture but also the composition. When these closed communities are finally broken by strong new invaders, for example during the course of conversion or revitalization, the individual's share of the culture may enter a rapidly changing period that leads to either impoverishment or enrichment.

The society as a whole resembles an archipelago. Its constituent islands are the separate minds of its members, which exchange culturgens far more frequently among themselves than they do with the units of neighboring societies. For purposes of analysis the society-archipelago can be treated as a single large insular unit. Ethnography therefore becomes the biogeography of archipelagoes in this special sense.

In order for these biogeographic analogies to be transformed into the real phenomena of human behavior, it is necessary to incorporate the actual processes of cognition and social interaction as independent variables that control the immigration and extinction of culturgens. In the sections to follow, we first recast relevant aspects of ethnography in island-biogeographic terms, then reexamine in this new context the cognitive processes that determine the discrimination and choice of culturgens. At the end we try to relate this approach to the important question, first raised in Chapter 1, of why human-level euculture has occurred so rarely in the history of life, and we ask whether the evolutionary route taken by the human species is the only one open to any organism.

Cultures as the Biotas of Archipelagoes

The basic island-biogeographic model, modified from the original form used for biological species to accommodate culturgens, is presented in Figure 7–1. Real islands and cultures are dynamic, evolving systems that rarely if ever occupy an exact steady state. Yet it remains true that on ecological time scales the total number of species changes relatively slowly on many islands. Similarly, cultures may pass through periods of vigorous innovation accompanied by culturgen turnover, but episodes during which *total* cultural diversity and complexity expand dramatically are rare. Thus the theory of equilibrium or steady-state island biogeography is an excellent starting point for this new branch of ethnology.

The space in which the culturgens accumulate is the minds of the members of the society, the equivalent of the archipelago in the original theory of island biogeography (see MacArthur and Wilson, 1967). The archipelago-society can be treated as one insular ensemble, in other words as though it were a single space open for occupation by competing culturgens. Societies that are high in culturgen innovation, or are in close contact with many surrounding societies, either through geographic proximity or more efficient channels of communication, will receive novel culturgens at a relatively high rate. However, this *immigration rate* can be expected eventually to decline as the number of adopted culturgens

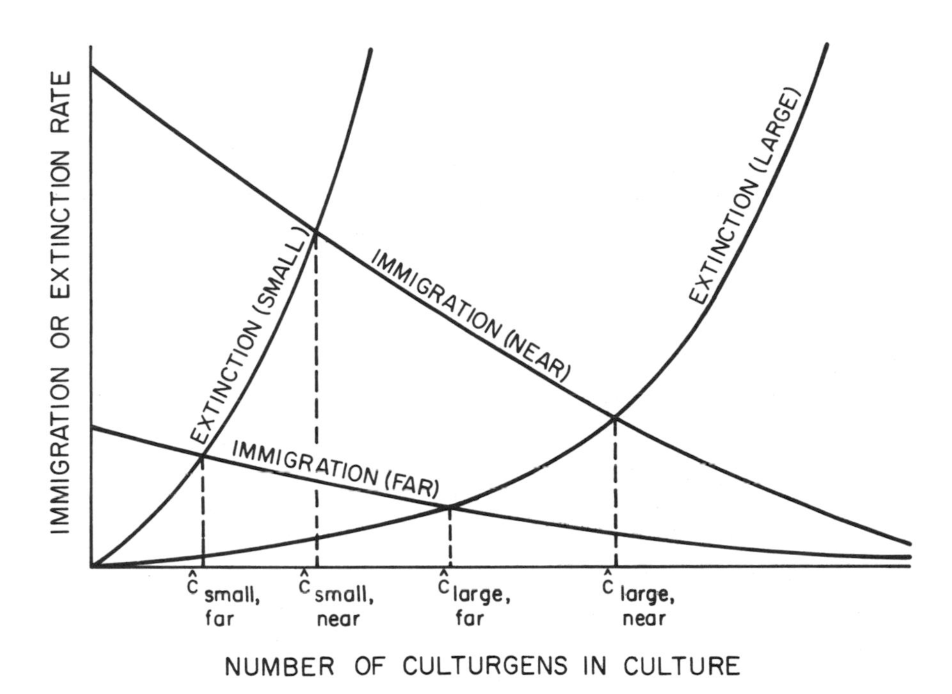

Figure 7–1 The island-biogeographic model applied to culturgen numbers possessed by entire societies. The predicted equilibrium numbers $\hat{c}$ increase in passing from societies that are small and isolated to those that are large and in close contact with neighboring societies.

grows, because the number still remaining to be adopted must shrink to a level where only a few probably less valued culturgens in the pool of surrounding cultures are still available for assimilation. In the absence of further information, we have represented the rate as a monotone declining function of c, the number of culturgens already assimilated. As the formulation in Figure 7–1 suggests, societies that are relatively isolated have lower immigration curves than those in close contact with surrounding societies.

The *extinction rate* of culturgens is the rate at which culturgens are lost entirely by the society or at least withdrawn from its active store. All other things being equal among the societies compared (that is, in the absence of substantial variation in societal organization that speeds communication or enhances the incentives to learn), large populations can be expected to retain culturgens longer, because the chance that every member of the society will discard a given culturgen is smaller when there are many people instead of only a few. Hence the extinction curve of large societies is lower than that of small societies. On the other hand, the extinction rate is not expected to be strongly influenced by the degree of isolation of the society from other societies.

A steady-state cultural diversity, defined as the number of active culturgens possessed by the society in periods when the number is neither increasing nor decreasing, is reached when the extinction rate equals the immigration rate. Note that the equilibrium exists in the *numbers* of culturgens only. Particular culturgens may be replacing others in quick succession, yet in a pattern that equalizes the extinction and immigration rates and hence does not alter the standing number of active culturgens possessed by the society through time.

Three relationships of potential importance for the social sciences have been discovered in the biogeography of organisms. The first is the *area effect,* which is simply the increase in the number of species of a given group (such as birds or flowering plants) with an increase in the space the species occupy. The relationship among islands of varying size but varying degrees of isolation from the mainland is

$$s = bA^z \qquad (7\text{-}1)$$

where s is the number of species on a particular island, A is the area of the island, and b and z are fitted constants that differ from one kind of organism to another, as from birds to flowering plants. The exponent z varies least; in most kinds of organisms it falls between 0.2 and 0.4 (MacArthur and Wilson, 1967). Preston (1962) showed that if the relative number of species that contain various numbers of organisms is lognormally distributed, and if there is some threshold number of individuals below which a species becomes extinct, then $z = 0.263$, a theoretical value that falls within the lower part of the range of estimates obtained from field studies.

The second biological phenomenon of probable relevance to cultural evolution is the *distance effect:* among islands or archipelagoes of equal size, the number of species at equilibrium decreases as isolation increases. The third is the *turnover principle:* the more rapid the growth in the number of species during the buildup to equilibrium, the higher the turnover rate at or near equilibrium, in other words the higher the number lost and also gained. In the special case where the immigration and extinction curves form straight lines (they are shown as concave in Figure 7–1), the turnover rate at equilibrium is

$$r = \hat{s}/\tau \qquad (7\text{-}2)$$

where r is the turnover rate in some unit of time chosen for convenience, $\hat{s}$ is the equilibrium number of species on the island, and τ is the time required to reach equilibrium from $s = 0$ (MacArthur and Wilson, 1967).

These basic principles of species biogeography can be applied to the evolution of cultural richness, but not in the exact form envisioned in the elementary models. Indeed, even in biological systems the basic model is relevant only in restricted conditions and must be modified in sometimes complex ways to fit other specialized conditions (see the review by Pielou, 1979). In the case of cultural evolution the modifications are certain to be more profound. Yet they are worth pursuing for the light they will shed on the mechanisms that create and maintain social complexity.

One principal alteration must be in the characterization of the immigration process. The immigration rates of culturgens cannot be regarded as largely independent of population size, unlike those of species. The gravity model of interaction, which is supported by empirical evidence from human geography (Haggett, 1972; Stephan, 1979), states that the social interaction between two localities is a direct function of the product of the sizes of their populations and an inverse function of the distance between them. Consequently, immigration curves can be expected to rise not only with proximity, as shown in Figure 7–1, but also with increasing population size (which would favor a rise in culturgen immigration because of increases in the rate of innovation as well).

In human studies it is important to make a distinction between diversity as measured by the number of culturgens within single cultures and diversity at a higher level measured as the number of cultures occupying an island or some other circumscribed land area. In the first case the society is the island and culturgens are the species; in the second case a literal landmass is the island and the cultures are the species. Terrell (1974, 1977), who was the first to apply biogeographic equilibrium theory to ethnography, took the second approach in analyzing the languages of the Solomon Islands. The number of languages found on particular islands is related to the landmass of the islands by the linear function

$$y = 0.0052x + 0.7812. \qquad (7\text{-}3)$$

The relationship suggests that as new areas become available, immigrants simply occupy them, form new tribal units, and eventually evolve distinguishable languages. The basic equilibrium model does not apply in this example, since there is no turnover of coexisting entities. On the other hand, Terrell showed that the equilibrium model can be used to characterize variation in the number of words shared by different villages. As the number of such cognate words increases, the loss ("extinction") rate of shared words increases, while the rate of borrowing decreases until a balance is struck between the two processes.

The other principal ethnographic measure, of culturgen diversity within individual societies, has not to our knowledge been examined in an explicit manner with reference to biogeographic theory. However, Carneiro (1967) has provided useful data relating the number of organizational traits in particular societies to the number of individuals constituting the societies. He considered only the presence or absence of organizational traits, such as craft specialization, nuclear family, taxation, service specialization, hierarchy of priests, and slavery. Technological and ideological qualities were ignored. Among the one hundred societies catalogued by Carneiro, ranging in complexity from hunter-gatherer bands to large tribes, the number of traits increased approximately as $0.6\ N^{0.6}$, where N is the size of the population (see Figure 7–2).

The comparison between species-area curves and cultural diversity curves is direct, because the areas of islands entered in the species-area estimates are generally regarded to be approximated by a linear function of the population sizes of the resident species. We are not aware of data relating population size to the size of the area occupied by the culture groups named in Figure 7–2. The simplest hypothesis is again that of uniform population density. Under this assumption the number of traits is proportional to $A^{0.6}$. The areal exponent 0.6 lies well above the range 0.2–0.4 occupied by the area-species curves of various kinds of organisms. However, human populations are not always so conveniently distributed. Marked variations in population density ρ characterize at least the more complex societies organized around population centers. Typical patterns relating ρ to the distance d from a center of the population are the negative exponential form

$$\rho = \rho_0 e^{-ad}, \qquad a \sim 1 \tag{7-4}$$

and the power law

$$\rho = \rho_0 d^{-a}, \qquad 2 \lesssim a \lesssim 20 \tag{7-5}$$

(Haggett, 1972). The linkage of cultural diversity to geographic area is more complex in these circumstances than in simple island biogeography. For example, given a single population center and the density function,

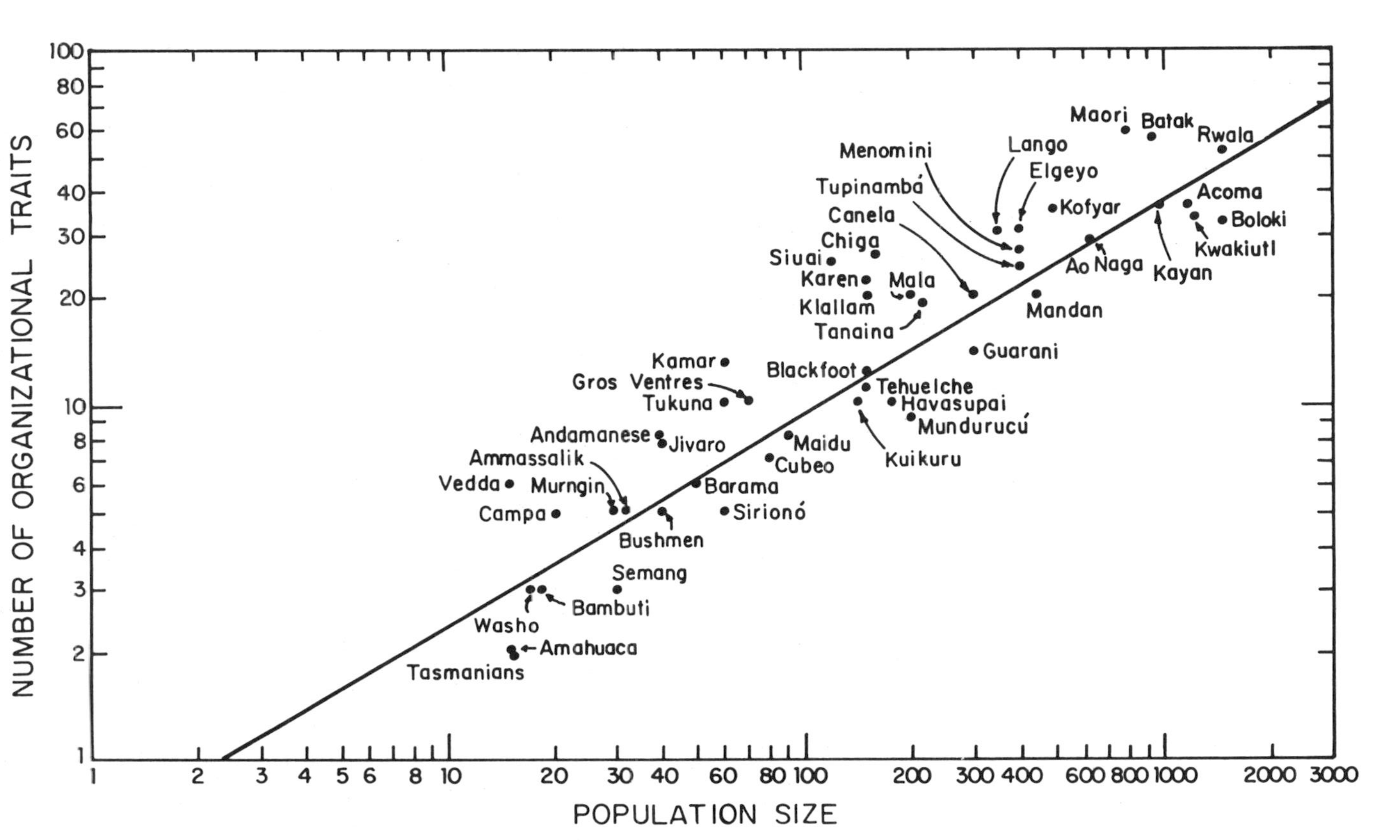

Figure 7–2 The relation of the number of organizational traits of cultures to the size of the population in the society. (Modified from Carneiro, 1967.)

Eq. (7-4), the population size N is related to area $A = \pi d^2$ by the expression

$$N = 2\pi\rho_0[1 - (1 + \pi^{-1/2}A^{1/2}) \exp(-\pi^{-1/2}A^{1/2})]. \tag{7-6}$$

For small centers, N varies as A^1 and the cultural diversity follows $A^{0.6}$; but for large A, N saturates to $2\pi\rho_0$ and the cultural diversity becomes virtually independent of the area. The areal diversity exponent z is thus a variable, with $0 \lesssim z \lesssim 0.6$. The precise effects of spatial patterning on the island-biogeographic exponents characteristic of real culture groups, such as those illustrated in Figure 7–2, remain unexplored.

On the other hand, there are marked differences between the assembly rules of groupings of species and groupings of culturgens. The most important is the existence of a stronger interdependence among organizational traits than occurs among species. As each organizational trait is added, it is more likely to favor and even to require the addition of others. Such mutualistic symbiosis also occurs among colonizing species but is much less common, and this circumstance alone may explain the lower z values of the area-species curves. As the size of human societies increases, new organizational traits are needed to continue the orderly conduct of the group. Once introduced, their extinction rate is lower and their persistence time correspondingly longer, just as an increase in the area of island habitats reduces the extinction rates of species of organisms living in them, lengthens their persistence time, and raises the overall equilibrium species diversity on the island. But in addition, the higher organizational traits do not necessarily exclude lower ones. Unlike interacting species of organisms, they tend to be added onto the behavioral repertory in a hierarchical and complementary manner and even to promote the adoption of still other culturgens that enhance communication. As a consequence the number of organizational traits, and probably other kinds of culturgens as well, can be expected to rise more quickly with an increase in population size than is the case in species of organisms.

Carneiro cites an instructive example of the details of hierarchical growth in the social structure of the Plains Indians. During most of the year the populations were broken into independent bands, each of which was loosely organized under a chief with limited authority. During the annual summer buffalo hunt, ten or more bands came together to form temporary aggregates of several thousand persons, and a more complex organization was adopted. The band chiefs organized themselves into a council and elected a paramount chief. Men's societies were activated to perform specialized functions; a prominent example was the temporary police force that kept order during the hunts, marches to new sites, and sun dances.

The distance effect predicted by the basic biogeographic model has not to our knowledge been considered systematically by ethnographers. However, the idea has been intuitively expressed by Jones (1977) in his analysis of the culture of the Tasmanian Aboriginals, who became extinct in the mid-1800s. Although information concerning these remarkable people is scarce and judgment difficult, Jones suggests that they were culturally depauperate in comparison with the Australian Aboriginals and most other hunter-gatherers. They apparently did not use bone points, boomerangs, or spear throwers. Although fish were readily available, the Tasmanians did not catch them for food. Also, the culture possessed few if any rituals involving large groups of people. Jones argues that the Tasmanians simply lost these various adaptations during their long isolation from the Australian mainland. In terms of the biogeographic model, this is the equivalent of saying that the extinction rate of culturgens remained the same as in populations of comparable size elsewhere, but owing to the reduction of the immigration rate of replacement culturgens, the total number of culturgens descended to a new, lower equilibrium level. This interpretation must be regarded as tentative. At least one anthropologist (Horton, 1979) has disputed Jones's explanation. He suggests that the culture was not as strongly depauperate as the largely anecdotal accounts from the early 1800s suggest. Even some of the reductions might have been special adaptations to the Tasmanian environment rather than a consequence of random loss.

However the particular case of the Tasmanians is eventually resolved, the distance effect and random cultural pauperization are potentially fruitful hypotheses if explored against the larger background of the models of island biogeography. They can be further strengthened by analysis of the kind of area-trait data assembled by Carneiro and by historical studies of appropriately chosen societies, especially those undergoing growth and major organizational changes.

Finally, biogeographic theory can be usefully adapted to focus attention on the survival rate of newly introduced culturgens. Most contemporary societies are bombarded by new culturgens that are competitive with the ones in predominant use. It would be useful to know the probability that a newly introduced culturgen adopted at first by only one person will persist and spread through at least part of the remainder of the society. How long will the culturgen be used by at least someone in the population? Models that conceptualize a "demography" of culturgens can provide answers to questions of this kind.

Let us envision a society of *N* persons as *N* receptacles, each of which can receive or lose a given culturgen. Suppose that at first just one person possesses the culturgen in active form. Within some interval of time—say, a day or a year—the culturgen can spread to a second person, or a third, or to the whole population of *N* members. During the same period that new individuals acquire the culturgen, others lose it, through death, forgetting, or conscious choice. The assimilation of the culturgen can be regarded as under the control of epigenetic rules that determine its rate of acquisition and rejection. To a lesser extent possession is under the direct control of genetic fitness, which affects whether the individual will die at an earlier or later date. The established results of island biogeography suggest that under such circumstances there may exist a critical population size beyond which the time required for the extinction of a culturgen becomes very long. Once past this threshold a sufficient number of refugia, in the form of individual minds, would communicate at a rate sufficient to guarantee the culturgen a minimum occupancy of one mind over substantial time periods. An example of this phenomenon is shown for the case of biological organisms in Figure 7–3. The takeoff phenomenon revealed by the model further suggests that the number of culturgens possessed by a society will increase steeply with an increase in the size of the society. It seems likely that the value of z will be even higher in many culturgen categories than that observed in the organizational traits (Figure 7–2).

A more exact characterization of culturgen survival and diversity must await the fitting of survivorship models to realistic assimilation functions. In this enterprise gene-culture theory will be greatly assisted by the sophisticated formal models of information diffusion and disease spread now under study in mathematical sociology and epidemiology (Coleman, 1964; Bartholomew, 1973; Bailey, 1975; and Hamblin et al., 1979, provide introductory treatments of these fields). Indeed, there is much in the literature of both disciplines to recommend vigorous application to island biogeography itself.

Culturgen Packing

With the delineation of culturgen immigration, extinction, and survival time, we can extend the biogeographic model one step further to examine culturgen packing. Cultural diversity depends on the number of cul-

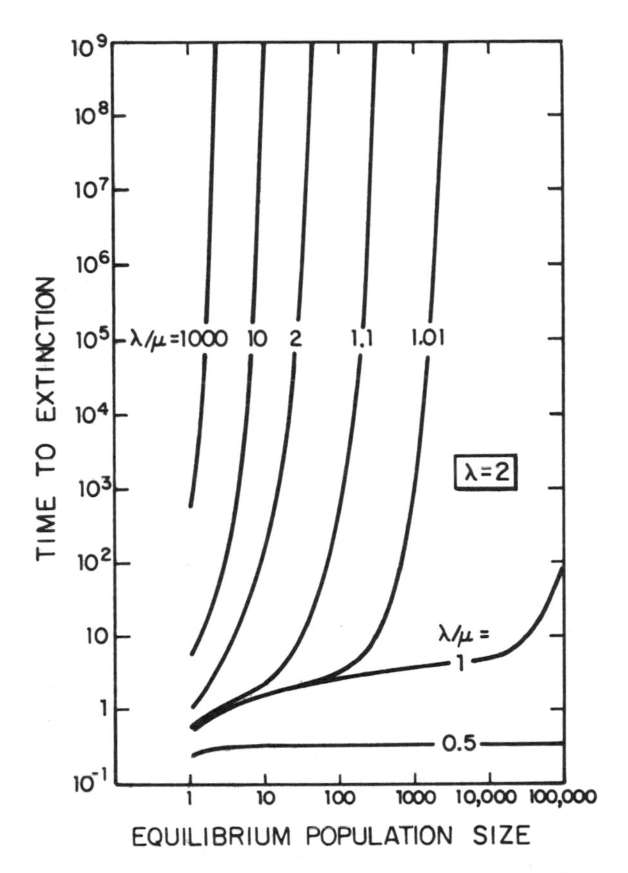

Figure 7–3 The mean survival time of a biological population given as a function of the birth rate λ, death rate μ, and the size of the population, in the case where $\lambda = 2$ per year. Similar thresholds to long survival times may exist for culturgens. (Modified from MacArthur and Wilson, 1967.)

turgens that can be incorporated into the mind. At low and intermediate levels of diversity, new culturgens should be added readily, because the small number already assimilated renders new culturgens easier to identify and less likely to be excluded by those already present. At high levels of diversity the likelihood of confusion and competitive interference is greater, especially among culturgens of the same category. We can imagine a level at which packing is complete. Cultural diversity will then have reached a dynamic equilibrium: as each new culturgen is added, an old one is dropped.

The condition can be characterized more exactly as follows. Let

$c_1, \ldots, c_M$ be a system of M culturgens. Suppose that there is a culturgen c_{M+1} that is initially at zero frequency, and δc_{M+1} is an event such as an innovation or importation that suddenly raises the frequency from zero to a small but finite level. If δc_{M+1} dies away to zero, or c_{M+1} displaces some other culturgen, we say that the system is fully packed. If in addition δc_{M+1} always dies away and no other culturgen is eliminated, the system is closed to invasion by c_{M+1}. On the other hand, if δc_{M+1} remains nonzero, and especially if it grows in magnitude, the system can be said to be partially packed and open to invasion by c_{M+1}.

Recall that culturgens can be classified into three categories. Artifacts, such as tools, dwelling places, and clothing, can be transmitted from one generation to the next. They will affect behavior even if, in the extreme imaginable case, they are unaccompanied by instructions. Behavior, including speech, the use of tools, and other forms of transmissible activity, constitutes the second principal category of culturgens. Finally, mentifacts (Huxley, 1958) are usefully distinguished as a third category, even though they blend imperceptibly into behavior at one end of the range of their variation.

Mentifacts play an especially powerful role in culturgen packing. They are the nearly pure creations of the mind, the reveries, fictions, and myths that have little connection with reality but take on a vigorous life of their own and can be transmitted from one generation to the next. Mentifacts are the most striking products of reification, the process that is a diagnostic feature of the human mind. Although far from being concrete entities, mentifacts occupy a central position in the life of every society. They often serve multiple functions through the differentiation of their meaning into several layers. Wilbert (1979) has provided an unusually clear example in the case of the Haburi saga, a myth of the Warao Indians. While accounting outwardly for the origin and ecology of the people, this legend permeates the culture in three distinct roles:

> To the Warao child the Haburi saga is a story crackling with adventure, detailing episodes of ogre chases, defeats, and triumphs. To the average adult, on the contrary, the episodic itinerary of the myth explains a terrestrial arrangement of resources and corresponding means of exploitation. To the sages of the tribe the myth conveys deep religious imports establishing as it does a link between mankind and the gods.

Constructions such as the Haburi saga extend to virtual infinity the number of culturgens that can be created. But of course only a limited set can be fitted together into a culture.

The potential growth of a culture depends not only on the culturgens that are active, in other words accepted and in current usage, but those that are kept in a passive state, meaning that they are remembered or at least recorded. The great divide in the evolution of cultural richness was the attainment of literacy. The number of passive culturgens that can be stored by preliterate societies is limited by the durability of unused artifacts and the memorization of oral traditions (recall the 10*J* Rule discussed in Chapter 3), but in literate societies it is theoretically almost unlimited. As the passive store grows in printed records and on film and tape, the immigration rate of culturgens into the active culture also inevitably rises. Information is retrieved and resynthesized, forgotten authors are revived and reinterpreted, and old theories are revitalized with the addition of new facts. A much higher culturgen equilibrium is reached as a result; we say that culture has been vastly enriched. For example, Indian tribes of California produced 3,000 to 6,000 artifact types, but the armed forces of the United States landed 500,000 artifact types at Casablanca in World War II (Steward, 1955). The ratio of passive to active culturgens also increases enormously, as well as the ratio of active culturgens at large in the society to those possessed by its individual members.

According to Marshack (1979), literacy of the most primitive kind began far back in Paleolithic times and had become relatively well developed some thirty-two thousand years ago. It took the form of scratches on ornaments, pieces of bone and clay, stones, and cave walls. The scratches were arranged by means of repeated motifs into descriptive classes such as meanders, fishlike images, and parallel lines. The patterns evidently transmitted records and messages. An increase in the abundance of the marks coincided with the invention of representational cave art in Europe, as well as with substantial increases in the number and stylistic design of artifacts in Europe and Africa (Isaac, 1972; Trinkhaus and Howells, 1979). The fossil record indicates that during the same period, human populations were expanding and *Homo sapiens sapiens* was in the process of replacing *H. sapiens neanderthalensis*. The symbols and representational art may have been used to register membership, status, and position, and to communicate among groups (Gamble, 1980).

What sets the limits of the packing process? We are particularly interested in the active culturgens that constitute living cultures. It is obvious that individual categories of culturgens are greatly enlarged when they become the objects of economic specialization. The Maori language recognizes 20 categories of greenstone used in axes, the Eskimos 100 categories of seal, some Amazon-Orinoco Amerindians 1,000 species of plants, and the Arabs 6,000 attributes of camels. Still, it is clear that there is an overall limit to the size of the active culture possessed by an individual or a society. As particular technologies, art forms, and institutions rise to prominence, others recede from practice and eventually from memory (Price, 1975). We suggest the operation of three constraining processes whose magnitudes are innate properties of the human brain: (1) discrimination and categorization of the stimuli that identify culturgens, (2) capacity and recall in long term memory, and (3) valuation of the multiple cues provided by each culturgen. The examination of these mental processes must be the first step toward the ultimate goal of characterizing the packing process and predicting cultural diversity from principles of cognition and behavior.

Discrimination and categorization. Culturgens present configurations of stimuli that are perceived and evaluated by a complex sequence of filtering and association in the central nervous system (recall Figure 3–4). There is some chance that during communication the same stimuli will be classified differently by various members of the society. The communication system can be characterized by the probability $P(c_m|s)$ that a culturgen transmitted with a signal attribute structure s will be assigned the classification c_m by the person perceiving it. The system can be designed according to an array of conceivable alternatives such as that depicted in Figure 7–4. At one extreme we can imagine razor-sharp transmission: each stimulus or combination of stimuli unerringly conveys the identification of the culturgen for which it is meant to be diagnostic. The second, more realistic alternative is the fuzzy logic recognized in recent experimental studies of cognition (Rosch, 1975; Rosch and Lloyd, 1978; Brown, 1978): the mind evaluates the stimulus and decides on the culturgen prototype closest to it, with an intermediate probability of making one choice as opposed to another in the boundary regions. The overlap curves we have drawn are also consistent with the evidence adduced by Shepard (1958a,b) and Getty and his associates (1979) that confusability between signals is a monotone decreasing function of the distance between the stimuli. Getty and coworkers give the following relation as most likely for their experimental preparations:

$$x_{ij} = e^{-aD_{ij}} \tag{7-7}$$

where x_{ij} is the frequency of confusion between two stimuli, D_{ij} is the interstimulus distance (which depends on the multidimensional scaling technique used), and a is the fitted "sensitivity" or discrimination factor.

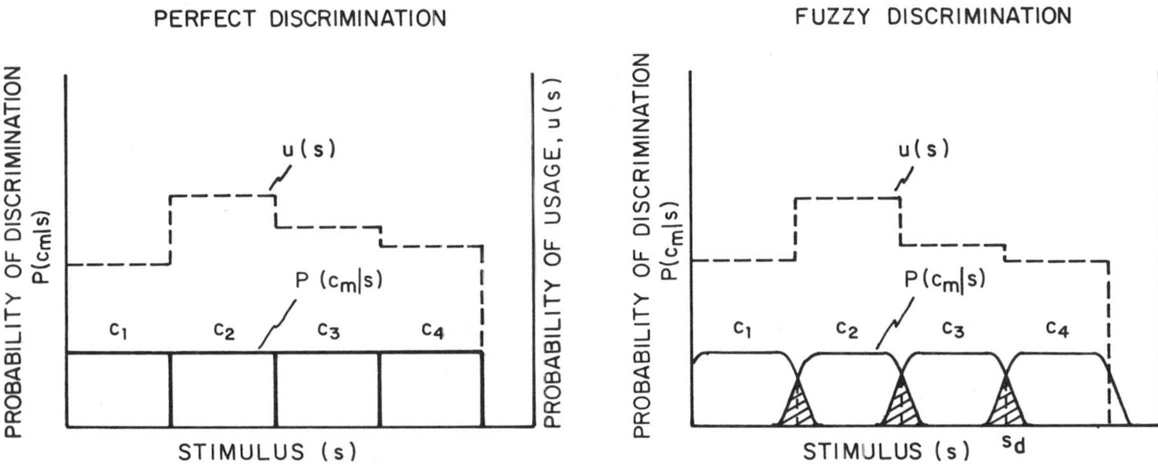

Figure 7–4 Alternatives in the design of a cognitive system that classifies culturgens. The left diagram presents one extreme in a continuous array of possibilities; the culturgens (c_m) are perfectly discriminated by means of sets of stimuli. In the right diagram ambiguity between the cells exists within a zone of overlap. In this fuzzy-logic case the zone is bell-shaped, with a midpoint of s_d. Usage bias curves, labeled $u(s)$, are superimposed to connect the measurement of discrimination to the remainder of gene-culture theory in the following way: once a culturgen has been identified, either correctly or incorrectly, it has a specific probability of being assimilated.

During the course of evolution the sensory systems and the brain have achieved certain intermediate degrees of discriminatory power. As we showed in Chapters 2 and 3, these levels vary greatly among the sensory modalities and categories of culturgens. The question of interest is, Do the levels represent evolutionary optima? If so, what are the determinants of the optima? The problem can be phrased more precisely as follows: Given that M culturgens must be recognized (M color categories or M edible plant species, for instance), what are the forms of the recognition function $P(c_m|s)$ that maximize genetic fitness subject to developmental and time-energy costs?

In order to illustrate the interplay of the design features, let us model the overlap region centered on s_d (see Figure 7–4) with a bell-shaped function, say of the Gaussian form

$$f(s) = e^{-(s-s_d)^2/\sigma^2}. \tag{7-8}$$

It is further reasonable to suppose that greater costs to the fitness of the individual are incurred when the signal is more ambiguous; in other words, the ambiguity costs are proportional to σ^α, where $\alpha > 0$. But a higher price is also exacted by the physiological development and maintenance of the supplementary cognitive apparatus needed to reduce ambiguity. This second cost can be regarded as proportional to some inverse power of σ, for example $\sigma^{-\beta}$, where $\beta > 0$.

Let B_0 represent the gross fitness benefit accruing to an information processing system with perfect culturgen discrimination; hence $\sigma = 0$. Then at finite ambiguity this performance is degraded to $B_0 - a\sigma^\alpha$. The development and maintenance costs must also be incorporated, yielding a net fitness contribution

$$B_{\text{net}} = B_0 - C(\sigma) \tag{7-9}$$

where $C(\sigma)$ is the cost function $a\sigma^\alpha + b\sigma^{-\beta}$. The two terms in the cost equation generate rising and falling curves respectively, as illustrated in Figure 7–5.

The optimal B_{net} then requires the minimum value of the cost function. This quantity, which defines the optimal ambiguity or culturgen cell overlap σ_0 for the system, occurs when

$$\sigma_0 = (b\beta/a\alpha)^{1/(\alpha+\beta)}. \tag{7-10}$$

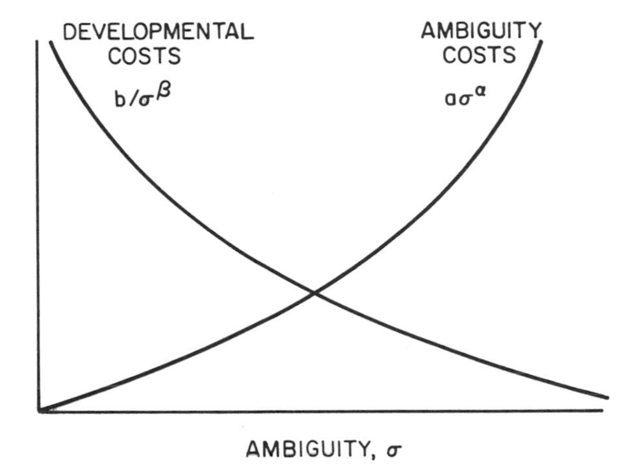

Figure 7–5 A model of the optimal ambiguity problem in the evolution of culturgen transmission. As ambiguity increases during the evolution of the cognitive system, the costs to the organisms rise with the penalties caused by the ambiguity, but they also fall because of the lower physiological expense of the neurosensory apparatus that tolerates the ambiguity.

Natural selection is expected to move the sensory and processing systems toward this level. If σ^* is the ambiguity at which developmental costs equal fitness costs, then

$$\sigma^* = (b/a)^{1/(\alpha+\beta)} \tag{7-11}$$

and

$$\begin{aligned} \sigma_0 &> \sigma^* \quad \text{if} \quad \alpha < \beta \\ \sigma_0 &= \sigma^* \quad \text{if} \quad \alpha = \beta \\ \sigma_0 &< \sigma^* \quad \text{if} \quad \alpha > \beta. \end{aligned} \tag{7-12}$$

A weakness of this first elementary illustration is the assumption that the optimum overlap is independent of d, the culturgen cell width. This is the equivalent of saying that the same amount of overlap is optimum for narrow cells (many culturgens) as for wide cells (few culturgens). But the ratio of overlap to cell width must become crucial near its upper limit. Even if the overlap were very small, ambiguity would still be intolerably high if the cell width were equally small. Consequently, the parameter of interest for many conceivable categories of culturgens is the ratio of overlap to width. Consider for purposes of clarity a simple model in which the culturgen categories are rectangular cells of width d and overlap l, as shown in Figure 7–6. A cost model suitable for illustrative purposes is

$$C(l) = al/d + b/l. \tag{7-13}$$

The optimal overlap l_0 at the packing density represented by d is then $(bd/a)^{1/2}$ and the minimal cost is

$$C_{\min} = C(l_0) = 2(ab/d)^{1/2}. \tag{7-14}$$

A consequence of special interest in such formulations is that $C_{\min}$ is d-dependent. Hence if there exists a maximum tolerable $C_{\min}$ for the system, there also exists a corresponding value of d and hence a maximum number of culturgens that can be packed into the cognitive system. Cultural evolution beyond this complexity limit must therefore depend on a division of learning among the group members.

An important feature of human cognition is that the amount of categorization attainable with different kinds of stimuli varies enormously. For example, luminance and loudness are perceived as continua and less often

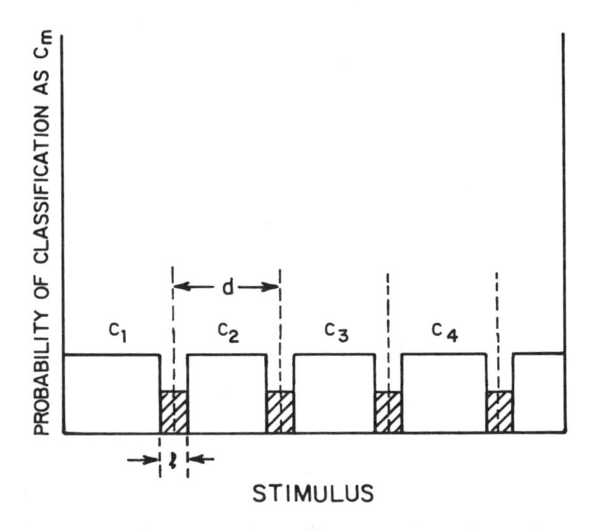

Figure 7–6 A system of rectangular culturgen categories, where the culturgen cells c_1–c_4 have width d and overlap l.

made the subjects of symbols and words, whereas hue and pitch are perceived in more nearly discrete categories and are the subjects of a large vocabulary. In addition, pitch discrimination forms the basis of a large part of the human communication system. We suggest that if natural selection has shaped human cognition, a proposition for which there is substantial evidence (see Chapters 1–3), it has moved discriminatory power toward optima in the various stimulus categories. Moreover, these ambiguity optima impose limits on the number of active culturgens that can be packed into the long term memory of individuals. As a consequence they have almost certainly constrained the evolution of cultures.

Capacity and recall in long term memory. Equally great variation occurs among the various sensory modalities with reference to learning capacity. Tastes and odors are memorized much more slowly than audiovisual stimuli but are retained over longer periods of time. Some classes of stimuli are commonly fashioned into metaphorical symbols, while others rarely serve more than a simple denotative function. Numbers and other signals expressed in audiovisual form, for example, are easily chunked into complex assemblages that facilitate recall. They contrast in this respect with melodies and odors, which resist mnemonic clustering.

Each of these properties of information processing, with its distinctly human parameters, exercises a deep influence on the number of culturgens that can be retained in long term memory and hence added to the active culture. At this early stage in the development of neurobiology and psychology it is only possible to speculate about the evolutionary history that led to the heterogeneous state of the cognitive system. The variation reflects the existence of the epigenetic rules, which we postulate to have been shaped by natural selection acting on their final consequences in social behavior. The important point is that the peculiarities of the human mind are not to be regarded as simply given, but rather as evolutionary products that have been tested through the relative adaptiveness of the social systems over many generations.

Valuation. The rates of culturgen immigration and abandonment depend on the strength of the emotional and semantic rewards induced by their identifying stimuli. But reinforcement of even the simplest culturgens is a complex phenomenon that entails multiple sensory modalities and differences in the values placed on various of the associated stimuli. It will be recalled that culturgens are constructed in long term memory by the polymer-like growth of nodes that are linked to one another. A single node is established, such as the concept "dog," and a network of associated nodes grows around it. The linkages are created by experience and learning. Studies employing the semantic differential technique have revealed that some of the linked nodes are purely denotative. In the case of the central word "dog," for example, they might include "mammal," "very hairy," and "moderately fast." Others conjure up emotional feeling, which insofar as they can be expressed in words perhaps include "very friendly" and "attractive." The knowledge structure associated with a concept spreads amoeba-like across the semantic landscape. Depending on individual experience, it can remain small or grow large. It can extend over a vast specialized domain ("carnassial teeth present," "territorial pheromone in male urine") that constitutes expertise, or it can be captured by the deep aversion of cynophobia that distorts or excludes many other of the possible associations. The knowledge structure can also be eroded or broken into fragments by the erasure of links in long term memory.

Although the growth of knowledge structures depends heavily on individual experience, it is not randomly experiential in origin; not all links can be established with equal ease. To say that epigenetic rules exist is the equivalent of saying that certain pairs of nodes are much more readily linked than others. Human beings are more prone to learn a color vocabulary based on the modal hue perceptions than they are to learn one based on wavelengths located between the modes. They are also far more likely to develop neutral or antipathetic feelings concerning brother-sister incest, as well as phobias against snakes, hostile associations between strangers and territory, and so forth. As a result the general forms of knowledge structures tend to converge across cultures.

A number of important questions are raised by these peculiarities in the cognitive process. The imagery of knowledge structures as polymers subject to regular constraints, like molecular polymers formed in obedience to the laws of physical chemistry, is a potentially useful aid in the development of a more vigorous learning theory. It would be important to know, for example, the maximum length of a chain of neutral or lightly laden concepts and their survival time in long term memory. Or the role that *topology* plays in the growth of the knowledge structures: does it matter whether the links form straight chains, circles, or tightly cross-bound polygons?

Such an inquiry can be extended into evolutionary time by examining the feedback of polymerization patterns to the valuation of the culturgens. It has been well established that the reinforcement of a denotative culturgen, in other words its linkage to emotional nodes, can be changed from neutral or negative to positive through linkage to a second, positively reinforced denotative culturgen (Brown and Herrnstein, 1975; Lindzey et al., 1975). In principle, continued linkage over a number of generations can genetically assimilate such secondary reinforcement by altering the epigenetic rules, and turn it into positive primary reinforcement.

Why Is Euculture So Rare?

A deep question in the biogeography of organisms is the frequency and raison d'être of exceptional species. During the four-billion-year history of life, a very few evolutionary innovations have required extremely complicated intermediate steps and occurred only once, yet the organisms achieving them became so enormously successful that they altered the environment in major ways. The amniotic egg of the ancestral reptiles is a familiar example. It permitted the vertebrates to remain away from the water throughout their life cycle and thus to become the first large-bodied animals to penetrate virtually every terrestrial habitat. A second, less familiar example is the cultivation of symbiotic fungi on freshly cut vegetation by leafcutter ants. The invention of this form of agriculture has made these insects the dominant herbivores in the New World tropics. It appears to have occurred in only one species, the ancestor of the attine genera *Acromyrmex* and *Atta,* among the millions of species of insects that have lived during the past 300 million years. A third and truly fundamental example is the origin of the eucaryotic cell, which characterizes all higher organisms from green algae to man. This complex unit apparently was created by the symbiotic incorporation of primitive procaryotic cells into a host procaryotic cell to form the mitochondria, chloroplasts, and cilia. For over two billion years life had existed exclusively in the form of bacteria, blue-green algae, and other single-celled procaryotic forms. The first eucaryotic cells launched new phyletic lines that culminated in large, multicellular organisms able to colonize every part of the sea, land, and air.

The geologically most recent example of a unique evolutionary product is euculture and the relatively enormous brain required to sustain it. In the course of achieving euculture the human species was the first to embark on gene-euculture coevolution. It invented reification, symbolization, and language, and magnified self-reflection until it could examine history and plan into the distant future. Through euculture mankind has literally altered the form of organic evolution. The time has come to try to answer the question we raised at the beginning of this book, that of why mankind is unique in this overwhelmingly important respect. Simply by comparison with the other singular events of evolution, euculture can be reasonably postulated to be reached by intermediate steps that are either improbable or else unstable and short-lived. Yet this interpretation produces a paradox. If mankind's demographic success can be regarded as indicative, the reification learning rules, together with their cultural manifestations, convey an enormous adaptive advantage. Then why have other large-bodied species—squids, eurypterids, fishes, reptiles, birds, and nonhuman mammals—not evolved repeatedly toward euculture?

Several hypotheses compatible with our knowledge of macroevolution can be advanced to resolve the paradox. One is that high cognition in animals, including the prehuman ancestors, has been of advantage in only a few very rare environmental niches. The scarcity of opportunity, viewed from this standpoint, imposes a bottleneck on the advance of cognitive evolution, confining it to certain ecologically specialized life forms. This explanation is contravened, however, by the extraordinarily diverse ecological adaptations possessed by the most neuropsychologically advanced nonhuman mammals. It is enough to list the chimpanzee (tropical forest and savanna: ground-foraging, omnivore), gibbon (tropical forest canopy: arboreal, vegetarian), elephant (tropical forest and savanna: terrestrial, vegetarian, gigantic), wolf (cold-temperate forest: terrestrial, carnivore), and toothed whales (marine: carnivores).

To cite this evidence is not to deny the existence of a special set of preadaptations favoring the origin of exceptional cognitive ability in man, of which penetration of a special niche can form a part. According to the prevailing "hunting hypothesis" (reviewed by Pfeiffer, 1969, and Wilson, 1975), the ancestral *Homo* were unique among all previous creatures in possessing the following combination of adaptations: bipedal locomotion, hands free and able to be used more fully in tool making, intelligent at the start (on a level comparable at least with modern chimpanzees), and partly carnivorous with access to the large hooved mammals of the African savanna. Because the early *Homo* were also relatively small and physically unimposing, a premium was placed on tool using and cooperative, closely coordinated behavior, which in turn led to a rapid increase in brain capacity and finally to the eucultural breakthrough. Nevertheless, this hypothesis, if correct, merely identifies the exceptional combination of anatomic preadaptations, ecological conditions, and historical circumstances that finally led one species to euculture. The fact remains that a relatively advanced form of cognition short of that required for euculture has arisen repeatedly in animal species in a great diversity of habitats; and we are left with the problem of why complicated brains, once attained, do not evolve more frequently into brains competent for euculture.

A second possible explanation for the rarity of the condition is based on the observation that the primitive species of *Homo,* including *H. habilis* and *H. erectus,* had an exceptionally diverse diet. Is it possible that ecologically generalized species require larger brains, and that the human species is merely the extremum of this rule? The answer appears to be no. The two hundred living nonhuman species of the order Primates display an enormous variation in the degrees and kinds of their ecological specialization (Napier and Napier, 1967, 1970; Eisenberg et al., 1972; Hill, 1972). The variation is equally great in both small-brained forms, such as the lemurs and other prosimians, and large-brained forms, such as the anthropoid apes. Among the latter can be counted the ecologically specialized orangutan (*Pongo pygmaeus*) and ecologically very generalized chimpanzee (*Pan troglodytes*). There appears to be little if any correlation across species between the degree of specialization and ecological success as measured in population density and breadth of geographic range. Even if a correlation does exist, it cannot be of sufficient magnitude by itself to account for the peculiar phylogenetic distribution of eucultural capacity.

In the absence of clear-cut ecological correlates, it seems appropriate to search as well for more internal constraints on the evolution of the highest levels of cognition and euculture. Like other macroevolutionary steps, the origin of euculture resembles the crossing of an activation threshold. The steps leading to the threshold become increasingly unlikely and short-lived, but once the threshold is crossed evolution accelerates. Progress is swift and certain. This imagery is not entirely contrived. The evolution of the human brain was in fact the fastest of any complex organ recorded in geologic history. Gene-euculture coevolution started when the *Homo* level was reached. Its outward appearance was that of an autocatalytic reaction, in which the products created hastened the process still further.

What intrinsic feature is likely to progressively stiffen resistance to neural evolution prior to the attainment of the breakthrough? Surely it is the cost of the eucultural brain measured in terms of factors such as energy and time; this cost increases with the capacity of the system, and in the absence of countervailing benefits acts as a severe load. Such a rise in cost with increased behavioral flexibility might at first seem unlikely. It must be the case that cognitive schemata precise enough to maintain high levels of selectivity cost a great deal in production and maintenance. An organism might increase its genetic fitness simply by disposing of this filter and becoming a generalist. But the exact opposite is the case. The organism deprived of automatic guides cannot choose blindly. It cannot be a tabula rasa. It must make the correct choice at each of many crucial moments, a capacity achievable only with a whole new array of cognitive machinery, including the neuronal circuits of memory, valuation, and decision making, and even abstract concept formation.

The distinction between the two strategies is nicely illustrated by the various adaptations that utilize olfactory behavior. Many insect species respond automatically to the odor of particular sex attractants, alarm substances, and other pheromones. The physiological requirements of this specificity are few. The male silkworm moth (*Bombyx mori*) makes a sexual response only to the chemical bombykol, and especially to one of the four geometric isomers of bombykol (*trans*-10-*cis*-12-hexadecadienol), a pheromone emitted by the female silkworm moth. This screening is entirely peripheral. The male catches the molecules on ten thousand specialized sensory hairs on each of his two feathery antennae. Each hair is innervated by one or two receptor cells that lead inward to the main antennal nerve and ultimately through connecting nerve cells to centers in the brain (Schneider, 1969). When we turn to mammalian olfactory behavior, as exemplified by territorial marking in the domestic cat, a whole new level of cognitive organization is encountered which is context specific. Cats do not just respond to the pheromones of other cats. They learn to recognize individual scent mixtures in the signposts of their rivals and to judge the whereabouts of the other animals by the degree to which the scent has faded. They then employ this information in deciding their own patterns of movement (Leyhausen, 1965).

The greater subtlety of olfactory discrimination and virtually every other category of behavior in the cat has been obtained at the price of an increase in the number of neurons of no less than four orders of magnitude, from 10^5 in the insect to 10^9 or more in the cat. The human brain, which may be the minimal instrument for reification and thus euculture, is even larger and more complex, containing approximately 10^{11} neurons connected by 10^{15} synapses (Crick, 1979). About 20 percent of the brain is devoted to speech and language (Jerison, 1975), which are principal channels of euculture. This portion comprises widely separated areas of the cortex that must cooperate swiftly, precisely, and continuously during most events of communication and conscious thought. In the course of fetal development the brain gains neurons at the rate of hundreds of thousands a minute. The possibilities for error are enormous during this growth and during later maintenance of the full-scale product. It is not surprising that neurologists and psychiatrists have compiled a long catalog of genetic and environmentally induced pathological syndromes. As Kety (1979) has said, "The wonder is that for most people the brain functions effectively and unceasingly for more than 60 years."

The attainment of high degrees of selectivity by flexible epigenetic rules incurs other costs. Judgments based on memory require long learning periods. In social species parental investment is an added requirement. Both forms of commitment reduce the immediate reproductive rate, because the parents must spend their time rearing the young instead of breeding. The loss must be compensated for by an increase in life span and higher juvenile survival rates.

As indicated in Figure 7–7, we postulate that the brain cost in genetic fitness rises as a monotone function of behavioral plasticity. In the formulation presented, the flexibility of an epigenetic rule is defined as the capacity of the rule to handle a variety of contingencies equally well. In other words, it measures the number and complexity of schemata, decision options, behavioral programs, and cognitive heuristics that the system can apply to different situations while maintaining a consistently high level of "correct choice." A value near MIN indicates a minimal flexibility and means that the species is genetically determined to choose only one specific culturgen. The closer the approach to MAX, the more flexible the response. Because the contingencies have to be evaluated by cognitive mechanisms, which require in turn the investment of neurons, time, and energy, the cost of the response is expected to rise with its flexibility—that is, with the precision with which an increasing number of contingencies are met.

On the other hand, the benefits of increased flexibility are not likely to be monotone in all cases. If they were, each behavioral category would tend to evolve either to complete flexibility or to complete specificity, as exemplified in Figure 7–7A and B. Although the evolution of behavioral repertories of most invertebrate species is dominated by the pattern of B, leading to rigid "instinctual" responses, some of the categories are intermediate in flexibility. These include the choosing and memorization of most sites, foraging pathways, and (in the case of social insects) the odor of the mother colony. In vertebrates, especially in man, intermediate degrees of flexibility in both social and nonsocial learning are commonplace. We can therefore reasonably postulate the existence of nonmonotone

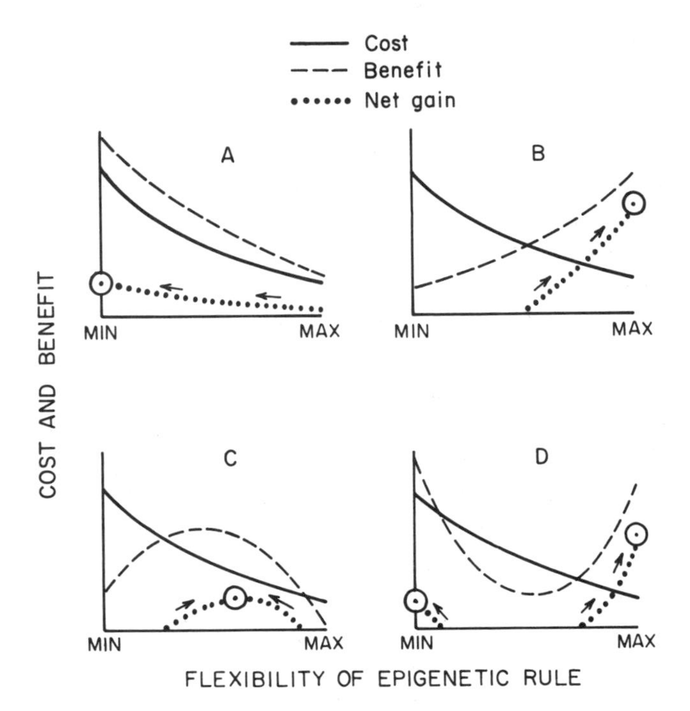

Figure 7–7 A hypothesis explaining the evolutionary rarity of human-level euculture. The cost in energy, developmental mistakes, and maintenance errors rises monotonically with epigenetic flexibility, that is, as the flexibility values increase from MIN toward MAX. As illustrated in *C* and *D*, any one of a variety of nonmonotone benefit functions can fix optima at higher levels of epigenetic flexibility.

benefit functions of the kind illustrated in Figure 7–7C and D. The shapes of these functions are likely to be particular to the social structure and environment of the species and perhaps can only be inferred from close empirical studies of the kind now routine in other topics of behavioral ecology and sociobiology.

We view *Homo* as an evolving genus that beat the odds. It overcame the resistance to advanced cognitive evolution by the cosmic good fortune of being in the right place at the right time. The ancestral species had certain preadaptations, including bipedalism, freed hands, a moderately complex brain by nonhuman standards, and a loose primate social organization, which were put to service in the exploitation of a rich food source on the tropical savanna. The combination added sufficient advantage to the congealing effects of reification (see Wilson, 1978, for a review) and to cognitive flexibility to outweigh their cost in natural selection. The eucultural threshold could at last be crossed.

Is Culture Necessary for Civilization?

In addressing the question of why euculture is so rare, we should expand the inquiry to ask whether the highest state it engenders, civilization, requires gene-culture coevolution. Does civilization even depend on culture? This is far from a frivolous question. Let us define civilization as a very advanced form of social existence, based on a written language, art, religion, and high technology. If we characterize it by these accomplishments and do not define it tautologically as the product of culture, we come up with an interesting new perception. Culture does not appear to be necessary for civilization. It is possible to imagine a world in which all of thought and behavior is anatomically preprogrammed in the brain, right down to the exact order of words used in complicated sentences. The use of language, construction and employment of tools, and economic transactions are context sensitive but predetermined in form. All that is learned are specific places and contingencies. A cantata may be sung at a festival, stirring feelings of deep pleasure through the audience, yet be entirely innate to the last note and inflection. The scientific report of a probe to a distant planet may be delivered at an international congress. The information is new, but the concepts and terminology used to describe it are genetically inherited and invariant.

It is possible in fact to envision three very different ways in which hypothetical brains can evolve a capacity for civilization. The first is by *complete hardwiring*, as suggested in the example just given. Conceivably the behavioral programs, semantic networks, and knowledge structures are already in place; such is the arrangement one would expect if termites had grown to immense size and produced a civilization. Or else there is a culture to be learned, but the transmission is of the pure genetic type (recall Figure 1–3), with individuals having the capacity to assimilate only one culturgen or a narrowly defined set of culturgens in each category. Complete genetic determinism can include a sensitivity to context, so that for every environmental contingency E there is one and only one behavior B. The totality of such a repertory gives the outward appearance of psychic flexibility and even "free will," but in fact the responses are not subject to modification by experience or reasoning.

A second mode of psychic development is of course *cultural transmission*. In pure cultural transmission, it will be recalled, the mind is not epigenetically biased or directed in any respect toward the culturgens it adopts; the genes have lost control of the behavioral repertory to the environment. In the much more likely intermediate case of gene-culture transmission, there is an innate bias in the selection of culturgens, and the genes and environment participate jointly in cognitive development. Looked at another way, the genes have accepted culture as a partner; they have off-loaded part of the information to another form of transmitting system. In human evolution, which is based on gene-culture transmission, social learning was a brilliant epigenetic coup that vastly extended the modes of operation and information capacity of the brain.

Finally, one can conceive of a level of programming intermediate between complete hardwiring and gene-culture transmission, driven by a *selectional,* rather than an *instructional,* mechanism. The procedure, which can be called a *schema activating system,* is still genetically fixed but permits a flexibility of response to the environment. The brain develops a large but finite number of neural assemblies that have the capacity to develop into cognitive schemata. When such a system operates in the gene-culture mode, a particular culturgen or class of culturgens triggers the maturation of one of the assemblies into its predestined schema, by means of cell growth and differentiation, synapse formation and modification, or the establishment of fields of electrogenic activity. Once the schema is chosen in this manner, the primordia of other related schemata are blocked. The phenomenon is an extreme form of the "recognizer neuron" activation hypothesized by Edelman (in Edelman and Mountcastle, 1978). Our purpose in reviewing it here is to point out that such a brain system, if it existed, would leave the genes firmly in control of knowledge. It could exclude social learning altogether, since endogenous triggers could activate specific schemata automatically.

Thus there are at least three conceivable kinds of brains capable of sustaining civilization, two of which can dispense with culture altogether. However, such a conception has little meaning and cannot be used to understand human evolution unless it is converted into a more rigorous theoretical formulation. To proceed let us first recognize a "civilization niche" of which mankind is the sole existing occupant of which we have knowledge. The formal problem can be stated precisely. We ask whether it is possible for genes to hardwire brains that contain a mature, eucultural complement of chunks, schemata, and knowledge structures. In other words, could the neural circuitry storing eucultural knowledge be laid down by a reasonable eucaryotic genome, coupled to an epigenetic system that does not rely on social learning? Or is it necessary for genes to invent culture in order to invade the civilization niche? In short, are genes and culture partners of convenience or partners of necessity? Was the human solution the only one possible?

The answers to all of these questions hinge on the *epigenetic compressibility* of the brain structures that in human civilization embody eucultural knowledge. Tissue structures with a high epigenetic compressibility can be built by relatively compact systems of epigenetic rules and small amounts of genetic code. Although the complete structural blueprint or description of such a finished tissue might be extremely large, the developmental mechanism that builds the tissue is fully described by a very small blueprint. In contrast, tissues that are low on the scale of epigenetic compressibility require lengths of genetic code and systems of epigenetic rules with a blueprint as complicated as that of the tissue itself. Highly regular structures built from repeating modules or subunits typify compressible systems. Systems whose components are assigned by lottery and thus manifest large amounts of idiosyncratic information are classic examples of incompressible systems.

Although kin networks, exchange relationships, mating and alliance systems, together with other forms of rule-based human behavior, often have simple and striking symmetry properties (see for example Lévi-Strauss, 1969a,b; Wallace, 1970), the compressibility of the associated neural circuitry is unknown. Given the absence both of data and of adequate theories of tissue epigenesis, we are obliged to rely on modeling techniques that are relatively weak.

In the analysis to follow we shall use information theory, as first suggested by Bremermann (1963) but now applied to a more thorough set of neural and cognitive structures. We caution the reader that the method ignores the possibility that the epigenetic blueprint may be much smaller than the structural blueprint. Indeed, the genome is not a description containing a microscopic picture of the finished organism; rather, it acts as a code that feeds and generates a rule-based assemblage of epigenetic constraints and developmental algorithms (see our overview of brain development in Chapter 6). Such rule-based epigenesis has great power over compressible structures, whereas our approach is most accurate for systems that are incompressible. Nevertheless, the arguments err on the side of overestimating the genetic requirements. Thus they are useful for mapping a first crude outer envelope of the evolutionary possibilities.

We shall show that if the epigenetic blueprint is indeed as complicated as the structural blueprint from which neurons and schemata can be read directly, it is not possible to fully program a human-sized brain and language structure with genes alone. It is possible, however, to prescribe some knowledge structures, as well as rules in the growth and differentiation of neurons, that reduce the amount of genetic specification required to some incomplete level. This amount cannot be exactly estimated, owing to the lack of an adequate developmental theory of brain structure.

The haploid chromosome complement of man contains approximately 2.9×10^9 nucleotide pairs (Dobzhansky et al., 1977), comprising on the order of a hundred thousand structural genes. At each nucleotide site there can exist any one of four bases, so that the information per nucleotide pair is $\log_2 4 = 2$ bits. An upper limit to the amount of information in the genome of one human being is therefore on the order of 10^{10} bits. The human genome is typical of most animals, especially the mammals. Many plants, as well as salamanders and some nonteleost fishes, have a haploid content between 10^{10} and 10^{11} nucleotide pairs. Much higher contents, say 10^{13} or 10^{14}, seem unlikely because of the difficulty of housing and maneuvering the requisite large bulk of nucleoprotein complexes within cells of conventional size.

There is no fundamental difficulty in specifying a versatile culturgen processor with the amount of information existing in human and animal genomes. With 10^9 bits, for example, it is possible to specify a processor that sorts 10^8 sensory signals into 10^8 culturgens. A comparable potential exists in the specification of epigenetic rules. In the special case of completely specific rules, which assign assimilation probabilities of 1.0 to the favored culturgen and 0 to the others, the number of culturgens that can be managed is in fact exactly equal to the number of nucleotides. Thus the genomic size, even when interpreted in the narrow information-theoretic sense, is not confining on the more stereotyped behavior evolved by most kinds of animals or even on the protocultures of more intelligent forms such as macaques and chimpanzees.

However, the genome does at first appear to be severely limiting in the specification of neural circuitry. The curves of Figure 7–8 suggest that a human-sized brain cannot be totally specified by a straightforward transfer of information from nucleotides to neurons. The problem is exacerbated by the existence of complex differences among the neurons. In most cases the cells integrate information received from many other specific neurons. The information subsequently emitted by the cell is not a simple on/off signal. It is enriched substantially by variation in the fre-

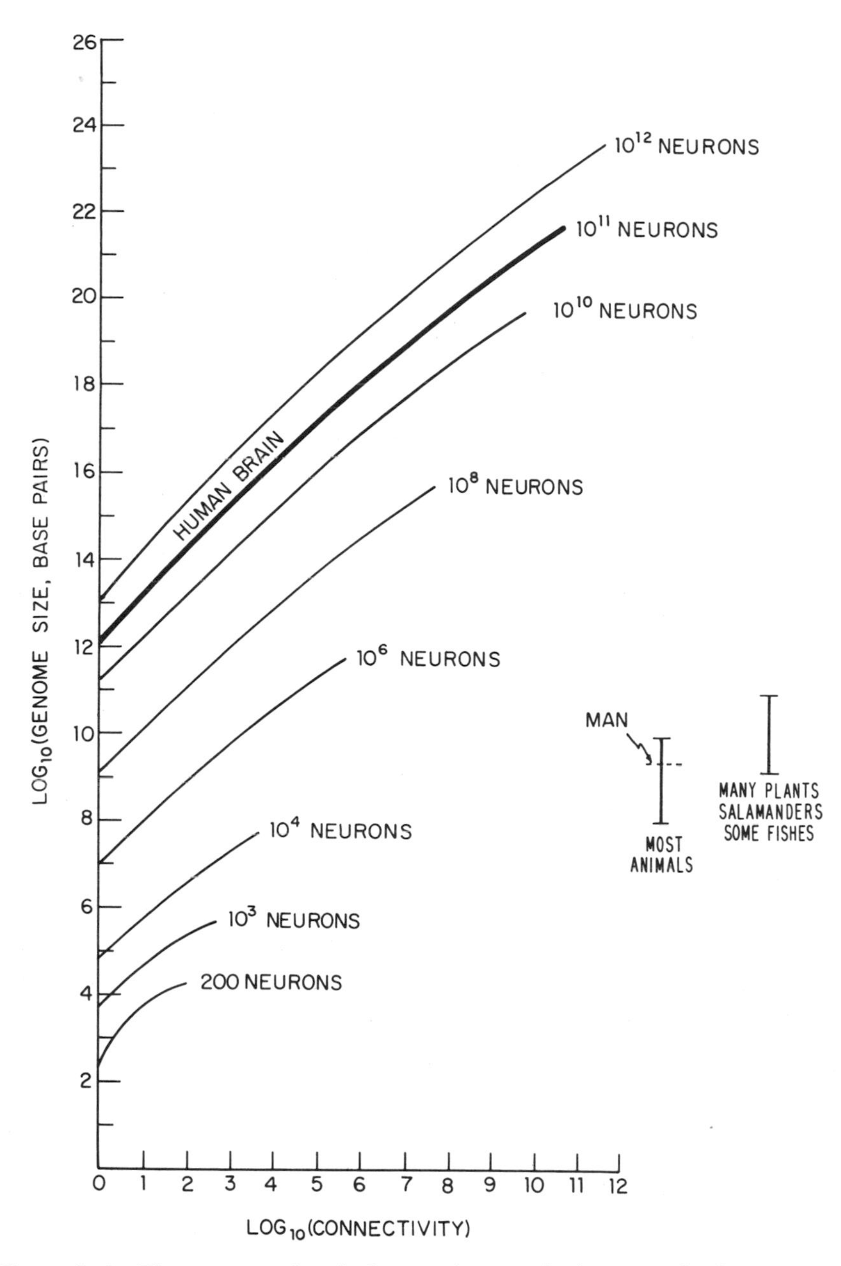

Figure 7–8 The genome size, in base pairs, required to contain the structural description of a hardwired brain structure with a specified number of neurons each having the same connectivity (number of neurons connected to the neuron under examination). The human nucleotide content and the amount of information needed to specify the human brain are indicated.

quency and duration of the action potential bursts, and by local processing based on passive waves. Thus the neuron is a unit that communicates with its neighbors in an individualistic manner.

As suggested in Figure 7–9, the size of the genome does not appear to preclude the full specification of hardwired knowledge structures. It is conceivable that a system could be constructed that matches the accomplishments of human long term memory. To see why this is true, consider that there are approximately 3.15×10^7 seconds in a year, so that during a generous life span of 100 years, 3.15×10^9 seconds at most are available to assimilate new information. At a minimum of 10 seconds per new con-

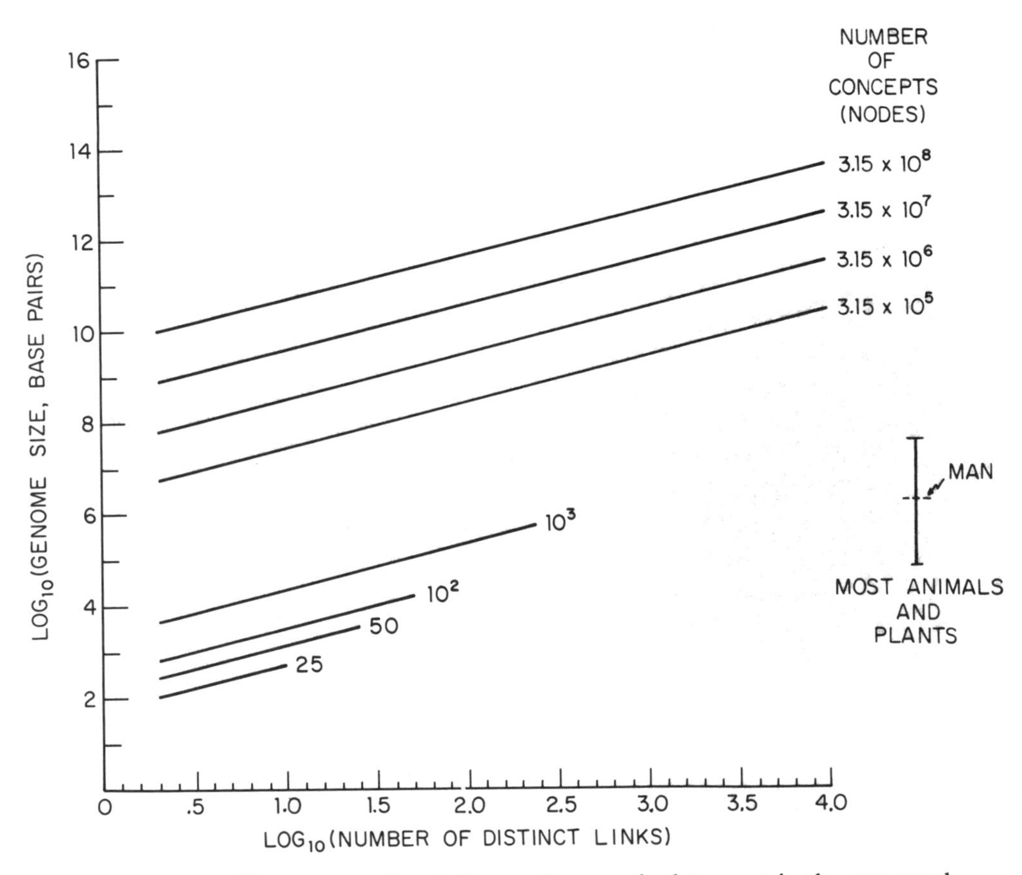

Figure 7–9 The genome size, in base pairs, required to contain the structural specification of a knowledge structure or semantic network consisting in *N* concepts. In this illustration each concept is joined to exactly *M* others by *M* distinct links, or relations of meaning, with no direct self-reference.

cept or concept chunk (see the description of the 10*J* Rule in Chapter 3), this is equivalent to a final knowledge structure in long term memory containing 3.15×10^8 chunks if the individual does nothing but learn new concepts. If 10 percent of the life span is devoted to learning, 3.15×10^7 chunks are assimilated. If a more modest 0.1 percent of the life span is used, 3.15×10^5 chunks are assimilated.

However, as shown in Figure 7–10, the grammatical and semantic structure of language at the human level does exceed the capacity of the genome to encode it. Although a very large number of symbols can be encoded separately, there is a strict limit to the length of the sentences that can be innately programmed in such a way that the truth value of each sentence is specified. For example, to possess a completely inborn vocabulary of 10,000 words and to speak in sentences of 10 words each would require a truly astronomical 10^{40} nucleotides, or 10^{16} kilograms of DNA, far more than the weight of the entire human species!

We have emphasized that these results concerning the constraints seemingly imposed by the size of the genome are nevertheless compromised by an assumption implicit in the analysis. This is the visualization of the genome as specifier of tissue structures, rather than as the ultimate source of epigenetic *rules* that guide development. It is known that neither genes nor the nucleotides constituting them specify neurons on a one-to-one basis. The epigenetic rules create fields and gradients within which neurons grow and proliferate (Bullock et al., 1977; M. Jacobson, 1978a). As we showed in Chapter 6, single mutations can alter entire zonal patterns and circuit structures entailing large populations of cells within the central nervous system.

The main remaining theoretical problem is therefore the epigenetic compressibility of the minds possessed by adult human beings. We need to know exactly how much specificity at the level of the neurons can be and need be condensed into the genes. Is it possible or necessary to accurately grow every one of the approximately 10^{11} neurons of the human brain with the information in 2.9×10^9 nucleotides? To do so would be to lower the information curve of the human brain to the level of only 10^6 neurons specified in a simple information-theoretic manner by the available nucleotides in the genome (see Figure 7–8).

Although this question cannot be answered fully at the present time, we can establish that the set of human-size brains attainable strictly on the basis of the human genome is vanishingly small. Consider the curve in Figure 7–8 that refers to the human brain. An information requirement equal to 10^{14} bits falls on the lower portion of this curve. There are $2^{10^{14}}$

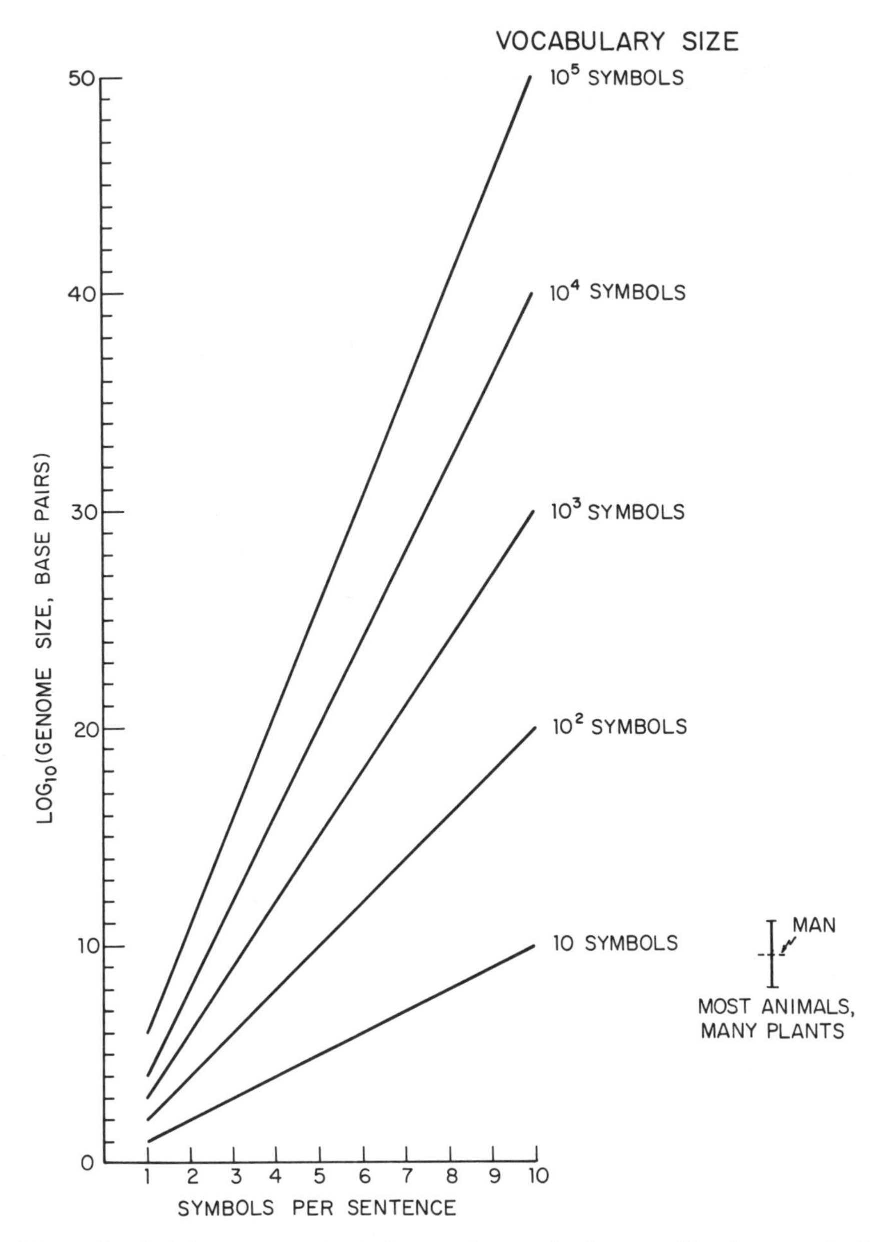

Figure 7–10 The genome size in base pairs required to specify a language built from N symbols (for example, words in a vocabulary) and joined in sentences containing M symbols each. Each sentence is classified into one or the other of the four truth values "meaningless," "undecidable," "false," and "true."

brains of the type dealt with in this diagram that have such an information requirement. Let us estimate what proportion of this set of brains can be built using genomes no larger than 10^{10} bits.

The following thought experiment will allow us to perform the calculation. A zygote is allowed to retain all of its intracellular machinery except its chromosomes; it therefore stands ready to transcribe any "model genomes" that are introduced into its nucleus. Obtain a large set of these zygotes. Into the first zygote inject a genome consisting of exactly one base pair. Into the next introduce a genome consisting of a different base pair. Continue in this manner until all genomes equal to one base pair have been used, one each to a cell. Repeat the implantation procedure for all genomes composed of two base pairs, three base pairs, and so on up to 10^9 base pairs.

In each case we allow development to proceed under normal physiological conditions but without social or observational learning. The end result is observed. Most often epigenesis will abort rapidly and completely, but occasionally a hardwired brain possessing a size and complexity in the human range may result. Our protocol allows us to enumerate all such possibilities.

At this point in the argument we introduce an overestimate that will serve to emphasize the startling nature of the epigenetic possibilities. Let us suppose that *each* "model genome," from one base pair to 10^9 base pairs, generates a 10^{14}-bit, hardwired brain. The number of such brains is

$$4 + 16 + \ldots + 2^{2\times 10^9} = \sum_{n=1}^{10^9} 2^{2n}.$$

The sum of this geometric series is $4 \cdot (4^{10^9} - 1)/3$, so the ratio of the number of compressible hardwired brains to the total number of 10^{14}-bit brains is *at most*

$$\frac{4}{3} \cdot \frac{(4^{10^9} - 1)}{2^{10^{14}}} \sim \frac{2^{2\times 10^9}}{2^{10^{14}}} \sim 2^{-10^{14}} \sim 10^{-3\times 10^{13}}$$

This number is almost inconceivably small. The length of the string of zeros that we would need to print it in the form 0.000 . . . 01 would stretch from the earth to the moon and back sixty times. Results of this nature are furthermore not peculiar to brains and genes. They appear to be part of the universal reciprocity between structural and algorithmic complexity. (Chaitin, 1975, Davis, 1978, and Simon, 1979, provide introductions to these subjects.)

The conclusion we draw from such numbers is that virtually no brains with size and structural complexity in the human range can be hardwired by human-sized genomes. In fact, a repeat of the preceding argument shows that almost all 10^{14}-bit brains require genomes at least (10^{14} – 10) bits in size, orders of magnitude beyond human limits. Thus in the universe of brains, those that are epigenetically compressible to any significant degree are very rare. The ratios are nevertheless nonzero, and the calculations do not preclude the existence of 10^{14}-bit brains that can be compressed into 10^{10} bits of genome. Likely cases, based on very regular and highly repetitious structures, are easy to imagine. But the numbers do indicate their scarcity and give some suggestion of both the extraordinary interplay of historical circumstances and the stringent developmental controls that would be necessary to achieve and maintain them.

Perhaps, then, a genome of the size possessed by human beings and other organisms could be large enough to encode a fully hardwired knowledge structure of the size acquired by learning during the human lifetime. But would such a knowledge structure function effectively in the human adaptive niche? We doubt it. A diagnostic feature of the human social condition is the idiosyncracy of the information that human beings use with skill. The sudden appearance of new opportunities, the identification of previously unsuspected rivals or allies, the accommodation to unexpected disasters, all exemplify the degree to which novelty and heterogeneity permeate the details of human knowledge structures. In a relatively static world with predictable features, innately programmed knowledge structures might pay off in fitness. Idiosyncracy is far less compressible. Its presence and exploitation call for epigenetic rules tuned to the underlying invariants and embedded features relevant to reproductive success, with the details free to vary.

We therefore tentatively conclude that genes and culture are partners of necessity, that a purely genetic, hardwired human civilization is conceivable in imagination but not in practice, as long as the genome is based on DNA. The degree to which dependence on culture can be reduced by incorporating epigenetic rules in the development of the brain cannot be estimated accurately at this time, because there is no adequately explicit theory of developmental biology addressed to cellular and organ differentiation (Ebert and Sussex, 1970; Wessells, 1977; M. Jacobson, 1978a). When such a formulation becomes available, it will be interesting to try to combine it with models of gene-culture coevolution to plot the most likely routes into euculture and the civilization niche, and to learn whether the human species has been "typical" with reference to the predicted trajectories.

Summary

The concept of equilibrium cultural diversity has been introduced. In order to evaluate steady states as well as to deduce rates of change at various nonequilibrium diversities, we have viewed the collective minds of a society as an "archipelago." Using the procedures of island biogeography, we envisage new culturgens being adopted and old ones lost or placed in passive store at rates that depend on the current culturgen diversity level (Figure 7–1). A regular increase in diversity with population size across societies, the area effect, can be predicted and documented. The decrease in diversity as a consequence of isolation, the distance effect, can also be projected, but data are inadequate to measure its magnitude.

The analogies to island biogeography also permit an evaluation of the turnover rate of culturgens, as well as estimates of the expected duration time of individual culturgens in various categories. It is possible to recognize not only similarities between culturgenic and biotic assemblages but also important differences, especially in the case of the organizational traits of societies. A fuller consideration of these differences is likely to enhance rather than diminish the value of biogeographic analysis in the study of cultural diversity.

Culturgen packing is another key property affecting the richness of cultures. The immigration and extinction rates of culturgens, and hence of cultural diversity within individual societies, are determined by the number of culturgens that can be packed into the same cognitive system. This property in turn is the result of three capacities in cognition: the ability to discriminate among culturgens belonging to particular categories, categorization and recall in long term memory, and valuation of the stimuli associated with each culturgen. Examination of these processes must be the first step toward the ultimate goal of characterizing the packing process. We have paid particular attention to discrimination, since it has been the best characterized in experimental psychology. Models are developed in which it is possible to estimate the optimum degrees of ambiguity in signal reception and the numbers of signals clustered into single culturgens.

Valuation is related to culturgen packing by considering the number of stimulus ensembles that can be linked together in stable configurations through value learning. The larger the linked ensembles, or "polymers," the greater the potential cultural diversity.

The study of diversity leads back to the question of the uniqueness of human-level euculture. We have considered simple hypotheses concerning ecological specialization only and rejected them on the basis of evidence from comparative zoology. We believe it necessary to seek internal constraints on the origin of euculture as well. From a priori considerations it appears that the cost of cognitive processing in terms of genetic fitness is a monotone rising function of cognitive and behavioral flexibility. If this conjecture is true, it follows that the benefit curves are nonmonotone (Figure 7–7). They are likely to vary greatly in form according to social structure and environment but to possess general properties that make the evolutionary pathways to advanced flexibility few and relatively inaccessible.

Finally, we have raised the theoretical question of whether the level of civilization created from human euculture might be duplicated entirely by genetic programming. In other words, we have explored whether genes and culture are partners of convenience or partners of necessity. Using a basic argument from information theory, we show that if the genome is regarded as a blueprint from which the structures of neural circuits and cognitive schemata are specified directly, it is not possible to program a human-sized brain and language structure solely with the amount of DNA found in human beings and other organisms. It is nevertheless possible to prescribe some knowledge structures, as well as rules in the growth and differentiation of neurons, that reduce the amount of genetic specification to some intermediate level. This amount cannot be exactly estimated at present, because of the lack of an adequate developmental theory of brain structure.

CHAPTER EIGHT

Gene-Culture Coevolution and Social Theory

The theory of gene-culture coevolution is an extension of sociobiology that creates an internally consistent network of causal explanation between biology and the social sciences. It is designed to include all cultural systems, from the protocultures of macaques and chimpanzees to the euculture of human beings, as well as forms of culture hitherto conceived only in the imagination. Hence we have spoken of the goal of a comparative social theory, within which the study of human behavior is embedded.

The success of human sociobiology, of which gene-culture theory is a part, will depend on its capacity to perform three services. First, it must derive rigorous propositions that are the unexplained axioms of other theories in the social sciences. Second, we require that it achieve a level of predictiveness and testability greater than that provided by other modes of explanation, or at least that it subsume the exact phenomenological models of disciplines such as economics and anthropology so as to make the underlying assumptions identical. Finally, it must suggest new questions and problems, as well as identify previously unknown parameters and laws to be woven into a network of verifiable explanation from genes through the mind to culture.

Epigenesis

The pivot of gene-culture theory is epigenesis. Traditional sociobiology has encountered difficulties in attempting to treat the transition from genes to culture as a "black box," where there are in fact three steps: from genes to epigenesis, from epigenesis to individual behavior, and from individual behavior to culture. The first two steps lie primarily within the domain of the brain sciences and psychology, the third step in population biology and the social sciences. The epigenetic rules channel the development of individual behavior according to the prescription of gene ensembles inherited by single organisms. The behavior of many individuals creates cultural patterns, which are characterized in elementary form as the ethnographic probability distributions. The behavior of individual members in particular cultural settings determines their survivorship and reproduction, hence their genetic fitness and the rate at which the gene ensembles spread or decline within the population. With appropriate techniques the reciprocating processes of gene-culture coevolution, gene-culture translation, and genetic feedback can be defined and their interactions measured.

We have established that no sharp line can be drawn between genetic and cultural evolution. Paradoxically, the distinction becomes even less clear and useful as the analysis of gene-culture coevolution gains in precision. Epigenetic rules can be rigid, resulting in an invariant single choice and a fully "genetic culture," or entirely unselective (at least across broad arrays of culturgens), resulting in what appears superficially to be a genetically liberated culture. Or they can be selective to an intermediate degree, the condition characterizing most categories of human cognition and behavior. Whatever the degree of selectivity, the epigenesis is prescribed by gene ensembles—even perfect indifference must be encoded genetically. Thus the link between genes and culture cannot be severed.

Learning

The theory of gene-culture coevolution can be used to examine nonsocial learning as a special, more primitive case. Even in the absence of contact with other members of the species, the individual processes information and chooses among alternative behaviors. Although its epigenetic rules do not impart any degree of sensitivity to choices made by other members of the population, they can operate in a manner otherwise basically similar to those of cultural species.

While the same principles underlie them all, epigenetic rules operating in cultural transmission presumably differ in many details from those in noncultural learning, and those that direct euculture differ from the rules of protoculture. To refer to the canalization of cultural evolution as biological is not to homologize it automatically with the development of animal behavior. The learning of language, reification, and disjunctive concept formation are biological in the sense of being subject to genetically underwritten epigenetic rules, but as far as we know these processes are distinctively human. Such learning can be analyzed by gene-culture theory, but not by direct comparison with animal species. The inseverable linkage between genes and culture does not also chain mankind to an animal level.

Complexity

The social sciences have been balkanized into a bewildering diversity of subdisciplines and schools of thought. If a single rationalization for this crippling state of affairs exists, it is the belief that human behavior is vastly more complex and subtle than anything encountered by the natural scientists. But human nature may be simpler and more transparent than we thought. If epigenetic rules are made the "molecular units" of learned behavior, the analysis of human behavior can be streamlined. Relatively elementary epigenetic rules generate greater diversity at the cultural level, while remaining obedient to complex genetic assemblages that prescribe relative simplicity in the sensory and neuronal responses. Under appropriate conditions simple rules can be employed to generate an arbitrary amount of detail in either direction.

Even the property of complexity can be turned to advantage. Vast integrative phenomena often emerge in the form of unexpectedly simple patterns, and large collections of units are more likely than small ones to be amenable to precise statistical description. The number of molecules in each cubic centimeter of air exceeds the entire population of human beings on earth by a factor of billions, yet physical theory can describe them with ease. Even when the systems have a history and the separate units behave in an idiosyncratic fashion, large numbers can often be handled with economy. Such historically constrained translation has been achieved, for example, in some of the central processes of population genetics, demography, and island biogeography where the units are genes, individuals, and species respectively. In Chapters 5 and 6 we showed how genetic heterogeneity and even limited forms of historicity can be built into the elementary models of gene-culture coevolution.

The crucial property of complexity per se is compressibility. A compressible system is one that can be described or regenerated by a set of rules or instructions much shorter than the shortest direct description of the system itself. It is further possible to recover the original system in detail. Much of science consists in codes that permit the folding of complex information into economical, easily retrievable forms. Science might even be defined, as Mach (1942) suggested, as "a minimal problem consisting of the completest presentation of facts with the least possible expenditure of thought." There is an additional reward. Time and again the search for codes has revealed a deep structure containing new phenomena and previously unsuspected connections to other systems. Many of the greatest advances have come when one discipline was mapped onto the axiomatic structure of another. Molecular biology, for example, has been united with quantum theory, Mendelian genetics with molecular biology, physiology with biochemistry, and population ecology with demography.

A worthy goal of the social sciences, then, is to discover the compressibility of human social phenomena. We suggest that in spite of its sometimes intimidating complexity much of the information is compressible, and that the codes we identify will reveal the deep structure of cultural evolution.

The Network of Causation

To this end let us consider the relationships between biology and the social sciences, beginning with the conception of gene-culture coevolution. Figure 8–1 presents the essential elements and processes in the sequence we envision. All of the steps have been separately documented, and with the exception of the reification rules their physical mechanisms are partially understood.

The brain is an organ whose construction is shaped by the genome of the individual human being. The information it receives about culturgens and other entities in the environment has already been filtered and organized to some extent by the peripheral sensory cells and coding interneurons. The cortex engenders consciousness, a reconstruction of reality consisting in the temporal sequencing of the neuronal signals that have encoded previous histories of sensory input. The sequencing can be rearranged arbitrarily to invent fiction and even to predict the future. The information is evaluated, and deliberated responses are keyed to it in accordance with the epigenetic rules. To some degree it is also translated into words and symbolic images under the influence of the reification learning rules. The channeling process is mediated in part by the limbic system, which serves as an "on-board computer" for assessing the ultimate genetic adaptiveness of contemplated actions. This process involves more than an ungeneralized will organized by elementary procedures of pure logic. It can be as automatic and complicated as the manner by which the cerebellum processes complex motor algorithms and feeds out-

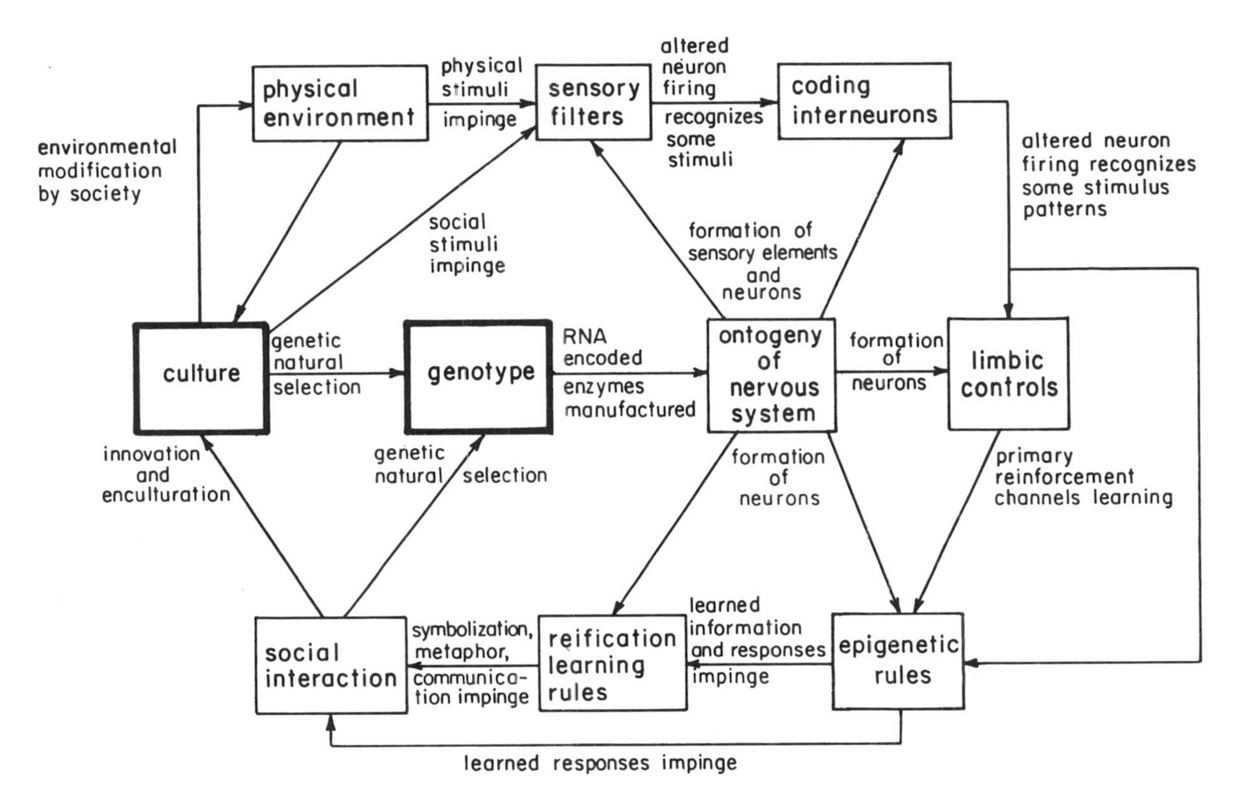

Figure 8–1 A diagram of gene-culture coevolution incorporating the essential relationships of the epigenetic rules.

put down the pyramidal tracts. The ultimate, evolutionary goals of the mind, toward which minute-by-minute problem solving is directed, reside in the epigenetic rules, and in that sense the core of both humanness and individuality are invested there rather than in the more purely cognitive and ratiocinating portions of the mind.

To complete this conception of gene-culture coevolution, the epigenetic rules interact with the signals received by the individual to convert that individual into a full cultural entity, a human being. Thus the mind is an active entity that shapes its own growth. The interaction of the members of the society create the culture, which together with the genotypes of the individuals and their relatives determine reproductive success. The degree of success in turn prescribes the gene frequencies over generations and thence the form of the epigenetic rules on which cultural evolution depends.

How is gene-culture coevolutionary theory related to the remainder of sociobiology? Traditional models from sociobiology have been concerned with neither the mechanics of the mind nor with the upward translation from genes to culture. Few of these models are in fact genetic at all. Instead, they apply basic principles of optimality in order to predict the behavior and the social patterns that should exist if human beings maximize their inclusive fitness. The predictions are rich and varied. Thus inbreeding avoidance is expected to be a general phenomenon, especially between siblings or other genetically close relatives. When males are strikingly larger or more brightly colored than females, the difference is usually part of a breeding strategy in which the males compete for many mates, while a close similarity between the sexes is ordinarily associated with monogamous bonding. When a size difference occurs, variation in its magnitude in comparing different species is a rising function of the average number of females inseminated by each successful male. Intraspecific aggression appears when other density-dependent factors of population growth regulation, such as predation and emigration, are absent or weak. An extremely common form of aggression is territorial defense, which is most likely to evolve around resources that are patchily distributed and predictable in occurrence.

Such responses are deduced as generalizations from the first principles of population biology (see for example Wilson, 1975; Chagnon and Irons, 1979; Symons, 1979; van den Berghe, 1979; Barlow and Silverberg, 1980; Lockard, 1980). They are evaluated by laboratory and field studies or, in the case of human beings, by cross-cultural ethnographic analyses. Traditional sociobiology is methodologically sound and has stimulated new forms of research. Yet it characterizes only the final behavioral product, which is to be prescribed in some manner by genes. The genetic foundation and the epigenetic mechanisms that give rise to these products are left largely unexplained.

In the case of insects and most nonhuman vertebrates, this vague conception of the linkage between genes and behavior perhaps leaves fewer mysteries. Most of the behavior is relatively invariant and species specific. We can safely assume that the pattern of territorial defense of an aeschnid dragonfly, consisting in a particular set of flight maneuvers over the margin of a pond, is genetically distinct from the defense of an agrionid damselfly, which maneuvers differently over the center of a marsh. But in order to have a real evolutionary theory of mind and culture, one must begin with genes and the mechanisms that the genes actually prescribe. In human beings the genes do not specify social behavior. They generate organic processes, which we have called epigenetic rules, that feed on culture to assemble the mind and channel its operation. Behavior is only one product of the mind as it deals with the contingencies of day-to-day existence. The mind chooses particular behavioral responses on the basis of knowledge structures that are the product of learning and of active reflection. Furthermore, there is no simple, one-to-one correspondence between culturgens and behavior. Although many culturgens are composed in part of overt behavior, they cannot be defined exclusively in terms of patterns of skeletal-muscular activity. They can be ultimately delimited as portions of knowledge structures. As such they contain fragments of information about culture, environment, and social action, and they constitute the raw material used by the epigenetic rules in the construction of the mind. This is the basis for our proposition that the epigenetic rules are the molecular units in the coevolution of genes and culture and the dividing line between the two levels of development—from genes to epigenesis and from epigenesis to culture.

The outcomes of the epigenetic rules of social behavior are abstractly characterized as the innate assimilation functions $v_{ij}(\xi)$, the rates at which individuals shift from one activity to another as a function of the surrounding culture. Fully instinctive patterns of valuation and decision constitute the special case in which there is no targeting of behavior plans and goal priorities by social or environmental learning. The gene-culture theory we propose recognizes that the values of $v_{ij}(\xi)$ in most higher vertebrates, and especially in human beings, are determined by a network of epigenetic rules operating on information created by learning and cognition. Instead of single fixed responses, the assimilation functions generate social patterns of behavior within which reproductive success is determined. The patterns evolve in response to particular selection pressures. Thus the key status of such phenomena as human incest avoidance, courtship practice, territoriality, optimum group size, kin toleration, and other basic properties of social organization is not altered, and it can be elucidated by a sociobiology that now incorporates gene-culture coevolutionary theory. We believe that this theory opens a new realm of sociobiology and through it introduces a mode of evolutionary analysis that will lead to a deeper and more precise understanding of human behavior.

Gene-Culture Theory and the Social Sciences

When cultural behavior is treated as an ultimate product of biology instead of an independent stratigraphic layer above it, the social sciences can be converted more readily into a continuous explanatory system. In fact, social behavior could prove to be substantially compressible. Although *Homo sapiens* is the most complex species on earth by a spectacular margin, it is probably far less complex and difficult to understand than contemporary social theory leads one to believe.

Consider ethnography itself. The rich details of individual cultures, viewed as closely as possible in the terms of the people who live them, are undeniably the essence of the social sciences. Goajiro pottery, beggars' argot, Balinese trances, women's fashions during the Restoration, and Tapirapé shaman training are all worthy of meticulous study in their own right. They deserve intense inquiry both at the emic level that interprets the feelings of the persons in the culture and at the etic level of the more detached, objective observer. This preoccupation with the particularities of culture is natural history, which is its own justification. But pure ethnography is often motivated by the world view of hermeneutics, which does require critical examination because of its implication for social analysis. In the original philological context, hermeneutics was the analysis of conflicting derivative texts to deduce the wording of the original manuscript. Now it is more broadly defined to be the set of epistemological and methodological devices by which the "true meaning" of the practitioner of a different culture can be clarified and rendered more congruent to a general consensus of what constitutes the truth (Bauman, 1978). Understanding of social phenomena grows by means of the "hermeneutic circle": rather than charting a straight course toward the most important and secure bodies of information, scholarship maneuvers through an endless recapitulation and reassessment of collective memories. During this process the social scientist is unavoidably engaged in a discourse with his own subject. In the extreme version of this philosophy of science, both observer and observed are considered to use the same resources, and truth consequently must be negotiated in the same manner as ordinary agreements.

The richness of ethnographic detail and the appeal of the hermeneutic circle as an alternative epistemology to conventional science tempt observers to emphasize phenomenology and induction. "We must descend into detail," Geertz (1973) wrote, "past the misleading tags, past the metaphysical types, past the empty similarities to grasp firmly the essential character of not only the various cultures but the various sorts of individuals within each culture if we wish to encounter humanity face to face." General principles can be achieved, but only after we have worked our way through the "terrifying complexity" of culture.

Such thick description, as Ryle (1971) called it, is necessary but not sufficient. We do not need to await the glacial approach of a pristine ethnology. Progress will be more rapid if the accumulating information is probed relentlessly with the aid of postulational-deductive models. The most successful models are likely to be based on premises from neurobiology, cognitive psychology, and population biology. Close historical parallels can be found in the emergence of modern ecology and biogeography. These subjects required a natural-history phase, in which species were examined "face to face," with a full appreciation of their idiosyncracies: plants in succession on the Indiana sand dunes, insects in the biomass pyramid of an English woodlot, birds migrating between Pacific islands, and so on. Nevertheless, progress resulted equally from postulational-deductive theory directed at both the general and the particular qualities of the species: for example, the convergence to species equilibria, general constraints on trophic levels, and the necessary conditions for species coexistence. Although most of the models have been applied only partially and still have limited accuracy, they are quantitative and are being improved steadily. Most significantly, they have stimulated bursts of new descriptive, "hermeneutic" research.

Many social scientists are fully committed to this dual procedure of scientific inquiry. They have sought the grail of a unifying postulational-deductive theory. Some of their most impressive attempts have been in the realm of the models relating climatic change, geography, and population growth to the rise of agriculture (Boserup, 1965; Flannery, 1965, 1973; Binford, 1968; Cohen, 1977) and to the origin of the state (Carneiro, 1970; Flannery, 1972; Stephens, 1979). It has been possible to quantify certain steps in these evolutionary processes and to compare them with optimum economic strategies deduced from a knowledge of local environments, as exemplified in the studies of Thomas (1971), Binford (1978), Gross et al. (1979), Hassan (1979), Howell (1979), Reidhead (1979), Kirch (1980), and Winterhalder and Smith (1981). Research of this nature is at the cutting edge of anthropology. It is supplemented by a relatively meager number of ethnographic correlations accumulated over the years, such as the relationship first noted by Steward (1955) between partrilineal, patrilocal, exogamous band organization with simple technology and limited or scattered food resources; and by broader "covering laws," such as the generalization by White (1949b) that culture evolves as a function of the amount of energy harnessed per capita and the efficiency of putting energy to work.

These enterprises are the expected first nomothetic reformulations of the natural-history data of the social sciences. They are parallel to phylogenetic analysis in biology—reconstruction, for example, of the stages of the emergence of the first tetrapods onto the land amid the drying of swamps, competition, and other environmental circumstances that might have impelled this adaptive shift. The generalizations contain elements of postulational-deductive theory. But they have fallen short in one crucial respect. They do not consider in any detail the development of individual motivation and hence cannot bridge psychology and biology. Motivation is conceived as a general, relatively unstructured force, and the form of translation has been left unspecified.

The need for a fuller sequence of causal explanation has been perceived by Marvin Harris (1979), who suggests an upward translation from what he calls the infrastructure (work patterns, demography, infant care, and comparable human properties) to the structure (division of labor, class structure, political organization, and so forth) and thence to the superstructure (art, music, rituals, sports, science). In Harris' view the psychobiological prime movers are very simple. They include the need to eat, especially higher-caloric food, the desire to expend less rather than more energy, the sex drives, and a need for love and affection. This particular formulation is said to be better than earlier explanations because it accounts for both differences and similarities. "For example," Harris writes, "the need to eat is a constant, but the quantities and kinds of foods that can be eaten vary in conformity with technology and habitat. Sex drives are universal, but their reproductive consequences vary in conformity with the technology of contraception, perinatal care, and the treatment of infants."

Such generalities, along with the models issuing from them, exemplify the real progress possible even with very primitive notions about the sources of human motivation. In their present form, however, such conceptions are not only incomplete but also seriously misleading. From analysis of the information available to us about human cognition and development, we have concluded that the human mind is far more structured than has been postulated by Harris and many others active in the theory of anthropology, economics, and sociology. This structure is based on epigenetic rules that are genetically prescribed, act on cultural information, are typically complex and selective, and cannot be inferred by unaided intuition. Gene-culture translation, comprising the feedforward from the individual to societal levels, is made explicit through a knowledge of these properties and generates verifiable predictions of differences among cultures in the form of statistical distributions. Genetic change in the epigenetic rules and assimilation functions, which conveys the feedback from the societal to the individual level, can be analyzed through the theory of population biology and sufficient information about the underlying cognitive processes.

It is a remarkable fact, with many ramifications, that the crucial processes of cognitive epigenesis have been explored by scientists working in three independent traditions. The neurobiologists and experimental psychologists have examined functions that are relatively simple and directly perceived. These functions, which include color perception, cognitive algebra, risk estimation, and comparable phenomena, are the ones most likely to be shared with other primate species and hence in that special sense to be considered more truly "biological." They are also more experimentally tractable. Structuralists, in contrast, have been largely concerned with the most complex, distinctively human processes. Chomsky, Brown, Lenneberg, and other psycholinguists have dealt with the basic qualities of language; Piaget with the stages of intellectual development that compose a "genetic epistemology"; and Lévi-Strauss, Barthes, Leach, and other ethnographic structuralists with the origins of ritual and myth. Finally, psychoanalysts have tried to characterize the ontogeny of even deeper, more elusive activities of the mind.

In spite of radical differences in the language used to describe them, these three modes of analysis are not fundamentally incompatible. All in effect attempt to define one or more categories of the rules of human cognition, somewhere along the chain of activities from perception through memory storage and recall to valuation and decision making. It is not surprising that the "hard" experimental approach of the neurobiologists and psychologists has not yet penetrated to deep grammar, myth, and other more subtle forms of mental activity. Structuralists and psychoanalysts are scientists endeavoring to approximate what we call the epigenetic rules of reification and symbolization. The phenomena they describe will eventually be investigated by more precise experimental techniques. Nevertheless, it would be a mistake to regard structuralism and psychoanalysis as independent theories or even as facets of a true theory in the ordinary sense employed in the natural sciences. They are simply descriptions, however deeply they cut. They do not yet formulate principles from a postulational-deductive framework. They have failed to suggest axioms that can be mapped onto the theoretical structure of other disciplines. They conceive of limited aspects of cognitive epigenesis in only one species, mankind, and by their nature cannot predict any form of measurable dynamics. Although fundamental to an understanding of the human mind, the phenomena of structuralism and psychoanalysis cannot be made part of science nor their role in shaping human history elucidated until they are incorporated into models of gene-culture coevolution.

The Explanation of History

This leads us to the question, Can human history be more deeply understood with real evolutionary theory? The ultimate triumph of both human sociobiology and the traditional social sciences would be to correctly explain and predict trends in cultural evolution on the basis of their own axioms. The most influential attempt to date has been that of Marx, who foretold the end of capitalism and the rise of a classless society. Modern Marxism has acquired many meanings beyond this original courageous forecast. It is the set of all radical critiques of capitalism, a prism, as Samuelson (1976) said, "through which mainstream economists can—to their own benefit—pass their analyses for unsparing audit." Marxism is still more significant as the emblem of revolutionary socialism. In this role it has become ideology, since a virtually unlimited number of propositions about historical process, true and untrue, Marxian and non-Marxian, could serve the same result, as long as the predicted end is socialism in a classless society.

The pure intellectual legacy of Marx is the method of analysis of social institutions. The modern Marxist seeks the deeper causes of historical change in economic process and class struggle; views contemporary societies as chaotic transitional states on the road to a more natural, harmonious order; and emphasizes the strictly material, dialectical nature of cultural evolution. Out of these basic conceptions has grown a rich mélange of textual analyses, apologetics, revisions, and specific applications to most disciplines within the sciences and humanities (Caute, 1967; Heilbroner, 1980). There is even a Marxian archaeology (Klejn, 1973), a Marxian psychology (Baran, 1959), a Marxian paleontology (Gould and Eldredge, 1977), and a Marxian biology (Rose and Rose, 1976).

Little agreement can be found among the adherents about what all this means. Present-day theoretical Marxism contains a vast array of exercises in exegesis and apologetics. For some scholars it is principally an exciting set of ideas to be developed; for others it is primarily the intellectual arm of revolutionary socialism. Neo-Marxism's central propositions are not advanced as hypotheses to be tested and possibly discarded. Rather, they are treated as core truths that must be actualized by those who accept them. Thus O'Laughlin (1975) spoke for most of her colleagues in making the following prescription: "To some extent then Marxist anthropology must be applied anthropology, the university and classroom a locus of political struggle, and praxis an essential aspect of verification." In other words, scholars should make Marxism come true.

In short, Marxism was a major development in intellectual history. It produced a mode of analysis that cut through the outer phenotypes of custom and institutions to examine the underlying economic forces. But is it a real theory of cultural evolution? Can it explain and perhaps even predict history with rigor and consistency? It clearly has not done so to the present time. Many socialist revolutions have occurred around the world, most under the banner of Marxism, but none has followed the script written by Marx himself. Their causes have been complex and diverse, entailing nationalism and ethnocentricity as much as class struggle. A closer examination shows that Marxism is not a scientific theory comparable in form to Darwinism and Mendelism, to take two other nineteenth-century creations. It is more like Lamarckism and orthogenesis, competing theories that correctly described certain outward features of evolutionary change but postulated erroneous underlying mechanisms. Darwinism and Mendelism led to accurate dynamical descriptions of heredity and evolution, but Lamarckism and orthogenesis did not and could not. In our opinion the key error of Marxism as a scientific theory of history is its tendency to conceive of human nature as relatively unstructured and largely or wholly the product of external socioeconomic forces.

Marx himself was equivocal on the issue of human nature. His collected writings contain statements that are often vague and contradictory, even mystical (Bloom, 1941; Wasserman, 1979), a circumstance that has resulted in endless controversy. Marx recognized the existence of a "generic man," with inborn drives characterizing the species as whole, as well as a "historical man," a more plastic portion of the mind that changes with the environment and hence with human activities. The components of generic man themselves changed erratically through the history of Marx's writings. In the 1844 Paris *Manuscripts* they included altruism, cooperation, empathy, and other specific qualities. Marx went so far as to speak of history itself as a part of natural history, and the inevitability of a union between the natural sciences and the science of man. Elsewhere, however, he spoke of history as a continuous transformation of human nature, concluding that "it is not the consciousness of men that determines their existence, but their social existence that determines their consciousness" (1971[1859]:21).

In fact, the writings of Marx contain no explicit characterization of mental heredity and development as corporeal processes. In this respect they differ fundamentally from the work of Darwin and Mendel. Followers of Marx were set free to develop their own image of human nature, and by and large they have chosen a biologically unstructured model that assigns maximum weight to the environment. In recent years this view has been contravened by the discovery of rich structure in the operation of the brain and in the development of social behavior, much of which has little to do with external socioeconomic forces. It is difficult to imagine a distinctly Marxian version of modern cognitive developmental psychology, human ethology, or population genetics. Incorporation of the findings of these disciplines into the Marxian conception of human nature would diffuse explanation to multifactorial models so decisively and shift emphasis of social theory so far from the isolated themes of class struggle and economic determinism as to bring Marxian scholarship close to the mainstream of Western social science and blend the two together. It would thus confront Marxian scholars with the exciting task of deepening their theories by means of a realistic, biologically fundamental picture of human nature. Indeed, we have noted that as Marxists such as Terray (1975), Godelier (1977), and Marvin Harris (1979) move away from philosophical critiques and reviews of intellectual history in order to address the substance of human behavior, their language and interpretation become remarkably similar to those of other anthropological theorists.

The original question can now be rephrased as follows: Will the social sciences, using all of the considerable resources at their disposal and designing ever more comprehensively multifactorial models, be able to explain history more fully and perhaps even to predict it with moderate accuracy? We believe that the answer is yes, at least on a very limited scale, in ways that can be enormously fruitful in scientific research and social planning. Part of the reason for our optimism is the newly analyzed implication of structure in the epigenetic rules. Because of these rules, the mind is a system that tends to organize itself into certain forms in preference to others, while the combined action of many minds seems to lead to the emergence of patterns of culture that are statistically predictable. This channeling action will be most conspicuous in those categories of cognition and behavior guided by epigenetic rules that are relatively inflexible and selective. An inflexible epigenetic rule is one that yields only a limited variety of usage bias curves in the face of environmental change. We need to remember that flexibility does not necessarily imply a lack of genetic determinism. A shift in bias curves can be orchestrated closely by the underlying epigenetic rules. Furthermore, each variant may prove to be adaptive in the particular environment that evokes it. Adaptiveness is only a hypothesis to be tested by economic and ecological studies. If upheld in a particular category, it can be explained more completely by reference to the full theory of genetic feedback in the evolution of epigenesis. In either case, whether the flexibility is adaptive or not, the epigenetic rules can be used to predict cultural patterns in the form of ethnographic probability distributions. When the epigenetic rules are both inflexible and selective, individuals favor a small fraction of culturgens strongly over others in *all* feasible environments, and as a consequence the cultural patterns they influence will change relatively little in the course of history.

Flexibility and selectivity are independent properties of mental development. It is possible for individuals to select one culturgen exclusively in one environment and then switch to the exclusive use of another culturgen in a second environment, thereby maintaining high specificity while revealing substantial flexibility. What matters is the pattern of choice and its consequences when translated upward to the societal level. We already know from studies on psychological development that the pattern varies greatly from one behavioral category to the next. Categories that are both inflexible and highly selective include basic color classification, sibling incest avoidance, many facial expressions used in nonverbal communication, and several properties of mother-infant bonding. Only with difficulty can individual development be deflected from the narrow channels along which the great majority of human beings travel. In most conceivable environments, and in the absence of a forceful attempt to produce other responses, these behaviors will persist as the norms of culture in most or all societies.

Our knowledge of epigenetic rules is rudimentary in the great majority of behavioral categories. Little can be said about aggression, religious beliefs, political institutions, and economic practice. Yet from a combination of research in the brain sciences and developmental psychology, aided by clues from ethnography, a more exact knowledge of antecedent developmental pathways and flexibility will eventually emerge. It should be possible to refine gene-culture translation models to explain both steady-state and dynamic cultural patterns in even the more subtle forms of behavior. Such translation can produce a useful retrodiction of ethnography—an inference about what has already occurred. In the case of epigenetic rules that are moderately flexible, the social scientist might also be able to predict short term changes in the forms of the ethnographic distributions.

To go beyond this elementary level, a great deal more empirical information is needed, as well as significant advances in the basic theory of gene-culture coevolution. The ontogeny of most forms of social behavior is not unilinear, as we depicted it in the first quantitative models of the more rigid categories. It is reticulate: development in one category of behavior affects development in other categories. Furthermore, the products sometimes blend together into confusing compounds. For example, it may well be that bigotry and group aggression stem from the interaction of several quite distinct behaviors, including the fear-of-strangers response, the proneness to associate with groups in the early stages of social play, and the intellectual tendency to dichotomize continua, including arrays of other human beings, into in-groups and out-groups. Although these elements are prominent in early development, they have not been studied in sufficient detail to measure their flexibility and possible interactions. Once reticulate development is more clearly documented, theory will be challenged to translate it into ethnographic distributions.

Moreover, the existence of mind means that human social evolution incorporates the assessment of history and of dimly perceived futures (Figure 8–2). An examination of the assessment process itself can reveal the full impact of cognitive epigenesis in the formation of cultures. A society that chooses to ignore the implications of the innate epigenetic rules will still navigate by them and at each moment of decision yield to their dictates by default. Economic policy, moral tenets, the practices of child rearing, and virtually every other social activity will continue to be guided by inner feelings whose origins are not examined. Such a society must

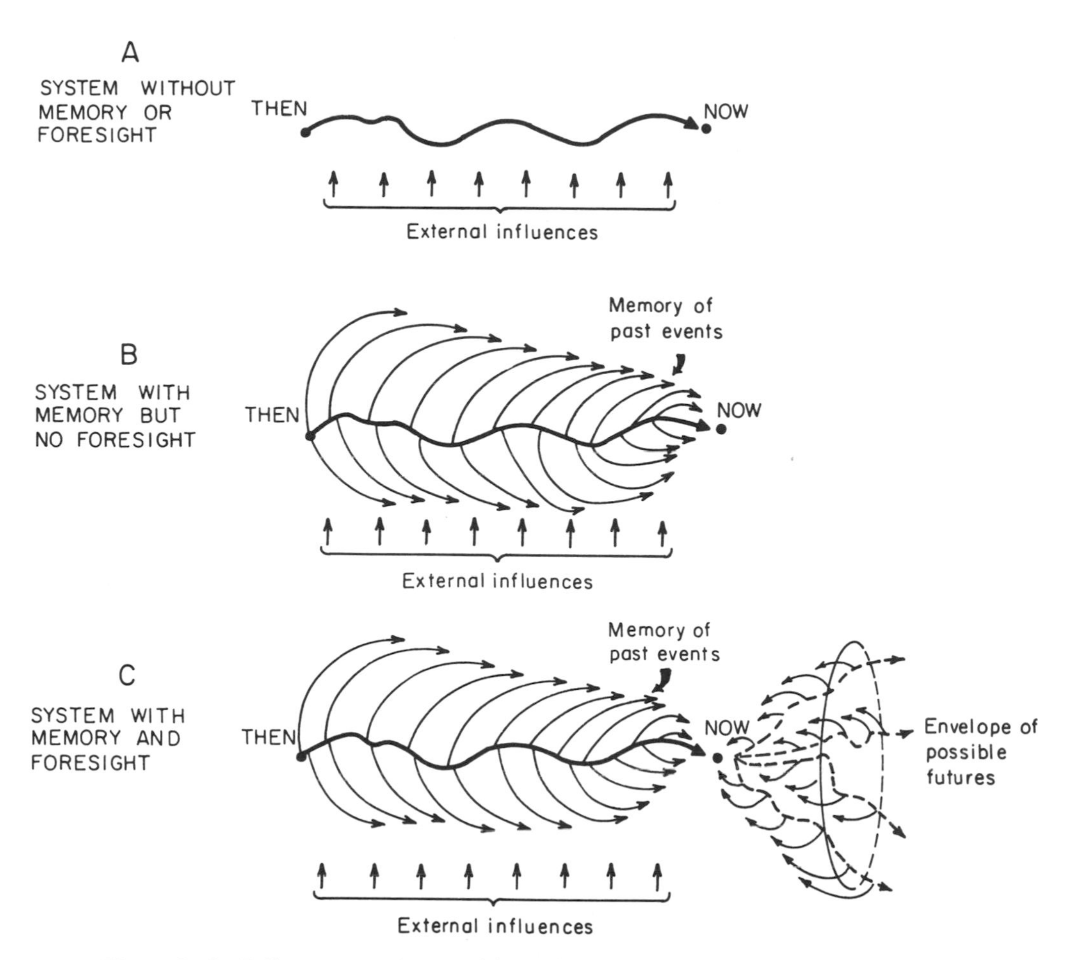

Figure 8–2 Influences on the cognitive epigenesis of an individual. *A*: In the simplest possible system, the development is affected by both internal and external events, but only from moment to moment. *B*: In more complex systems, it is influenced also by the memory of past events. *C*: In human development, the epigenetic trajectory is affected not only by external events and memory but also by anticipation of future events. In all cases the epigenetic rules act to structure and constrain the envelope of possible futures of both individual cognition and the histories of particular societies.

consult but cannot effectively challenge the oracle residing within the epigenetic rules. It will continue to live by the "conscience" of its members and by "God's will." Such an archaic procedure might lead in the most direct and untroubled manner to a stable and thoroughly benevolent culture. More likely, it will perpetuate conflict and relentlessly drag humanity along what is at best a tortuous and dangerous path. On the other hand, the deep scientific study of the epigenetic rules will call the oracle to account and translate its commands into a precise language that can be understood and debated. Societies that know human nature in this way might well be more likely to agree on universal goals within the constraints of that nature. And although they cannot escape the inborn rules of epigenesis, and indeed would attempt to do so at the risk of losing the very essence of humanness, societies can employ knowledge of the rules to guide individual behavior and cultural evolution to the ends upon which they agree.

Then, having accomplished so much, can we also hope to extend our understanding forward into future evolutionary time, when the epigenetic rules themselves and even the physical basis of mind may be altered? Dyson (1979) has written:

> Looking at the past history of life, we see that it takes about 10^6 years to evolve a new species, 10^7 years to evolve a new genus, 10^8 years to evolve a class, 10^9 years to evolve a phylum, and less than 10^{10} years to evolve all the way from the primaeval slime to Homo Sapiens. If life continues in this fashion in the future, it is impossible to set any limit to the variety of physical forms that life may assume. What changes could occur on the next 10^{10} years to rival the changes of the past? It is conceivable that in another 10^{10} years life could evolve away from flesh and blood and become embodied in an interstellar black cloud or in a sentient computer.

Nevertheless, the history of our own era can be explained more deeply and more rigorously with the aid of biological theory. And across a time span of a few generations, within the epigenetically constrained behavioral categories, the prediction of history is a worthwhile venture that can at the very least spur a deeper analysis of existing societies and their past history. We should keep in mind that most of the wondrous inventions of science and technology serve in practice as enabling mechanisms to achieve territorial defense, communication of tribal ritual, sexual bonding, and other ancient sociobiologic functions. Curiosity, even the artistic impulse itself, might also fill such a role.

These are some of the possibilities. The real prime movers, which permit and sustain such steps in cognitive evolution, cannot yet be identified with confidence. An exploration of such domains at the level of cognitive psychology and the neurosciences will almost certainly uncover new phenomena that can be translated into short term historical predictions with results beyond our present capacity to imagine. For evolutionary time intervals, gene-culture theory predicts a coevolution of epigenetic rules and cultural forms. In model worlds that to varying degrees approximate the real one, the envelope of possible futures can thereby be estimated and analyzed (Figure 8–3).

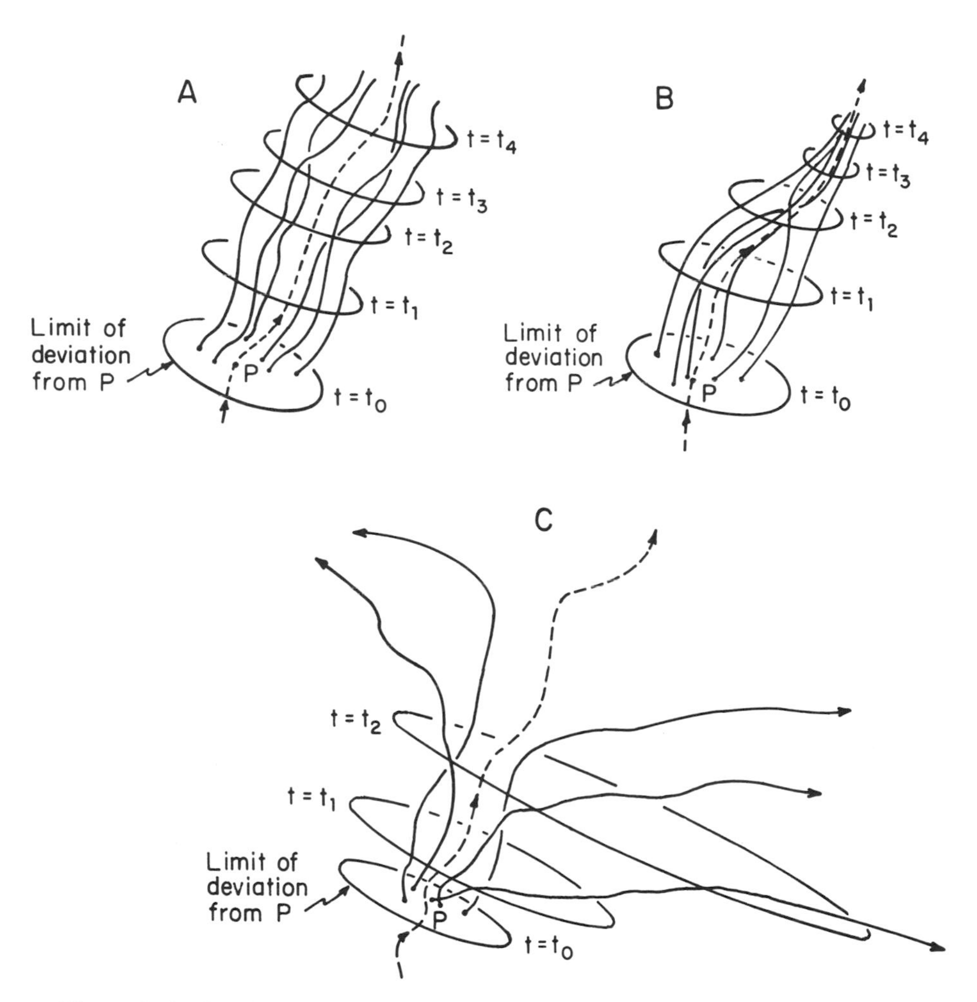

Figure 8–3 Parallel, convergent, and divergent gene-culture histories. *P* is the path or trajectory sought, while the disk about *P* models the maximum accuracy that can be achieved because of *observational* uncertainties about the system's structure. In the divergent evolution shown in *C*, the cultures are extremely sensitive to both initial and boundary conditions. Accurate long term predictions about single histories are then impossible. This circumstance can arise even for very simple, completely deterministic systems. For background on such in-principle limitations, see Arnold and Avez (1968), Ford and Lunsford (1970), Ford (1974), Cohen (1976), May (1976), and May and Oster (1976).

The decomposition of social behavior into objective functional units, the discovery of new and sometimes surprising epigenetic rules, the measure of human genetic diversity, the tracing of the microevolution of the features of behavioral epigenesis, the paleobiological reconstruction of the origin of culture, the retrodiction of ethnography and connection of the results to new and more powerful covering laws in economics and sociology, perhaps even a sighting down the world-tube of possible future histories—these are the activities that will come increasingly to occupy the social sciences as the links between biology and the study of culture are more powerfully forged.

GLOSSARY

BIBLIOGRAPHY

INDEX

Glossary

The theory of gene-culture coevolution draws on many disciplines, including a few, such as neurobiology and cultural anthropology, that have rarely been juxtaposed in the past. For this reason we have stocked the following glossary with relatively elementary terms, in addition to those that are narrowly technical or else used by us in a specialized manner.

Adaptation In evolutionary biology, a particular structure, physiological process, or behavior that equips an organism to survive and reproduce. Also, the evolutionary process that leads to the formation of such a trait. (See also *Natural selection.*)

Adaptive landscape The array of genetic fitnesses of all of the genotypes that can be assembled by a population. (See also *Gene-culture adaptive landscape.*)

Afferent neuron A neuron that carries impulses to the central nervous system. More generally, a neuron that carries signals toward a localized region of processing anywhere in the nervous system. (Compare with *Efferent neuron.*)

Allele A particular form of a gene, distinguishable in its nucleotide sequence from other forms or alleles of the same gene, hence occupying the same locus in the chromosomes.

Alpha waves Unsynchronized brain waves that accompany a state of wakeful relaxation but cease when a relaxed subject is aroused.

Amino acid The basic building unit of proteins: an organic compound of the general formula H_2N—CHR—COOH, where R is any one of twenty or more different side chains.

Amplification In molecular biology, the cellular mechanism for increasing the dosage of specific genes by making numerous DNA copies of the gene. In sociobiology, the magnification of the effects of the epigenetic rules during the translation of individual behavior to societal patterns, especially ethnographic curves.

Anthropology The study of man, with special reference to evolutionary history, racial differentiation, and cultural diversity.

Artifact Something made by human effort or intervention.

Assimilation function The relation between the probability of adopting a culturgen or switching from one culturgen to another (or the rate at which such changes are made) and the proportion of other members of the society who possess the culturgen. For example, the probability may rise linearly as others adopt the culturgen, or not change at all.

Australopithecine Relating to the "man-apes," primates of the genus *Australopithecus,* which were primitive forms that lived as recently as one million years ago and probably included among the earlier species the ancestors of modern man (genus *Homo*).

Autonomic nervous system The system that controls such involuntary "housekeeping" functions as digestion, breathing, and the release of glandular secretions.

Axon A portion of the neuron that carries action potentials, usually away from the cell body of the neuron; the axon is generally the longest segment of the cell. (Contrast with *Dendrite*.)

Behavioral biology The scientific study of the biological basis of all aspects of behavior, including neurobiology, ethology, and sociobiology.

Bias curve See *Usage bias curve*.

Biogeography The scientific study of the geographic distribution of organisms. (See also *Island biogeography*.)

c_i The symbol used to denote a culturgen.

Canalization The developmental process that leads to the production of phenotypes near the species norm even when the growing organism is stressed by unusual environments.

Central nervous system That part of the nervous system most concentrated and centrally located, such as the brain and spinal cord of man and other vertebrate animals.

Cerebellum A structure of the brain, lying beneath and behind the forebrain, which coordinates sensory information and motor movement.

Cerebrum The enlarged anterior portion of the brain, consisting of forebrain, midbrain, and their derivatives.

Chromosome A complex, often rodlike structure found in the nucleus of a cell, bearing part of the basic genetic units (genes) of the cell.

Chunk A group of linked symbols in long term memory that can be designated by a single "chunk symbol." The chunking process appears to be automatic and unconscious, although it can be aided by conscious attention to the relationship involved. Thus in the node-link graphs often used to pictorialize the manner in which information is stored in long term memory, a chunk symbol can be represented by a node together with its links. A chunk in short term memory may be either the activated chunk of long term memory or a symbol that identifies such a chunk. Some psychologists also identify as chunks those patterns of stimuli that activate chunk nodes. (See also *Concept, Node, Symbol.*)

Chunking The grouping together of several symbols by the human brain to facilitate learning and long term memory. The recall of a single symbol then suffices to recover the entire assemblage.

Cistron The genetic unit of function in DNA, considered loosely equivalent to the gene. Usually each cistron contains the information needed to prescribe the formation of a single polypeptide chain.

Codon A "triplet" of three nucleotides in DNA, messenger RNA, and transfer RNA that directs the placement of a particular amino acid into a polypeptide chain.

Coevolution An evolutionary change "in a trait of the individuals in one population in response to a trait of the individuals of a second population, followed by an evolutionary response by the second population to the change in the first" (Janzen, 1980). We have extended this biological term to include the reciprocal effects of genetic and cultural evolution (see *Gene-culture coevolution*). More generally, diachronic changes in two or more interacting objects or systems.

Cognition As broadly defined by many psychologists (see for example Lindzey et al., 1975) and used in this book, the inner mental processes of perception, memory, and information processing by which the individual acquires information, makes plans, and solves problems. Cognition in the narrow sense is often considered to entail at least some conscious thought and to be uniquely human, but the possibility of conscious experiences on the part of the more intelligent animals cannot be discounted (see Griffin, 1976).

Cognitive algebra The quantitative rules governing valuation and decision making.

Cognitive psychology The scientific study of cognition.

Concept A primitive element of semantic memory. In the human mind *concept symbols* are often linked to symbols that can generate words or phrases. Concepts appear to be low-level generalizations built up automatically from repeated exposure to situations that have properties in common. They acquire *meaning* not only from these referent episodes but also, perhaps, from *propositions* and other more complex semantic entities in which they occur. In cognition a concept can be used as a fuzzy-logical function that classifies events or percepts in the individual's world into various degrees of membership in the concept's category. Although most if not all real-life concepts have fuzzy edges, at least some contain symbols that refer to "ideal examples" or *prototypes* of the concepts. (See also *Knowledge structure, Node, Proposition, Prototype, Symbol.*)

Cortex In neurobiology, a clearly differentiated zone of cells and synapses, usually gray matter composing a morphologically outer layer of the brain, in particular the cerebrum (where the cortex is a major locus of association and intelligent behavior) and the cerebellum.

Cultural evolution Any change in culturally transmitted artifacts, behavior, institutions, or mental concepts within or across generations. (See also *Evolution, Gene-culture coevolution.*)

Culture The sum of all of the artifacts, behavior, institutions, and mental concepts transmitted by learning among members of a society, and the holistic patterns they form. In human beings the culture of each society is characterized by some traits (culturgens) that are general to the whole species and others that are idiosyncratic. The transmission also entails cognition, which loads the traits with meaning and typically but not invariably labels them with words and other symbols that are then manipulated in language to create complex new messages. (See also *Culturgen, Euculture, Gene-culture coevolution, Protoculture.*)

Culturgen (pronounced "kul' tur jen") The basic unit of culture. A relatively homogeneous set of artifacts, behaviors, or mentifacts (mental constructs having little or no direct correspondence to reality) that either shares without exception one or more attribute states selected for their functional importance or at least shares a consistently recurrent range of such attribute states within a given polythetic set. Culturgens can be mapped into node-link structures in long term memory and in many instances can be treated as identical with them.

Culturgen assimilation The process whereby culturgens are invented and adopted in behavioral categories whose epigenetic rules are permissive toward the culturgens. (Compare with *Genetic assimilation.*)

Cytoarchitecture The arrangement of cells in tissues and organs.

Darwinism The theory of evolution by natural selection, as formulated by Charles Darwin (1809–1882). The modern, vastly expanded version is often referred to as neo-Darwinism and sometimes as the modern synthesis of evolutionary theory.

Demography The scientific study of birth rates, mortality, age distribution, and related quantitative properties of populations. (See also *Population biology.*)

Dendrite A twig-like portion of a neuron characterized by rapid tapering and frequent branching. The dendrite serves as a principal receptive and integrative area of most neurons. (Contrast with *Axon.*)

Deoxyribonucleic acid See *DNA*.

Determinism A fixed cause-effect relationship between known or at least knowable variables. In the philosophy of science, the expression usually implies that a process of interest is set by a relatively few, narrowly construed variables. Thus ''genetic determinism'' means to many that behavior is rigidly constrained by the genes, while ''cultural determinism'' means that it depends almost entirely on the the particularities of the surrounding culture.

Deterministic In mathematics, referring to a relationship between two or more variables, without taking into account the effect of chance on the outcome of particular cases. (Contrast with *Stochastic.*)

Developmental psychology In the original sense, the scientific study of the development of mental operations and cognition in children; the field now encompasses all of the later stages of life as well.

Diallelic Pertaining to two alleles situated on the same locus.

Differentiation In biology, the process whereby originally similar cells follow different developmental pathways.

Diploid With reference to a cell or to an organism, having a chromosome complement consisting of two copies (called homologues) of each chromosome. A diploid cell or organism usually arises as the result of the union of two sex cells, each bearing just one copy of each chromosome. Thus, the two homologues in each chromosome pair in a diploid cell are of separate origin, one derived from the mother and the other from the father. (Contrast with *Haploid.*)

Dishabituation The reversal of habituation: the restoration of an original level of response because of a shift in some quality of the stimulus eliciting it.

DNA (*deoxyribonucleic acid*) The basic hereditary material of all organisms; a nucleic acid polymer incorporating the sugar deoxyribose. In higher organisms, including animals, the great bulk of DNA is located within the chromosomes.

Dominance In behavioral biology, the physical domination of some members of a group by other members, usually in relatively orderly and long-lasting patterns (called dominance hierarchies). In genetics, the ability of one allelic form of a gene to determine the phenotype of a heterozygous individual, in which the homologous chromosome carries a different allele. For example, if *A* and *a* are two allelic forms of a gene, *A* is said to be dominant to *a* if *AA* diploids and *Aa* diploids are phenotypically identical (or nearly so) and are distinguishable from *aa* diploids. The *a* allele then is said to be recessive.

Drosophila A genus of fruit flies widely used in biological studies, especially genetics.

Ecology The scientific study of the interaction of organisms with their environment.

Efferent neuron A neuron that carries impulses away from the central nervous system. More generally, a neuron that carries signals away from one localized neural region of information processing and into another.

Enculturation The transmission of a particular culture, especially to young members of the society. Some authors make a distinction between socialization, regarded as the development of social behavior basic to every normal human being, and enculturation, the act of learning one culture in all its uniqueness and particularity (see for example Mead, 1963). However, in this work we have used the two terms interchangeably. (See also *Socialization*.)

Epigene A gene affecting a particular developmental process under study.

Epigenesis The processes of interaction between genes and the environment that ultimately result in the distinctive anatomical, physiological, cognitive, and behavioral traits of the organism. Epigenetic events occur from the moment that RNA is transcribed from DNA, then forward through all phases of development to the final assembly of tissues and cognition itself; the interacting environment first is composed entirely of the cell medium but then expands until—in the case of human beings especially—it includes all aspects of culture.

Epigenetic rule Any regularity during epigenesis that channels the development of an anatomical, physiological, cognitive, or behavioral trait in a particular direction. Epigenetic rules are ultimately genetic in basis, in the sense that their particular nature depends on the DNA developmental blueprint. They occur at all stages of development, from protein assembly through the complex events of organ construction to learning. Some epigenetic rules are inflexible, with the final phenotype being buffered from all but the most drastic environmental changes. Others permit a flexible response to the environment; yet even these may be invariant, in the sense that each possible response in the array is matched to one environmental cue or a set of cues through the operation of special control mechanisms. In cognitive development, the epigenetic rules are expressed in any one of the many processes of perception and cognition to influence the form of learning and the transmission of culturgens. The *primary epigenetic rules* of mental development are based upon the more automatic processes that lead from sensory filtering to perception (Chapter 2). The *secondary epigenetic rules* affect information displayed in the perceptual fields and include the channeling of memory, emotional response, decision making, and ultimately the *usage bias curves* (see also Chapter 3).

Epigenotype The set of genes affecting a particular developmental process under study.

Epistasis The nonadditive interaction of genes on different loci, especially the suppression of one set by another.

Ethnographic curve The array of probabilities that a society will possess particular cultural patterns. In the present analysis of gene-culture coevolution, cultural patterns are defined as the various proportions of individuals in the society that possess one active culturgen as opposed to another. The curve can be based on many societies observed concurrently or on one or a few societies observed repetitively through time.

Ethnography The description of the cultures of individual societies and the comparison of cultures.

Ethnology The scientific study of cultures, of which ethnography is the more primary, descriptive part.

Ethology The biological study of whole patterns of animal behavior in natural environments, stressing the analysis of adaptation and the evolution of the patterns. It is distinguished from sociobiology, which is the study of the biological basis of social behavior in particular, and especially of the properties of entire societies. (See also *Sociobiology*.)

Euculture The most advanced form of culture, in which individuals not only teach and learn information but also conceptualize much of it into concrete entities that can be more readily labeled by symbols and handled by language. (See also *Protoculture, Reification*.)

Evolution Any gradual change. Genetic or organic evolution, often referred to as evolution for short, is any genetic change in organisms from generation to generation; or, more strictly, a change in gene frequencies within populations from generation to generation. Cultural evolution is an alteration in culturally transmitted artifacts, behavior, institutions, or concepts within or across generations. As purely conceived, cultural evolution does not necessarily entail any form of genetic change, but when the two modes of change are coupled in a reciprocally interacting manner, the total process is referred to as gene-culture coevolution.

Evolutionary biology The collective disciplines of biology that treat the evolutionary process and the characteristics of populations of organisms, as well as ecology, behavior, and systematics. (See also *Population biology*.)

Fitness In evolutionary biology, differential representation in later generations. (See *Gene-culture fitness, Genetic fitness.*)

Frequency curve A curve plotted on a graph to display a particular frequency distribution.

Frequency distribution The array of numbers of individuals possessing different values of some variable quantity; for example, the numbers of animals of different ages (the "age distribution") or the numbers of societies possessing various proportions of culturgens (the "ethnographic distribution").

Fuzzy logic The probabilistic classification of continuously varying qualities, such as voice onset time in phoneme formation, into overlapping categories. (See also *Prototype.*)

Gamete The mature sexual reproductive cell: the egg or the sperm.

Gene The basic unit of heredity. Often defined as the rough equivalent of the cistron, the segment of DNA that carries information prescribing a single polypeptide.

Gene-culture adaptive landscape The array of gene-culture fitnesses of all of the combinations of genotypes and culturgens that can be assembled by a population.

Gene-culture coevolution The coupled evolution of genes and culture. More precisely, any change in gene frequencies that alters culturgen frequencies in such a way that the culturgen changes alter the gene frequencies as well. (See also *Coevolution.*)

Gene-culture coevolutionary theory The theory treating the full circuit of the coevolution process; sometimes referred to as gene-culture theory (GCT) for brevity.

Gene-culture fitness The contribution to the next generation of one particular combination of genotype and culturgens relative to the contribution of other such combinations.By definition, the differential contribution leads eventually to the prevalence of the genotypes with the highest fitnesses. (Compare with *Genetic fitness.*)

Gene-culture theory (GCT) See *Gene-culture coevolutionary theory.*

Gene-culture translation The effect of epigenetic rules of individual development on social patterns.

Gene-culture transmission The transmission of culturgens in which the choices are not all equiprobable. (Compare with *Pure cultural transmission* and *Pure genetic transmission.*)

Gene flow The exchange of genes between different species (an extreme case referred to as hybridization) or between different populations, caused by the migration of individuals or the long-distance dispersal of gametes.

Gene pool All the genes (hence, hereditary material) in a population.

Genetic assimilation A change in the frequency of genes making it more likely that individuals will develop a previously rare trait. In a typical sequence the phenotype appears as a result of exposure to an unusual environment; the bearers of the trait prove superior at survival and reproduction; as a consequence, there is an increase in frequency of genes that promote the appearance of the trait even in normal environments. (Compare with *Culturgen assimilation.*)

Genetic drift Evolution (change in gene proportions) by chance processes alone.

Genetic evolution See *Evolution.*

Genetic fitness The contribution to the next generation of one genotype relative to the contribution of other genotypes, averaged over all the culturgens possessed by individuals of various genotypes (hence in effect without reference to culturgens). By definition, the differential contribution leads eventually to the prevalence of the genotypes with the highest fitnesses. (Compare with *Gene-culture fitness.*)

Genome The complete genetic constitution of an organism.

Genotype The genetic constitution of an individual organism, designated with reference to either a single trait or a set of traits. (Contrast with *Phenotype.*)

Genus (plural: *genera*) A group of related, similar species. Examples include *Canis* (wolves, domestic dogs, and their close relatives) and *Homo* (primitive man, including *Homo habilis* and *H. erectus,* and modern man, *H. sapiens*).

Group Any set of organisms, belonging to the same species, that remain together for a period of time while interacting with one another to a distinctly greater degree than with other conspecific organisms.

Group selection Selection that operates on two or more members of a lineage group as a unit. Defined broadly, group selection includes both kin selection and selection of entire populations. (See also *Kin selection.*)

Habituation The simplest form of learning, in which an animal presented with a stimulus without reward or punishment eventually ceases to respond. (See also *Dishabituation.*)

Haploid Having a chromosome complement consisting of just one copy of each chromosome. Sex cells are typically haploid. (Contrast with *Diploid.*)

Heritability The fraction of variation of a trait within a population—more precisely, the fraction of its variance, which is the statistical measure—due to heredity as opposed to environmental influences. A heritability score of one means that all the variation is genetic in basis; a heritability score of zero means that all the variation is the result of the environment.

Heterarchy A mixed-level hierarchy: a hierarchy-like system of two or more levels of units, with the properties of the higher levels affecting the lower in some degree, but with additional activity in the lower units feeding back to some extent to influence the higher level. The units at the higher level are composed of the lower-level units. For example, individual behavior generates institutions and other macrocultural patterns, which then influence individual decisions.

Heterozygous Referring to a diploid organism having different alleles of a given gene on the pair of homologous chromosomes carrying the gene.

Hierarchy A system whose components are organized in such a way that two or more natural levels of system-into-subsystem decompositions can be distinguished. (See also *Heterarchy*.)

Holism The method of explanation of complex systems, and the philosophy motivating it, that includes not just the properties of the components of the system but also the patterns and even the history of their relationships and the manner in which they are integrated to cause the system to function as a distinct, superordinate entity. Extreme holism eschews any meaningful connections between superordinate entities and their components. (Contrast with *Reductionism*.)

Homeostasis The maintenance of a steady state, especially a developmental, physiological, or social steady state, by means of self-regulation through internal feedback responses.

Homo The genus of true men, including several extinct forms (*Homo habilis, H. erectus, H. sapiens neanderthalensis*) as well as contemporary man (*Homo sapiens sapiens*), who were or are primates characterized by completely erect stature, bipedal locomotion, reduced dentition, and above all an enlarged brain size.

Homozygous Referring to a diploid organism possessing identical alleles of a given gene on both homologous chromosomes. An organism can be a homozygote with respect to one gene and at the same time a heterozygote with respect to another gene.

Imprinting A rigid form of learning, usually induced during a brief, "sensitive" period of the life cycle, in which an animal comes to make a particular response to only one other animal or object.

Inbreeding The mating of kin. The degree of inbreeding is measured by the fraction of genes that will be identical owing to common descent.

Incest Sexual relations between closely related persons.

Inclusive fitness The sum of an individual's own fitness plus all its influence on fitness in its relatives other than direct descendants; hence the total effect of kin selection with reference to an individual.

Interlocus interactions The various modes in which genes on different loci affect one another's activity. (See for example *Epistasis*.)

Interneuron A neuron that connects other neurons to each other and is neither sensory nor primarily efferent to the periphery. Same as internuncial neuron.

Invertebrate Any animal, such as a snail or insect, whose nerve cord is not enclosed in a backbone of bony segments, or vertebrae. (Contrast with *Vertebrate*.)

Island biogeography The study of the distribution of organisms on islands and similar isolated habitat patches. The theory of island biogeography is based principally on models of immigration, extinction, and the attainment of equilibrium in the numbers of species inhabiting islands.

Kin selection The change in gene frequencies due to one or more individuals favoring or disfavoring the survival and reproduction of relatives (other than offspring) who possess the same genes by common descent.

Knowledge structure Any set of linked nodes in long term memory, from a single node to a schema to the entire contents of long term memory. Some knowledge structures can be mapped onto culturgens and in many cases are identical with them.

Lamarckism The theory of evolution by acquired characteristics, as propounded by Jean Baptiste de Lamarck (1744–1829). The belief that traits acquired by the activities of individuals are passed directly to the offspring—in contradiction to Darwinism, which holds that offspring receive only genetic traits unaltered by the environment and that evolution occurs by the greater success of individuals inheriting these traits, in other words by natural selection. (See also *Darwinism*.)

Learning Behavior that is modified according to an organism's experience.

Learning rule An epigenetic rule that directly affects the learning of particular culturgens. Given appropriate methods, the learning rule can be evaluated by means of a usage bias curve.

Leash principle The principle deduced from natural selection theory that epigenetic rules will tend to evolve in such a way as to make individuals choose certain culturgens over others; in other words, "the genes hold culture on a leash." (See also *Epigenetic rule*.)

Life cycle The entire life span of an organism from the moment of fertilization (or asexual creation) to the time it reproduces in turn.

Limbic region (*limbic system*) The higher brain structures that control or at least serve as a principal conduit for the emotional and motivational aspects of behavior. Anatomically, the limbic region is a ring of structures around the anterior end of the brain stem and interhemispheric commissures that includes the hypothalamus, anterior thalamus, amygdala, septal nuclei, hippocampal formation, and other cellular complexes.

Link A pictorial representation of the association between symbols in the brain.

Linkage In genetics, the association between alleles on different loci of the same chromosome that results in their nonrandom assortment during the formation of the sex cells. In psychology, an association between two nodes in long term memory.

Locus (plural: *loci*) The location of a gene on the chromosome.

Logistic A particular form of change, such as growth of a population or the spread of a culturgen, that first accelerates and then slows steadily as the process approaches its upper limit.

Long term memory (*secondary memory*) The storage system of the brain in which information is retained for long periods of time, often for the life of the individual. (Contrast with *Short term memory*.)

Major gene A gene that individually accounts for a large part of the variation of a trait under consideration, as opposed to a polygene. The effects of a major gene can be altered to some degree by modifier genes. (See also *Polygenes*.)

Mammal Any animal of the class Mammalia, such as a wolf or a human being, characterized by the production of milk by the female mammary glands and the possession of hair for body covering.

Maximin strategy The procedure in economic and other forms of behavior leading to the maximum return under the worst possible circumstances, rather than the maximum return under the most favorable circumstances.

Meaning The pattern of relationships between one symbol and all others.

Mentifacts Mental constructs having no direct correspondence to real objects, people, or events.

Microevolution A small amount of evolutionary change, consisting in minor alterations in gene proportions, chromosome structure, or chromosome numbers. (A larger amount of change would be referred to as macroevolution or simply as evolution.)

Modifier gene A gene that alters the effect of a major gene in a relatively minor way. (See also *Major gene*.)

Monotonic function A function that always increases or decreases; thus in the case of a monotonic increasing function, if $x < y$ then $f(x) < f(y)$ for all paired values of x and y.

Multidimensional scaling analysis The measurement of the differences of objects or processes in such a way that the multiple traits used to distinguish them can be combined and mapped, often onto a plane, for more convenient examination and analysis. Same as multiple-dimensional scaling analysis.

Multifactorial inheritance The control of variation in a trait by genes situated on more than one locus; the term usually implies further that the genes are polygenes.

Mutant An allele that has recently been created by mutation or at least is rare enough in the population to be at the frequency level expected from recent mutation activity. Also, an individual organism carrying such an allele.

Mutation In the broad sense, any discontinuous change in the genetic constitution of an organism. In the narrow sense, the word refers usually to a "point mutation," a change along a very narrow portion of the nucleic acid sequence.

Natural selection The differential contribution to the next generation by individuals of different genetic types but belonging to the same population. This is the basic mechanism proposed by Darwin and is generally regarded today as the main guiding force in evolution.

Neurobiology (neurosciences) The scientific study of the nervous system in the broadest sense of its anatomy, development, and physiology. (Compare with *Neurophysiology*.)

Neuron A nerve cell.

Neurophysiology The scientific study of the nervous system, especially the physiological processes by which it functions.

Neurotransmitter A substance, such as serotonin, that is released at synapses and mediates the transmission of nerve impulses.

Node An abstract representation of a symbol, or elementary memory unit in the brain. The symbol structures of the brain are often pictorialized as graphs or *node-link diagrams*. The nodes of the diagram represent the symbols and the links represent the relations, or patterns of activation, among the symbols. (See also *Concept, Knowledge structure, Symbol*.)

Node-link structure A set of nodes associated in long term memory. When the structure is well defined and its elements strongly linked, it comprises a *Schema*.

Nucleic acid A long-chain alternating polymer of deoxyribose or ribose sugar rings and phosphate groups, with organic bases (adenine, thymine or uracil, guanine, cytosine) as side chains. DNA and RNA, the basic genetic coding materials, are nucleic acids.

Nucleotide The basic chemical unit in a nucleic acid polymer, as in the genetic coding materials DNA and RNA.

Ontogeny The development of a single organism through the course of its life history.

Organism Any living creature.

Orthogonal In the mathematics of vectors and multidimensional spaces, two vectors are orthogonal when they have a zero inner product. The two vectors are then perpendicular to each other in the space. In a restricted literary sense, orthogonality is the independent action of two or more variables with respect to a third variable.

Paleontology The scientific study of fossils and all aspects of extinct life.

Penetrance In gene-culture theory, the number of culturgens of a particular category that various members of the population are able to incorporate. (See also *Selectivity*.)

Perception The psychological process of receiving and recognizing cues from the environment. The environment includes both the outside world and the internal physiological state of the organism. (See also *Cognition*.)

Perceptual field The configurations formed within the associative centers of the brain that correspond to information received through the sense organs. Thus the organization of light frequencies in the form of discrete hues represents a basic portion of the brain's perceptual field.

Phasic reentry The transitory reentry of impulses in patterns that are otherwise dormant. When the pattern in the conscious centers of the brain is sequenced in time, the result can be the reconstruction of the memory of a past event.

Phenocopy An individual of a scarce, aberrant kind, which superficially resembles a mutant (hence the individual "copies" the phenotype of the mutant). The exceptional trait marking the phenocopy is usually induced by environmental stress during development. An example is the crossveinless wing condition of *Drosophila* fruit flies, which is induced by heat shock administered during the pupal stage.

Phenotype The observable properties of a trait or set of traits of an individual. Phenotypic features develop under the combined influences of the genetic constitution of the individual and its environment. (Contrast with *Genotype;* see also *Epigenesis.*)

Phoneme The smallest possible subdivision of a word that can still be distinguished as a discrete sound. Single phoneme changes generally alter the meaning of a spoken word.

Physiology The scientific study of the functions of organisms and of individual organs, tissues, and cells of which they are composed. In its broadest sense, physiology also encompasses most of molecular biology and biochemistry.

Pleiotropism The control of more than one phenotypic characteristic, for example eye color, courtship behavior, or size, by the same gene or set of genes.

Polygenes Broadly defined, genes affecting the same trait but arrayed on two or more chromosomal loci. More narrowly defined, polygenes not only are scattered over multiple loci but are also roughly equal in the degree of their influence on the phenotype.

Population A set of organisms belonging to the same species and occupying a clearly delimited space at the same time. Often a particular population is coterminous with a society, especially if the society forms a closed or partially closed breeding system.

Population biology The scientific study of the biological properties peculiar to populations, including their distribution, ecology, growth, age characteristics, and genetic structure. (See also *Demography, Population ecology,* and *Population genetics*—all of which are subdisciplines of population biology.)

Population ecology The scientific study of the distinctive relationships between whole populations of organisms and their environments, covering such topics as the determinants of mortality and fertility, the rate of population growth, and the processes of competition and symbiosis.

Population genetics The scientific study of the distinctive properties of heredity viewed at the level of populations, including genetic diversity and the changes in gene frequencies that comprise evolution.

Primate Any member of the order Primates, such as a lemur, monkey, ape, or man.

Primiparous Bearing young for the first time.

Primitive Referring to a trait that appeared first in evolution and gave rise to other, more "advanced" traits. The primitive traits are often but not always less complex than the advanced ones.

Principle of parsimony The conjecture that epigenetic rules tend to evolve to the lowest level of genetic specification that will suffice; it follows that special physiological devices in cognition will decline when the need for specificity is reduced or eliminated. (See also *Transparency principle*.)

Proposition An element of semantic memory that joins two or more symbols functioning as "arguments" through a symbol functioning as a relational term. Thus in "John hit Sam," "John" and "Sam" are the arguments and "hit" is the relational term. Proposition nodes are often linked to nodes that can generate clauses or sentences in the individual's language. (See also *Concept, Node, Symbol*.)

Protoculture A form of culture, found in a few higher animals, in which information is transmitted by imitation and even by teaching, but in which reification and symbolization do not occur. (Contrast with *Euculture*.)

Prototype A schema or symbol in long term memory used as a referent by the brain during perception and classification of new sets of stimuli. (See also *Concept, Schema, Symbol*.)

Psychobiology The scientific study of the biological basis of behavior, especially at the levels of whole patterns of behavior and their physiological controls.

Psycholinguistics The scientific study of the processes of the mind that produce and interpret language.

Psychology The scientific study of behavior, with emphasis on the traits and capacities that are peculiar to human beings, or at least most highly developed in them.

Pupa The inactive developmental stage of the holometabolous insects, such as flies and wasps, during which maturation into the final adult form is completed.

Pure cultural transmission The transmission of culturgens in which all of the choices are equiprobable. (Compare with *Pure genetic transmission* and *Gene-culture transmission*.)

Pure genetic transmission In the theory of gene-culture coevolution, the extreme case in which the transmission of culturgens is limited to a single choice. (Compare with *Pure cultural transmission* and *Gene-culture transmission*.)

Receptor See *Sensory receptor*.

Recessive In genetics, referring to an allele the phenotype of which is suppressed when the allele occurs in combination with a dominant allele.

Reductionism Oversimplification in the explanation of a complex system owing to the attempt to account for the system solely on the basis of the properties of its components. Usually ascribed by social scientists to biologists, by biologists to chemists, and by chemists to physicists. (Contrast with *Holism*.)

Reification The mental activity in which hazily perceived and relatively intangible phenomena, such as complex arrays of objects or activities, are given a factitiously concrete form, simplified, and labeled with words or other symbols. In Marxian literature, the word is used more narrowly to mean the objectification of concepts and institutions in a way that removes them from the power of individual thought and change.

Reification learning rule An epigenetic rule operating in the formation and labeling of categories of reified phenomena.

RNA (*ribonucleic acid*) The basic material used in the copying and translating of genetic information (encoded by DNA) during the production of proteins; a nucleic acid polymer using the sugar ribose.

Schema (plural: *schemata*) A substantial, often functional fragment of long term memory. *Schema* is a frequently used but vaguely defined term of cognitive and developmental psychology. It has two general meanings. The first, originating with Bartlett (1932), is a large fragment of the knowledge or symbol structure in the mind. The individual refers to schemata in reflection and decision making. With Piaget (1952) the word schema, and in particular the expression sensorimotor schema, acquired a very explicit connotation of a plan of action and of knowledge that relates input stimuli to decisions that activate behavioral responses. (See also *Concept, Knowledge structure, Node, Symbol.*)

Selection pressure Any process in the environment that results in natural selection. For example, food shortage, the activity of a predator, or competition for a mate from other members of the same sex provide selection pressures by causing individuals of various genetic types to survive to different average ages, to reproduce at different rates, or both.

Selectivity In gene-culture theory, the strength of the tendency for individuals to use one or a small set of culturgens in preference to others that are available.

Semantic differential technique A method for measuring the association of words or symbols with points along scales of various meaning; for example, the word *dog* can be placed somewhere between friendly-hostile, hot-cold, and so forth.

Semantic memory The network of symbols contained in human long term memory. (See *Symbol.*)

Semantics The study of meaning.

Sensory receptor A cell, tissue, or organ specialized for detecting physical signals (such as light or sound) and translating them into coded responses that can then be "read" by interneurons and transferred to the central nervous system.

Short term memory The storage system in which information is retained for at most about fifteen seconds without rehearsal and in amounts not exceeding about seven symbols. (See also *Long term memory*.)

Social contagion The spreading of a successful culturgen through a society by means of personal contact.

Socialization The total modification of behavior in an individual because of the interaction of that individual with other members of the society, including its parents. Some authors make a distinction between socialization, regarded as the development of social behavior basic to every normal human being, and enculturation, the act of learning one culture in all its uniqueness and particularity (see for example Mead, 1963). In this work we have used the two terms interchangeably.

Society A group of individuals belonging to the same species and organized in a cooperative manner. The diagnostic criterion of a society is reciprocal communication of a cooperative nature, extending beyond sexual activity alone. (See Wilson, 1975.)

Sociobiology The systematic study of the biological basis of all forms of social behavior, including sexual behavior and parent-offspring interaction, in all kinds of organisms.

Sociology Broadly defined, the general study of human societies at all levels of political organization.

Species The basic lower unit of classification, composed of a population or series of populations of closely related and similar organisms. The more narrowly defined "biological species" is composed of individuals capable of interbreeding freely with one another but not with members of other species.

Steady state An apparently unchanging condition that results from the balanced synthesis and degradation of components within the system, or alternatively from the balance in the arrival of new components into the system and the departure of old ones to the outside.

Stochastic Referring to the properties of mathematical probability. A stochastic model takes into account variations in outcome that result from chance. (Compare with *Deterministic*.)

Supergene A group of linked genes that control a distinctive set of characteristics and hence act approximately like a single major gene.

Symbiosis The living together of two or more species or other dissimilar sets of organisms in a prolonged and intimate ecological relationship.

Symbol A term used in two ways, both of which connote a basic unit in cognition. In discussions of computational-theoretic aspects of cognition, *symbol* refers to the elementary units manipulated by information processing systems. The physical basis of a symbol in the brain is thought to be a network or cluster of neurons. In this sense a memory trace or engram would be a symbol. Cognition creates both new symbols and new relationships between preexisting symbols. Thus human memory is said to consist in *symbol structures* and to operate associatively; that is, the activation of a symbol activates associated symbols, bringing them into the sphere of conscious awareness. The term *symbol* is also used in a less specialized manner in much of anthropology, linguistics, psychology, and psychoanalysis to denote some element or feature used in communication. It is not completely arbitrary in form but is chosen to be freighted with meaning and significance for the particular culture that incorporates it.

Synapse A functional connection between neurons or between neurons and muscle fibers at neuromuscular junctions.

Synaptogenesis Development at the cellular level leading to the creation of synapses.

Teaching rule An epigenetic rule that directly affects the teaching of particular culturgens.

Territory An area occupied more or less exclusively by an organism or group of organisms by means of repulsion through overt defense or aggressive display.

Threshold A state of a system from which the system can pass to other states possessing very different properties.

Tissue An aggregate of cells related in function and organized to form part of an organ. For example, the conjunctival epithelium forms part of the eye.

Transcription The synthesis of RNA, which uses one strand of DNA as the template, and hence the first step in the "decoding" of DNA that leads to the prescription and production of proteins.

Translation The synthesis of a protein using the information encoded in RNA.

Transparency principle The conjecture that the more the relation between a category of behavioral response and genetic fitness depends on environmental circumstances, the more clearly the conscious mind perceives the relation and the more flexible the response.

u_i The symbol used to denote either innate bias, the probability of choosing one culturgen over another at the start of the learning process and hence in the absence of sensitivity to choices made by others, or the probability of making such a choice at each decision point after the initial learning has occurred, whether influenced by others or not.

$u_{ij}(\xi)$ The symbol used to denote the transition probability from culturgen i to culturgen j of an individual at a decision point when a proportion ξ of the other members of the society are using culturgen i.

Usage See *Usage of culturgen.*

Usage bias curve A curve that displays the probabilities that an organism will use one or the other of various culturgens within a culturgen category, given that it possesses a particular genotype and lives in a particular environment.

Usage of culturgen In the broad sense we have employed throughout, any response to a culturgen: either the initial perception and learning of the culturgen, or a relative preference for it during reflection and valuation, or a decision to employ it.

Usage pattern The proportion of individuals in a population that use one culturgen as opposed to another.

$v_{ij}(\xi)$ The rate of transition by an individual from culturgen i to culturgen j. The term $v_{ij}(\xi)$ denotes one form of the assimilation function.

Variance The most commonly used statistical measure of variation (dispersion) of a trait within a population. It is the mean squared deviation of all individuals in a sample from the sample mean.

Vertebrate An animal having a vertebral column (backbone); in particular, a fish, amphibian, reptile, bird, or mammal. (Contrast with *Invertebrate.*)

W The symbol used to denote genetic fitness.

Wild type In genetics, the standard or reference type. Deviants from this type are referred to as mutant, whether they have arisen recently by mutation or merely occur at relatively low frequencies in populations. (See also *Mutant.*)

ξ The lowercase Greek letter xi, an order parameter used to denote the frequency of occurrence of a culturgen in a society; the society can range from a state in which no members possess the culturgen ($\xi = -1$) to the state in which all possess it ($\xi = +1$). (See also *Assimilation function.*)

Zoology The scientific study of animals.

Zygote The cell created by the union of two gametes, in which the nuclei are also fused. The earliest stage of the diploid generation.

Bibliography

Abelson, R. P. 1973. The structure of belief systems. In R. C. Schank and K. M. Colby, eds., *Computer models of thought and language,* pp. 287–339. W. H. Freeman, San Francisco.

Abramowitz, M., and Irene A. Stegun, eds. 1965. *Handbook of mathematical functions.* Dover, New York. xiv + 1046 pp.

Ajzen, I., and M. Fishbein. 1977. Attitude-behavior relations: a theoretical analysis and review of empirical research. *Psychological Bulletin,* 84(5): 888–918.

Alcock, J. 1969. Observational learning in three species of birds. *Ibis,* 111(3): 308–321.

Alcock, J. 1979. *Animal behavior,* 2nd ed. Sinauer Associates, Sunderland, Mass. xii + 532 pp.

Alexander, R. D. 1971. The search for an evolutionary philosophy of man. *Proceedings of the Royal Society of Victoria,* 84(1): 99–120.

Alexander, R. D. 1979a. *Darwinism and human affairs.* University of Washington Press, Seattle. xxiv + 317 pp.

Alexander, R. D. 1979b. Evolution and culture. In N. A. Chagnon and W. Irons, eds., *Evolutionary biology and human social behavior: an anthropological perspective,* pp. 59–78. Duxbury Press, North Scituate, Mass.

Alexander, R. D., J. L. Hoogland, R. D. Howard, Katherine M. Noonan, and P. W. Sherman. 1979. Sexual dimorphisms and breeding systems in pinnipeds, ungulates, primates, and humans. In N. A. Chagnon and W. Irons, eds., *Evolutionary biology and human social behavior: an anthropological perspective,* pp. 402–435. Duxbury Press, North Scituate, Mass.

Allen, Elizabeth, et al. (35 authors, constituting the Sociobiology Study Group of Science for the People). 1976. Sociobiology—another biological determinism. *BioScience,* 26(3): 182, 184–186.

Altmann, S. A. 1968. Sociobiology of rhesus monkeys: III, the basic communication network. *Behaviour,* 32(1–3): 17–32.

Ambrose, J. A. 1969. Cited by J. Bowlby, *Attachment and loss,* vol. 1, *Attachment,* see pp. 293–294. Basic Books, New York. xx + 428 pp.

Ammerman, A. J., and L. L. Cavalli-Sforza. 1973. A population model for the diffusion of early farming in Europe. In C. Renfrew, ed., *The explanation of cul-*

ture change: models in prehistory, pp. 343–357. University of Pittsburgh Press, Pittsburgh.

Amoore, J. E. 1977. Specific anosmia and the concept of primary odors. *Chemical Senses and Flavor*, 2: 267–281.

Anderson, N. H. 1979. Algebraic rules in psychological measurement. *American Scientist*, 67(5): 555–563.

Arehart-Treichel, Joan. 1978. The pituitary's powerful protein. *Science News*, 114(22): 374–375, 381.

Argyle, M., and M. Cook. 1976. *Gaze and mutual gaze*. Cambridge University Press, Cambridge. xii + 210 pp.

Arnold, V. I., and A. Avez. 1968. *Ergodic problems of classical mechanics*. W. A. Benjamin, New York. x + 286 pp.

Asch, S. E. 1951. Effects of group pressure upon the modification and distortion of judgments. In H. Guetzkow, ed., *Groups, leadership, and men*, pp. 177–190. Carnegie Press, Carnegie Institute of Technology, Pittsburgh.

Ashton, G. C., J. J. Polovina, and S. G. Vandenberg. 1979. Segregation analysis of family data for 15 tests of cognitive ability. *Behavior Genetics*, 9(5): 329–347.

Atkinson, R. C., G. H. Bower, and E. J. Crothers. 1965. *An introduction to mathematical learning theory*. John Wiley, New York. xiv + 429 pp.

Bailey, N. T. 1975. *The mathematical theory of infectious diseases and its applications*, 2nd ed. (First edition published in 1957 under the title *The mathematical theory of epidemics*.) Griffin, London. xvi + 413 pp.

Banister, E. W. 1979. The perception of effort: an inductive approach. *European Journal of Applied Physiology and Occupational Physiology*, 41(2): 141–150.

Banker, G. A., and W. M. Cowan. 1977. Rat hippocampal neurons in dispersed cell culture. *Brain Research*, 126(3): 397–425.

Baran, P. A. 1959. Marxism and psychoanalysis. *Monthly Review*, 11(6): 186–200.

Barker, J. S. F. 1979. Inter-locus interactions: a review of experimental evidence. *Theoretical Population Biology*, 16(3): 323–346.

Barlow, G. W., and J. Silverberg, eds. 1980. *Sociobiology: beyond nature/nurture?* (AAAS Selected Symposium 35). Westview Press, Boulder, Colo. xxvi + 627 pp.

Barry, H. III, Margaret K. Bacon, and I. L. Child. 1957. A cross-cultural survey of some sex differences in socialization. *Journal of Abnormal and Social Psychology*, 55(3): 327–332.

Bartholomew, D. J. 1973. *Stochastic models for social processes*, 2nd ed. John Wiley, New York. xii + 411 pp.

Bartlett, F. C. 1932. *Remembering: a study in experimental and social psychology*. Cambridge University Press, Cambridge. x + 317 pp.

Bateson, P. P. G. 1976. Rules and reciprocity in behavioural development. In P. P. G. Bateson and R. A. Hinde, eds., *Growing points in ethology*, pp. 401–421. Cambridge University Press, Cambridge.

Bauman, Z. 1978. *Hermeneutics and social science*. Columbia University Press, New York. 263 pp.

Bearison, D. J. 1979. Sex-linked patterns of socialization. *Sex Roles*, 5(1): 11–18.

Beauchamp, G. K., O. Maller, and J. G. Rogers, Jr. 1977. Flavor preferences in cats (*Felis catus* and *Panthera* sp.). *Journal of Comparative and Physiological Psychology,* 91(5): 1118–1127.

Beck, B. B. 1980. *Animal tool behavior: the use and manufacture of tools by animals.* Garland, New York. xvi + 307 pp.

Becker, G. S. 1976. *The economic approach to human behavior.* University of Chicago Press, Chicago. iv + 314 pp.

Beets, M. G. J. 1979. Informational deficiencies and preference in human chemoreception. In J. H. A. Kroeze, ed., *Preference behaviour and chemoreception,* pp. 23–37. International Retrieval, London.

Bentler, P. M., and G. Speckart. 1979. Models of attitude-behavior relations. *Psychological Review,* 86(5): 452–464.

Bentley, D., and M. Konishi. 1978. Neural control of behavior. *Annual Review of Neuroscience,* 1: 35–59.

Benzer, S. 1973. Genetic dissection of behavior. *Scientific American,* 229(6): 24–37.

Berelson, B., and G. A. Steiner. 1964. *Human behavior: an inventory of scientific findings.* Harcourt, Brace, & World, New York. xxiv + 712 pp.

Berger, P. L., and T. Luckmann. 1966. *The social construction of reality: a treatise in the sociology of knowledge.* Doubleday, Garden City, N.Y. x + 203 pp.

Berlin, B., and P. Kay. 1969. *Basic color terms: their universality and evolution.* University of California Press, Berkeley. xii + 178 pp.

Berlyne, D. E. 1971. *Aesthetics and psychobiology.* Appleton-Century-Crofts, New York. xiv + 336 pp.

Bielicki, T., and Z. Welon. 1971. The operation of natural selection on human head form in an East European population. In C. J. Bajema, ed., *Natural selection in human populations,* pp. 92–102. John Wiley, New York.

Biesele, Megan. 1978. Sapience and scarce resources: communication systems of the !Kung and other foragers. *Social Science Information,* 17(6): 921–947.

Binford, L. R. 1968. Post-Pleistocene adaptations. In Sally R. and L. R. Binford, eds., *New perspectives in archaeology,* pp. 313–341. Aldine Publishing Co., Chicago.

Binford, L. R. 1978. *Nunamiut ethnoarchaeology.* Academic Press, New York. xiv + 509 pp.

Bitterman, M. E. 1975. The comparative analysis of learning. *Science,* 188(4189): 699–709.

Bloom, S. F. 1941. *The world of nations: a study of the national implications in the work of Karl Marx.* Columbia University Press, New York. viii + 255 pp.

Blum, H. F. 1963. On the origin and evolution of human culture. *American Scientist,* 51(1): 32–47.

Blumberg, B. S., and Jana E. Hesser. 1975. Anthropology and infectious disease. In A. Damon, ed., *Physiological anthropology,* pp. 260–294. Oxford University Press, New York.

Blurton Jones, N. G., and M. J. Konner. 1973. Sex differences in behaviour of London and Bushman children. In R. P. Michael and J. H. Crook, eds., *Comparative ecology and behaviour of primates,* pp. 689–750. Academic Press, London.

Boddy, J. 1978. *Brain systems and psychological concepts*. John Wiley, New York. xiv + 461 pp.

Boehm, C. 1978. Rational preselection from hamadryas to *Homo sapiens:* the place of decisions in adaptive process. *American Anthropologist,* 80(2): 265–296.

Bohman, M. 1978. Some genetic aspects of alcoholism and criminality: a population of adoptees. *Archives of General Psychiatry,* 35(3): 269–276.

Bolton, R. 1978. Black, white, and red all over: the riddle of color term salience. *Ethnology,* 17(3): 287–311.

Bolton, R., and Diane Crisp. 1979. Color terms in folk tales: a cross cultural study. *Behavior Science Research* (Human Relations Area Files, New Haven, Conn.), 14(4): 231–253.

Bonner, J. T. 1980. *The evolution of culture in animals*. Princeton University Press, Princeton, N.J. x + 216 pp.

Bornstein, M. H. 1973. Color vision and color naming: a psychophysiological hypothesis of cultural difference. *Psychological Bulletin,* 80(4): 257–285.

Bornstein, M. H. 1979. Perceptual development: stability and change in feature perception. In M. H. Bornstein and W. Kessen, eds., *Psychological development from infancy: image to intention,* pp. 37–81. Lawrence Erlbaum Associates, Hillsdale, N.J.

Bornstein, M. H., W. Kessen, and Sally Weiskopf. 1976. The categories of hue in infancy. *Science,* 191(4223): 201–202.

Bortz, J. 1978. Psychologische Ästhetikforschung—Bestandsaufnahme und Kritik. *Psychologische Beiträge,* 20(4): 481–508.

Boserup, Ester. 1965. *The conditions of agricultural growth*. Aldine Publishing Co., Chicago. 124 pp.

Boulding, K. E. 1978. *Ecodynamics: a new theory of societal evolution*. Sage Publications, Beverly Hills, Calif. 368 pp.

Bowers, R. B. 1978a. Statistical dynamic models of social systems: the general theory. *Behavioral Science,* 23(2): 109–119.

Bowers, R. B. 1978b. Statistical dynamic models of social systems: discontinuity and conflict. *Behavioral Science,* 23(2): 120–129.

Box, G. P., and G. M. Jenkins. 1970. *Time series analysis*. Holden-Day, San Francisco. xx + 553 pp.

Boyd, R., and P. J. Richerson, 1976. A simple dual inheritance model of the conflict between social and biological evolution. *Zygon,* 11(3): 254–262.

Brainerd, C. J. 1978. The stage question in cognitive-developmental theory. *The Behavioral and Brain Sciences* 1(2): 173–213.

Brainerd, C. J. 1979. Markovian interpretations of conservation learning. *Psychological Review,* 86(3): 181–213.

Braitenberg, V. 1977. *On the texture of brains*. Springer-Verlag, New York. x + 127 pp.

Bremermann, H. J. 1963. Limits of genetic control. *IEEE Transactions on Military Electronics,* MIL-7(2 and 3): 200–205.

Brian, M. V. 1979. Caste differentiation and division of labor. In H. R. Hermann, ed., *Social insects,* vol. 1, pp. 121–222. Academic Press, New York.

Brown, R. 1973. *A first language: the early stages*. Harvard University Press, Cambridge, Mass. xxii + 437 pp.

Brown, R. 1978. A new paradigm of reference. In G. A. Miller and E. Lenneberg, eds., *Psychology and biology of language and thought: essays in honor of Eric Lenneberg,* pp. 151–166. Academic Press, New York.

Brown, R., and R. J. Herrnstein. 1975. *Psychology*. Little, Brown & Co., Boston. xviii + 762 pp.

Bullock, T. H., R. Orkand, and A. Grinnell. 1977. *Introduction to nervous systems*. W. H. Freeman, San Francisco. xvi + 559 pp.

Buser, P. A., and A. Rougeul-Buser, eds., 1978. *Cerebral correlates of conscious experience*. North-Holland Publishing Co., Amsterdam. x + 364 pp.

Bush, R. R., and F. Mosteller. 1955. *Stochastic models for learning*. John Wiley, New York. xvi + 365 pp.

Buys, C. J., and K. L. Larson. 1979. Human sympathy groups. *Psychological Reports,* 45(2): 547–553.

Cain, W. S. 1979. To know with the nose: keys to odor identification. *Science,* 203(4379): 467–470.

Campbell, D. T. 1965. Variation and selective retention in socio-cultural evolution. In H. R. Barringer, G. I. Blanksten, and R. W. Mack, eds., *Social change in developing areas,* pp. 19–49. Schenkman Publishing Co., Cambridge, Mass.

Campbell, D. T. 1975. On the conflicts between biological and social evolution and between psychology and moral tradition. *American Psychologist,* 30(12): 1103–1126.

Carneiro, R. L. 1967. On the relationship between size of population and complexity of social organization. *Southwestern Journal of Anthropology,* 23(3): 234–243.

Carneiro, R. L. 1970. A theory of the origin of the state. *Science,* 169(3947): 733–738.

Carter-Saltzman, Louise. 1980. Biological and sociocultural effects on handedness: comparison between biological and adoptive families. *Science,* 209(4462): 1263–1265.

Cassirer, E. 1944. *An essay on man*. Yale University Press, New Haven, Conn. xii + 237 pp.

Cassirer, E. 1946. *Language and myth*. Dover, New York. xii + 103 pp.

Caute, D., ed. 1967. *Essential writings of Karl Marx*. Collier Books, Macmillan, New York. 254 pp.

Cavalli-Sforza, L. L. 1971. Similarities and dissimilarities of sociocultural and biological evolution. In F. R. Hodson, D. G. Kendall, and P. Tautu, eds., *Mathematics in the archaeological sciences,* pp. 535–541. Edinburgh University Press, Edinburgh.

Caviness, V. S., Jr., and P. Rakic. 1978. Mechanisms of cortical development: a view from mutations in mice. *Annual Review of Neuroscience,* 1: 297–326.

Chagnon, N. A. 1976. Fission in an Amazonian tribe. *The Sciences,* 16(1): 14–18.

Chagnon, N. A. 1977. *Yanomamö: the fierce people,* 2nd ed. Holt, Rinehart & Winston, New York. xvi + 174 pp.

Chagnon, N. A. 1979. Is reproductive success equal in egalitarian societies? In N. A. Chagnon and W. Irons, eds., *Evolutionary biology and human social behavior: an anthropological perspective*, pp. 374–401. Duxbury Press, North Scituate, Mass.

Chagnon, N. A., and R. B. Hames. 1979. Protein deficiency and tribal warfare in Amazonia: new data. *Science*, 203(4383): 910–913.

Chagnon, N. A., and W. Irons, eds. 1979. *Evolutionary biology and human social behavior: an anthropological perspective*. Duxbury Press, North Scituate, Mass. xvi + 623 pp.

Chaitin, G. J. 1975. Randomness and mathematical proof. *Scientific American*, 232(5): 47–52.

Changeux, J.-P., and A. Danchin. 1976. Selective stabilisation of developing synapses as a mechanism for the specification of neuronal networks. *Nature*, 264(5588): 705–712.

Chevalier-Skolnikoff, Suzanne. 1977. A Piagetian model for describing and comparing socialization in monkey, ape, and human infants. In Suzanne Chevalier-Skolnikoff and F. E. Poirier, eds., *Primate bio-social development: biological, social, and ecological determinants*, pp. 159–187. Garland, New York.

Chiva, M. 1979. Comment la personne se construit en mangeant. *Communications* (École des Hautes Études en Sciences Sociales—Centre d'Études Transdisciplinaires, Paris), 31: 107–118.

Chomsky, N. 1972. *Language and mind*, enlarged ed. Harcourt Brace Jovanovich, New York. xii + 194 pp.

Clarke, D. L. 1978. *Analytical archaeology*, 2nd ed., rev. by B. Chapman. Columbia University Press, New York. xxii + 526 pp.

Cloak, F. T., Jr. 1975. Is a cultural ethology possible? *Human Ecology*, 3(3): 161–182.

Cloninger, C. R., J. Rice, and T. Reich. 1979a. Multifactorial inheritance with cultural transmission and assortative mating: II, a general model of combined polygenic and cultural inheritance. *American Journal of Human Genetics*, 31(2): 176–198.

Cloninger, C. R., J. Rice, and T. Reich. 1979b. Multifactorial inheritance with cultural transmission and assortative mating: III, family structure and the analysis of separation experiments. *American Journal of Human Genetics*, 31(3): 366–388.

Cohen, J. E. 1971. *Casual groups of monkeys and men: stochastic models of elemental social systems*. Harvard University Press, Cambridge, Mass. xiv + 175 pp.

Cohen, J. E. 1976. Irreproducible results and the breeding of pigs (or nondegenerate limit variables in biology). *BioScience*, 26(6): 391–394.

Cohen, M. N. 1977. *The food crisis in prehistory: overpopulation and the origins of agriculture*. Yale University Press, New Haven, Conn. x + 341 pp.

Colby, K. M. 1973. Simulations of belief systems. In R. C. Schank and K. M. Colby, eds., *Computer models of thought and language*, pp. 251–286. W. H. Freeman, San Francisco.

Colby, K. M. 1978. Mind models: an overview of current work. *Mathematical Biosciences*, 39(3/4): 159–185.

Coleman, J. S. 1964. *Introduction to mathematical sociology*. Free Press, New York. xvi + 554 pp.

Coleman, J. S. 1973. *The mathematics of collective action.* Aldine Publishing Co., Chicago. x + 191 pp.

Collins, A. M., and Elizabeth F. Loftus. 1975. A spreading-activation theory of semantic processing. *Psychological Review,* 82(6): 407–428.

Collins, A. M., and M. R. Quillian. 1969. Retrieval time from semantic memory. *Journal of Verbal Learning and Verbal Behavior,* 8(2): 240–247.

Comings, D. E. 1979. Pc 1 Duarte, a common polymorphism of a human brain protein, and its relationship to depressive disease and multiple sclerosis. *Nature,* 277(5691): 28–32.

Connolly, K. 1973. Factors influencing the learning of manual skills by young children. In R. A. Hinde and Joan Stevenson-Hinde, eds., *Constraints on learning: limitations and predispositions,* pp. 337–369. Academic Press, New York.

Connolly, K., and J. Elliott. 1972. The evolution and ontogeny of hand function. In N. Blurton Jones, ed., *Ethological studies of child behaviour,* pp. 329–383. Cambridge University Press, Cambridge.

Coss, R. G. 1972. Eye-like schemata: their effect on behaviour. Ph.D. dissertation, University of Reading, England. (Cited by Eibl-Eibesfeldt, 1979.)

Cowan, W. M. 1979. Selection and control in neurogenesis. In F. O. Schmitt and F. G. Worden, eds., *The neurosciences: fourth study program,* pp. 59–79. MIT Press, Cambridge, Mass.

Crick, F. H. C. 1979. Thinking about the brain. *Scientific American,* 241(3): 219–232.

Crow, J. F., and M. Kimura. 1970. *An introduction to population genetics theory.* Harper & Row, New York. xiv + 591 pp.

Dahlberg, G. 1947. *Mathematical models for population genetics.* S. Karger, New York. 182 pp.

Daly, M., and Margo Wilson. 1978. *Sex, evolution, and behavior.* Duxbury Press, North Scituate, Mass. xii + 387 pp.

Davenport, W. 1960. Jamaican fishing: a game theory analysis. In S. W. Mintz, comp., *Papers in Caribbean anthropology* (Yale University Publications in Anthropology, nos. 57–64), no. 59, 11 pp. Department of Anthropology, Yale University, New Haven, Conn.

Davidson, G. 1977. Teaching and learning in an aboriginal community. *Developing Education,* 4(4): 2–8.

Davis, Clara M. 1928. Self selection of diet by newly weaned infants. *American Journal of Diseases of Children,* 36(4): 651–679.

Davis, M. 1978. What is a computation? In L. A. Steen, ed., *Mathematics today. Twelve informal essays,* pp. 241–267. Springer-Verlag, New York.

Davis, R. G. 1979. Olfactory perceptual space models compared by quantitative methods. *Chemical Senses and Flavour,* 4(1): 21–33.

Dawkins, R. 1976a. *The selfish gene.* Oxford University Press, New York. viii + 224 pp.

Dawkins, R. 1976b. Hierarchical organisation: a candidate principle for ethology. In P. P. G. Bateson and R. A. Hinde, eds., *Growing points in ethology,* pp. 7–54. Cambridge University Press, Cambridge.

Dawkins, R. 1980. Good strategy or evolutionarily stable strategy? In G. W.

Barlow and J. Silverberg, eds., *Sociobiology: beyond nature/nurture?,* pp. 331–367. Westview Press, Boulder, Colo.

DeCasper, A. J., and W. P. Fifer. 1980. Of human bonding: newborns prefer their mothers' voices. *Science,* 208(4448): 1174–1176.

Derynck, R., Jean Content, E. DeClercq, G. Volckaert, J. Tavernier, R. Devos, and W. Fiers. 1980. Isolation and structure of a human fibroblast interferon gene. *Nature,* 285(5766): 542–547.

De Soto, C. B. 1960. Learning a social structure. *Journal of Abnormal and Social Psychology,* 60(3): 417–421.

Dobzhansky, T. 1970. *Genetics of the evolutionary process.* Columbia University Press, New York. xiv + 505 pp.

Dobzhansky, T., H. Levene, and B. Spassky. 1972. Effects of selection and migration on geotactic behaviour of *Drosophila,* III. *Proceedings of the Royal Society,* ser. B, 180: 21–41.

Dobzhansky, T., F. J. Ayala, G. L. Stebbins, and J. W. Valentine. 1977. *Evolution.* W. H. Freeman, San Francisco. xvi + 572 pp.

Dodd, S. C. 1955. Diffusion is predictable: testing probability models for laws of interaction. *American Sociological Review,* 20(4): 392–401.

Doran, J. E., and F. R. Hodson. 1975. *Mathematics and computers in archaeology.* Harvard University Press, Cambridge, Mass. xii + 381 pp.

Douglas, Mary. 1979. Accounting for taste. *Psychology Today,* 13(2): 44–51.

Draper, Patricia. 1976. Social and economic constraints on child life among the !Kung. In R. B. Lee and I. DeVore, eds., *Kalahari hunter-gatherers: studies of the !Kung San and their neighbors,* pp. 199–217. Harvard University Press, Cambridge, Mass.

Drew, Jean S., W. T. London, E. D. Lustbader, Jana E. Hesser, and B. S. Blumberg. 1978. Hepatitis B virus and sex ratio of offspring. *Science,* 201(4357): 687–692.

Durham, W. H. 1976. The adaptive significance of cultural behavior. *Human Ecology,* 4(2): 89–121.

Durham, W. H. 1978. The coevolution of human biology and culture. In N. Blurton Jones and V. Reynolds, eds., *Human adaptation and behavior,* pp. 11–32. Halsted Press, Wiley, New York.

Durham, W. H. 1979. Toward a coevolutionary theory of human biology and culture. In N. A. Chagnon and W. Irons, eds., *Evolutionary biology and human social behavior: an anthropological perspective,* pp. 39–59. Duxbury Press, North Scituate, Mass.

Dyson, F. J. 1979. Time without end: physics and biology in an open universe. *Reviews of Modern Physics,* 51(3): 447–460.

Dyson-Hudson, Rada, and E. A. Smith. 1978. Human territoriality: an ecological reassessment. *American Anthropologist,* 80(1): 21–41.

Ebert, J. D., and I. M. Sussex. 1970. *Interacting systems in development,* 2nd ed. Holt, Rinehart & Winston, New York. xii + 338 pp.

Edelman, G. M., and V. B. Mountcastle. 1978. *The mindful brain: cortical organization and the group-selective theory of higher brain function.* MIT Press, Cambridge, Mass. 100 pp.

Edwards, A. W. F. 1977. *Foundations of mathematical genetics*. Cambridge University Press, New York. viii + 119 pp.

Ehrman, Lee, and P. A. Parsons. 1976. *The genetics of behavior*. Sinauer Associates, Sunderland, Mass. viii + 390 pp.

Eibl-Eibesfeldt, I. 1975. *Ethology: the biology of behavior*, 2nd ed. Holt, Rinehart & Winston, New York. xiv + 625 pp.

Eibl-Eibesfeldt, I. 1979. Human ethology: concepts and implications for the sciences of man. *The Behavioral and Brain Sciences*, 2(1): 1–57.

Eilers, Rebecca E. 1977. Context-sensitive perception of naturally produced stop and fricative consonants. *Journal of the Acoustical Society of America*, 61(5): 1321–1336.

Eilers, Rebecca E., and F. D. Minifie. 1975. Fricative discrimination in early infancy. *Journal of Speech and Hearing Research*, 18(1): 158–167.

Eilers, Rebecca E., W. R. Wilson, and J. M. Moore. 1977. Developmental changes in speech discrimination in infants. *Journal of Speech and Hearing Research*, 20(4): 766–780.

Eimas, P. D., E. R. Siqueland, P. Jusczyk, and J. Vigorito. 1971. Speech perception in infants. *Science*, 171(3968): 303–306.

Eisenberg, J. F., N. A. Muckenhirn, and R. Rudran. 1972. The relation between ecology and social structure in primates. *Science*, 176(4037): 863–874.

Ekman, G. 1954. Dimensions of color vision. *Journal of Psychology*, 38(2nd half): 467–474.

Ekman, P. 1973. Cross-cultural studies of facial expression. In P. Ekman, ed., *Darwin and facial expression: a century of research in review*, pp. 169–222. Academic Press, New York.

Ember, M. 1975. On the origin and extension of the incest taboo. *Behavior Science Research* (Human Relations Area Files, New Haven, Conn.), 10(4): 249–281.

Emlen, J. M. 1967. On the importance of cultural and biological determinants in human behavior. *American Anthropologist*, 69(5): 513–514.

Emlen, S. T. 1976. An alternative case for sociobiology. *Science*, 192(4241): 736–738.

Emlen, S. T. 1980. Ecological determinism and sociobiology. In G. W. Barlow and J. Silverberg, eds., *Sociobiology: beyond nature/nurture?*, pp. 125–150. Westview Press, Boulder, Colo.

Engen, T. 1974. Method and theory in the study of odor preferences. In A. Turk, J. W. Johnson, and D. G. Moulton, eds., *Human responses to environmental odors*, pp. 121–141. Academic Press, New York.

Engen, T. 1979. The origin of preferences in taste and smell. In J. H. A. Kroeze, ed., *Preference behaviour and chemoreception*, pp. 263–273. Information Retrieval, London.

Engen, T., and B. M. Ross. 1973. Long-term memory of odors with and without verbal descriptions. *Journal of Experimental Psychology*, 100(2): 221–227.

Eysenck, H. J. 1968. An experimental study of aesthetic preference for polygonal figures. *Journal of General Psychology*, 79(1st half): 3–17.

Fagan, J. F. III. 1979. The origins of facial pattern recognition. In M. H. Bornstein and W. Kessen, eds., *Psychological development from infancy: image to intention,* pp. 83–113. John Wiley, New York.

Fagen, R. 1981. *Animal play behavior.* Oxford University Press, New York. xvii + 684 pp.

Falconer, D. S. 1960. *Introduction to quantitative genetics.* Ronald Press, New York. x + 365 pp.

Fantz, R. L. 1963. Pattern vision in newborn infants. *Science,* 140(3564): 296–297.

Fantz, R. L., J. F. Fagan III, and S. B. Miranda. 1975. Early visual selectivity: as a function of pattern variables, previous exposure, age from birth and conception, and expected cognitive deficit. In L. B. Cohen and P. Salapatek, eds., *Infant perception: from sensation to cognition,* vol. 1: *Basic visual processes,* pp. 249–345. Academic Press, New York.

Farley, J. D. 1976. Phylogenetic adaptations and the genetics of psychosis. *Acta Psychiatrica Scandinavica,* 53(3): 173–192.

Feldman, M. W., and L. L. Cavalli-Sforza. 1976. Cultural and biological evolutionary processes, selection for a trait under complex transmission. *Theoretical Population Biology,* 9(2): 238–259.

Feldman, M. W., and L. L. Cavalli-Sforza. 1977. The evolution of continuous variation: II, complex transmission and assortative mating. *Theoretical Population Biology,* 11(2): 161–181.

Feldman, M. W., and L. L. Cavalli-Sforza. 1979. Aspects of variance and covariance analysis with cultural inheritance. *Theoretical Population Biology,* 15(3): 276–307.

Feller, W. 1958. *An introduction to probability theory and its applications,* 2nd ed., vol. 1. John Wiley, New York, xvi + 461 pp.

Finger, S. 1975. Child-holding patterns in Western art. *Child Development,* 46(1): 267–271.

Fishbein, H. D. 1976. *Evolution, development, and children's learning.* Goodyear, Pacific Palisades, Calif. xx + 332 pp.

Fishbein, M., and I. Ajzen. 1975. *Belief, attitude, intention and behaviour: an introduction to theory and research.* Addison-Wesley, Reading, Mass. xii + 578 pp.

Flannery, K. V. 1965. The ecology of early food production in Mesopotamia. *Science,* 147(3663): 1247–1256.

Flannery, K. V. 1972. The cultural evolution of civilizations. *Annual Review of Ecology and Systematics,* 3: 399–426.

Flannery, K. V. 1973. The origins of agriculture. *Annual Review of Anthropology,* 2: 271–310.

Ford, E. B. 1971. *Ecological genetics,* 3rd ed. Chapman and Hall, London. xx + 410 pp.

Ford, J. 1974. Stochastic behavior in nonlinear oscillator systems. In W. C. Schieve and J. S. Turner, eds., *Lectures in statistical physics: lecture notes in physics,* vol. 28, pp. 204–247. Springer-Verlag, New York.

Ford, J., and G. H. Lunsford. 1970. Stochastic behavior of resonant nearly linear oscillator systems in the limit of zero nonlinear coupling. *Physical Review A,* 1(1): 59–70.

Fox, R. 1980. *The red lamp of incest.* E. P. Dutton, New York, xiv + 271 pp.

Freedman, D. A. 1979. The sensory deprivations: an approach to the study of the emergence of affects and the capacity for object relations. *Bulletin of the Menninger Clinic,* 43(1): 29–68.

Freedman, D. G. 1974. *Human infancy: an evolutionary perspective.* Lawrence Erlbaum Associates, Hillsdale, N.J. xii + 212 pp.

Freedman, D. G. 1979. *Human sociobiology.* Free Press, Macmillan Co., New York. 188 pp.

Freedman, J. L., D. O. Soars, and J. M. Carlsmith. 1978. *Social psychology,* 3rd ed. Prentice-Hall, Englewood Cliffs, N.J. xiv + 628 pp.

Fry, R. E. 1979. The economics of pottery at Tikal, Guatemala: models of exchange for serving vessels. *American Antiquity,* 44(3): 494–512.

Futuyma, D. J. 1979. *Evolutionary biology.* Sinauer Asociates, Sunderland, Mass. xii + 565 pp.

Gajdusek, D. C. 1970. Physiological and psychological characteristics of stone age man. *Science and Technology,* 33(6): 26–62.

Gajdusek, D. C. 1977. Unconventional viruses and the origin and disappearance of kuru. In W. Odelberg, ed., *Les Prix Nobel en 1976* (Nobel Foundation), pp. 167–216. P. A. Norstedt and Söner, Stockholm.

Gamble, C. 1980. Information exchange in the Palaeolithic. *Nature,* 283(5747): 522–523.

García-Bellido, A. 1975. Genetic control of wing disc development in *Drosophila.* In S. Brenner, ed., *Cell patterning* (Ciba Foundation Symposium 29, new series), pp. 161–182. Elsevier, New York.

Geertz, C. 1966. Religion as a cultural system. In M. P. Banton, ed., *Anthropological approaches to the study of religion,* pp. 1–46. Tavistock, London.

Geertz, C. 1973. *The interpretation of cultures: selected essays by Clifford Geertz.* Basic Books, New York. x + 470 pp.

Geschwind, N. 1979. Specializations of the human brain. *Scientific American,* 241(3): 180–199.

Getty, D. J., J. A. Swets, J. B. Swets, and D. M. Green. 1979. On the prediction of confusion matrices from similarity judgment. *Perception and Psychophysics,* 26(1): 1–19.

Ghysen, A., and J. Richelle. 1979. Determination of sensory bristles and pattern formation in *Drosophila:* II, the achaete-scute locus. *Developmental Biology,* 70(2): 438–452.

Girgus, Joan S., S. Coren, and R. Fraenkel. 1975. Levels of perceptual processing in the development of visual illusions. *Developmental Psychology,* 11(3): 268–273.

Godelier, M. 1975. Modes of production, kinship, and demographic structures. In M. Block, ed., *Marxist analyses and social anthropology,* pp. 3–27. John Wiley, New York.

Godelier, M. 1977. *Perspectives in Marxist anthropology.* Cambridge University Press, New York. vi + 243 pp.

Goel, N. S., and N. Richter-Dyn. 1974. *Stochastic models in biology.* Academic Press, New York. x + 269 pp.

Goldman, P. S., and P. T. Rakic. 1979. Impact of the outside world upon the developing primate brain: perspective from neurobiology. *Bulletin of the Menninger Clinic,* 43(1): 20–28.

Goleman, D., and Sherida Bush. 1977. The liberation of sexual fantasy. *Psychology Today,* 11(5): 48–53, 104–107.

Goodall, Jane. 1965. Chimpanzees of the Gombe Stream Reserve. In I. DeVore, ed., *Primate behavior: field studies of monkeys and apes,* pp. 425–473. Holt, Rinehart & Winston, New York.

Goodman, C. S. 1978. Isogenic grasshoppers: genetic variability in the morphology of identified neurons. *Journal of Comparative Neurology,* 182(4): 681–705.

Görtz, R. 1976. On the solution of Markovian master equations. *Journal of Physics A: Mathematical and General,* 9(7): 1089–1092.

Gould, P. R. 1963. Man against his environment: a game theoretic framework. *Annals of the Association of American Geographers,* 53(3): 290–297.

Gould, S. J., and N. Eldredge. 1977. Punctuated equilibria: the tempo and mode of evolution reconsidered. *Paleobiology,* 3(2): 115–151.

Gray, P. H. 1958. Theory and evidence of imprinting in human infants. *Journal of Psychology,* 46: 155–166.

Greenberg, J. 1978. The brain: holding the secrets of behavior. *Science News,* 114(22): 363–364, 366.

Greene, L. S. 1974. Physical growth and development, neurological maturation, and behavioral functioning in two Ecuadorean Andean communities in which goiter is endemic: II, PTC taste sensitivity and neurological maturation. (Unpublished manuscript cited by P. Rozin, 1976.)

Greeno, J. G. 1974. Representation of learning as discrete transition in a finite state space. In D. H. Krantz, R. C. Atkinson, R. D. Luce, and P. Suppes, eds., *Contemporary developments in mathematical psychology,* vol. 1: *Learning, memory and thinking,* pp. 1–43. W. H. Freeman, San Francisco.

Gregg, L. W., and H. A. Simon. 1967. Process models and stochastic theories of simple concept formation. *Journal of Mathematical Psychology,* 4(2): 246–276.

Grewal, T., T. Gopaldas, P. Hartenberger, I. Ramakrishnan, and G. Ramachandran. 1973. Influence of sugar and flavour on the acceptability of instant CSM: trials on young children from an urban orphanage. *Journal of Food Science and Technology,* 10(4): 149–152.

Griffin, D. R. 1976. *The question of animal awareness: evolutionary continuity of mental experience.* Rockefeller University Press, New York. viii + 135 pp.

Griliches, Z. 1957. Hybrid corn: an exploration in the economics of technological change. *Econometrica,* 25(4): 501–522.

Gross, D. R., G. Eiten, Nancy M. Flowers, Francisca M. Leoi, Madeline L. Ritter, and D. W. Werner. 1979. Ecology and acculturation among native peoples of central Brazil. *Science,* 206(4422): 1043–1050.

Grossberg, S. 1978. A theory of human memory: self-organization and performance of sensory-motor codes, maps, and plans. *Progress in Theoretical Biology,* 5: 233–374.

Guttman, L. 1954. A new approach to factor analysis: the radex. In P. F. Lazarsfeld, ed., *Mathematical thinking in the social sciences,* pp. 258–348. Free Press, Glencoe, Ill.

Haaf, R. A., and R. Q. Bell. 1967. A facial dimension in visual discrimination by human infants. *Child Development,* 38(3): 893–899.

Hage, P. 1976. Structural balance and clustering in bushmen kinship relations. *Behavioral Science,* 21(1): 36–47.

Haggett, P. 1965. *Locational analysis in human geography.* St. Martin's Press, New York. xii + 339 pp.

Haggett, P. 1972. *Geography: a modern synthesis.* Harper & Row, New York. xx + 483 pp.

Haken, H. 1975. Cooperative phenomena in systems far from thermal equilibrium and in nonphysical systems. *Reviews of Modern Physics,* 47(1): 67–122.

Haken, H. 1977. *Synergetics.* Springer-Verlag, New York. xii + 325 pp.

Haldane, J. B. S., and S. D. Jayakar. 1963. Polymorphism due to selection of varying direction. *Journal of Genetics,* 58(2): 237–242.

Hall, J. C., and R. J. Greenspan. 1979. Genetic analysis of *Drosophila* neurobiology. *Annual Review of Genetics,* 13: 127–195.

Hallpike, C. R. 1979. *The foundations of primitive thought.* Clarendon Press, Oxford. xiv + 516 pp.

Hamblin, R. L., J. L. L. Miller, and D. E. Saxton. 1979. Modeling use diffusion. *Social Forces,* 57(3): 799–811.

Hamburg, M. 1977. *Statistical analysis for decision making,* 2nd ed. Harcourt Brace Jovanovich, New York. xiv + 801 pp.

Hames, R. 1979. A comparison of the efficiencies of the shotgun and the bow in Neotropical forest hunting. *Human Ecology,* 7(3): 219–252.

Hamilton, W. D. 1964. The genetical evolution of social behaviour, I and II. *Journal of Theoretical Biology,* 7(1): 1–52.

Hansen, Judith F. 1979. *Sociocultural perspectives on human learning.* Prentice-Hall, Englewood Cliffs, N.J. viii + 280 pp.

Hardyck, C., and L. F. Petrinovich. 1977. Left-handedness. *Psychological Bulletin,* 84(3): 385–404.

Harkness, Sara. 1973. Universal aspects of learning color codes: a study in two cultures. *Ethos,* 1(2): 175–200.

Harlow, H. F., M. K. Harlow, R. O. Dodsworth, and G. L. Arling. 1966. Maternal behavior of rhesus monkeys deprived of mothering and peer associations in infancy. *Proceedings of the American Philosophical Society,* 110(1): 58–66.

Harris, Adrienne. 1979. Recent findings on infant socialization from North American research. *International Social Sciences Journal,* 31(3): 415–428.

Harris, M. 1968. *The rise of anthropological theory.* Thomas Y. Crowell, New York. x + 806 pp.

Harris, M. 1979. *Cultural materialism: the struggle for a science of culture.* Random House, New York. xii + 381 pp.

Harrison, G. A., ed. 1977. *Population structure and human variation.* Cambridge University Press, New York. xviii + 342 pp.

Hartl, D. L. 1980. *Principles of population genetics.* Sinauer Associates, Sunderland, Mass. xvi + 488 pp.

Haslerud, G. M. 1938. The effect of movement of stimulus objects upon avoidance reactions in chimpanzees. *Journal of Comparative Psychology,* 25(3): 507–528.

Hassan, F. A. 1979. Demography and archaeology. *Annual Review of Anthropology,* 8: 137–160.

Hatch, E. 1973. *Theories of man and culture.* Columbia University Press, New York. xii + 384 pp.

Heilbroner, R. L. 1980. *Marxism: for and against.* W. W. Norton, New York. 186 pp.

Herrnstein, R. J. 1971. I. Q. *Atlantic Monthly,* 228(3): 43–64.

Hershenson, M., H. Munsinger, and W. Kessen. 1965. Preference for shapes of intermediate variability in the newborn human. *Science,* 147(3658): 630–631.

Hess, E. H. 1973. *Imprnting: early experience and the developmental psychobiology of attachment.* Van Nostrand Reinhold, New York. xvi + 472 pp.

Heston, L. L., and J. Shields. 1968. Homosexuality in twins: a family study and a registry study. *Archives of General Psychiatry,* 18(2): 149–160.

Hiernaux, J. 1977. Long-term biological effects of human migration from the African savanna to the equatorial forest: a case study of human adaptation to a hot and wet climate. In G. A. Harrison, ed., *Population structure and human variation,* pp. 187–217. Cambridge University Press, Cambridge.

Hill, J. 1978. The origin of sociocultural evolution. *Journal of Social and Biological Structures,* 1(4): 377–386.

Hill, W. C. O. 1972. *Evolutionary biology of the primates.* Academic Press, New York. x + 233 pp.

Hillman, D. E. 1979. Neuronal shape parameters and substructures as a basis of neuronal form. In F. O. Schmitt and F. G. Worden, eds., *The neurosciences: fourth study program,* pp. 477–498. MIT Press, Cambridge, Mass.

Hinde, R. A. 1970. *Animal behaviour: a synthesis of ecology and comparative psychology,* 2nd ed. McGraw-Hill Book Co., New York. xvi + 876 pp.

Hinde, R. A., and Yvette Spencer-Booth. 1969. The effect of social companions on mother-infant relations in rhesus monkeys. In D. Morris, ed., *Primate ethology: essays on the socio-sexual behavior of apes and monkeys,* pp. 343–364. Aldine Publishing Co., Chicago.

Hirshleifer, J. 1977. Economics from a biological viewpoint. *Journal of Law and Economics,* 20(1): 1–52.

Hirshleifer, J. 1978. Natural economy versus political economy. *Journal of Social and Biological Structures,* 1(4): 319–337.

Hoagland, H. 1964. Science and the new humanism. *Science,* 143(3602): 111–114.

Hofstadter, D. R. 1979. *Gödel, Escher, Bach: an eternal golden braid.* Basic Books, New York. xxii + 777 pp.

Holt, L. E., and J. Howland. 1939. *Holt's diseases of infancy and childhood,* 11th ed. Rev. by L. E. Holt, Jr., and R. McIntosh. Appleton-Century, New York. xxviii + 1421 pp.

Horr, D. A. 1977. Orang-utan maturation: growing up in a female world. In Suzanne Chevalier-Skolnikoff and F. E. Poirier, eds., *Primate bio-social development: biological, social, and ecological determinants,* pp. 289–321. Garland, New York.

Horsthemke, W., and L. Brenig. 1977. Non-linear Fokker-Planck equation as an asymptotic representation of the master equation. *Zeitschrift für Physik B,* 27: 341–348.

Horton, D. R. 1979. Tasmanian adaptation. *Mankind,* 12(1): 28–34.

Howard, R. A. 1960. *Dynamic programming and Markov process.* MIT Press, Cambridge, Mass. viii + 136 pp.

Howard, R. A. 1971a. *Dynamic probabilistic systems,* vol. 1: *Markov models.* John Wiley, New York. xviii + 576 pp.

Howard, R. A. 1971b. *Dynamic probabilistic systems,* vol 2: *Semi-Markov and decision processes.* John Wiley, New York. xviii + 533 pp.

Howell, Nancy. 1979. *Demography of the Dobe !Kung.* Academic Press, New York. xxii + 389 pp.

Hubel, D. H., T. N. Wiesel, and S. LeVay. 1977. Plasticity of ocular dominance columns in monkey striate cortex. *Philosophical Transactions of the Royal Society of London* (Biology), 278(961): 377–409.

Hubert, Helen B., R. R. Fabsitz, M. Feinlab, and K. S. Brown. 1980. Olfactory sensitivity in humans: genetic versus environmental control. *Science,* 20(4444): 607–609.

Hutchins, E. 1980. Culture and inference: a Trobriand case study. Harvard University Press, Cambridge, Mass. viii + 144 pages.

Huxley, J. S. 1958. Cultural process and evolution. In Anne Roe and G. G. Simpson, eds., *Behavior and evolution,* pp. 437–454. Yale University Press, New Haven, Conn.

Huxley, J. S. 1962. Evolution: biological and human. *Nature,* 196(4851): 203–204.

Irons, W. 1979. Cultural and biological success. In N. A. Chagnon and W. Irons, eds., *Evolutionary biology and human social behavior: an anthropological perspective,* pp. 257–272. Duxbury Press, North Scituate, Mass.

Isaac, G. L. 1972. Chronology and the tempo of cultural change during the Pleistocene. In W. W. Bishop and J. A. Miller, eds., *Calibration of hominid evolution,* pp. 381–430. University of Toronto Press, Toronto.

Jacobson, A. G. 1978. Some forces that shape the nervous system. *Zoon,* 6: 13–21.

Jacobson, M. 1978a. *Developmental neurobiology,* 2nd ed. Plenum Press, New York. xii + 562 pp.

Jacobson, M. 1978b. Clonal origins of the central nervous system: towards a developmental neuroanatomy. *Zoon,* 6: 149–156.

Janzen, D. H. 1980. What is coevolution? *Evolution,* 34(3): 611–612.

Jardine, N., and R. Sibson. 1971. *Mathematical taxonomy.* John Wiley, New York. xviii + 286 pp.

Jenkins, J. J., W. A. Russell, and G. J. Suci. 1958. An atlas of semantic profiles for 360 words. *American Journal of Psychology,* 71(4): 688–699.

Jenni, D. A., and Mary A. Jenni, 1976. Carrying behavior in humans: analysis of sex differences. *Science,* 194(4267): 859–860.

Jerison, H. J. 1975. Fossil evidence of the evolution of the human brain. *Annual Review of Anthropology,* 4: 27–58.

Jinks, J. L. 1979. The biometrical approach to quantitative variation. In J. N. Thompson, Jr., and J. M. Thoday, eds., *Quantitative genetic variation,* pp. 81–109. Academic Press, New York.

Jirari, Carolyn G. 1970. Form perception, innate form preference, and visually mediated head-turning in the human neonate. Ph.D. dissertation, University of Chicago. (Cited by Hess, 1973, and Freedman, 1974.)

Jochim, M. A. 1976. *Hunter-gatherer subsistence and settlement: a predictive model.* Academic Press, New York. xviii + 206 pp.

Johnson, E. G. 1977. The development of color knowledge in preschool children. *Child Development,* 48(1): 308–311.

Johnston, F. E., and H. Selby. 1978. *Anthropology: the biosocial view.* W. C. Brown, Dubuque, Iowa. xiv + 625 pp.

Jones, J. S., B. H. Leith, and P. Rawlings. 1977. Polymorphism in *Cepaea:* a problem with too many solutions? *Annual Review of Ecology and Systematics,* 8: 109–143.

Jones, R. 1977. The Tasmanian paradox. In R. V. S. Wright, ed., *Stone tools as cultural markers: change, evolution, and complexity,* pp. 189–204. Australian Institute of Aboriginal Studies, Canberra.

Kaffman, M. 1977. Sexual standards and behavior of the kibbutz adolescent. *American Journal of Orthopsychiatry,* 47(2): 207–217.

Kagan, J. 1970. The determinants of attention in the infant. *American Scientist,* 58(3): 298–306.

Kaplan, D., and R. A. Manners. 1972. *Culture theory.* Prentice-Hall, Englewood Cliffs, N.J. xii + 212 pp.

Karlin, S. 1979a. Models of multifactorial inheritance: I, multivariate formulations and basic convergence results. *Theoretical Population Biology,* 15(3): 308–355.

Karlin, S. 1979b. Models of multifactorial inheritance: II, the covariance structure for a scalar phenotype under selective assortment and sex-dependent symmetric parental-transmission. *Theoretical Population Biology,* 15(3): 356–393.

Karlin, S. 1979c. Models of multifactorial inheritance: III, calculation of covariance of relatives under selective assortative mating. *Theoretical Population Biology,* 15(3): 394–423.

Karlin, S. 1979d. Models of multifactorial inheritance: IV, asymmetric transmission for a scalar phenotype. *Theoretical Population Biology,* 15(3): 424–438.

Karlin, S. 1980a. Models of multifactorial inheritance. V. Linear assortative mating as against selective (nonlinear) assortative mating. *Theoretical Population Biology,* 17(3): 255–275.

Karlin, S. 1980b. Models of multifactorial inheritance. VI. Formulas and properties of the vector phenotype equilibrium covariance matrix. *Theoretical Population Biology,* 17(3): 276–297.

Katz, S. H. 1980. Fava bean consumption and biocultural fitness. *Annales de Sociologie,* in press.

Katz, S. H., M. L. Hediger, and L. A. Valleroy. 1974. Traditional maize processing techniques in the New World. *Science,* 184(4138): 765–773.

Kay, P. 1975. Synchronic variability and diachronic change in basic color terms. *Language in Society,* 4(3): 257–270.

Keene, A. S. 1979. Economic optimization models and the study of hunter-gatherer subsistence settlement systems. In C. Renfrew and K. L. Cooke, eds., *Transformations: mathematical approaches to culture change,* pp. 369–404. Academic Press, New York.

Keene, A. S. 1981. Optimal foraging in a nonmarginal environment: a model of prehistoric subsistence strategies in Michigan. In B. P. Winterhalder and E. A. Smith, eds., *Hunter-gatherer foraging strategies: ethnographic and archaeological analyses.* University of Chicago Press, Chicago, forthcoming.

Kemeny, J. G., and J. L. Snell. 1962. *Mathematical models in the social sciences.* Blaisdell Publishing Co., Waltham, Mass. viii + 145 pp.

Kennell, J. H., and M. H. Klaus. 1979. Early mother-infant contact: effects on the mother and infant. *Bulletin of the Menninger Clinic,* 43(1): 69–78.

Kety, S. S. 1979. Disorders of the human brain. *Scientific American,* 241(3): 202–214.

Kirch, P. V. 1980. Polynesian prehistory: cultural adaptation in island ecosystems. *American Scientist,* 68(1): 39–48.

Klaus, M. H., R. Jerauld, Nancy C. Kreger, W. McAlpine, M. Steffa, and J. H. Kennell. 1972. Maternal attachment: importance of the first post-partum days. *New England Journal of Medicine,* 286(9): 460–463.

Klejn, L. S. 1973. Marxism, the systemic approach, and archaeology. In C. Renfrew, ed., *The explanation of culture change: models in prehistory,* pp. 691–710. University of Pittsburgh Press, Pittsburgh.

Konner, M. J. 1972. Aspects of the developmental ethology of a foraging people. In N. G. Blurton Jones, ed., *Ethological studies of child behaviour,* pp. 285–304. Cambridge University Press, Cambridge.

Konner, M. J. 1977. Quoted in J. Greenberg, The brain and emotions. *Science News,* 112(5): 74–75.

Kosslyn, S. M. 1980. *Image and mind.* Harvard University Press, Cambridge, Mass. xviii + 500 pp.

Kovach, J. K. 1980. Mendelian units of inheritance control color preferences in quail chicks (*Coturnix coturnix japonica*). *Science,* 207(4430): 549–551.

Kraut, R. E., and R. E. Johnston. 1979. Social and emotional messages of smiling: an ethological approach. *Journal of Personality and Social Psychology,* 37(9): 1539–1553.

Kretchmer, N. 1972. Lactose and lactase. *Scientific American,* 227(4): 70–78.

Kronenfeld, D., and H. W. Decker. 1979. Structuralism. *Annual Review of Anthropology,* 8: 503–541.

Langer, Susanne K. 1967. *Mind: an essay on human feeling,* vol. 1. Johns Hopkins University Press, Baltimore. xxii + 487 pp.

Langer, Susanne K. 1972. *Mind: an essay on human feeling,* vol. 2. Johns Hopkins University Press, Baltimore. xii + 400 pp.

Larkin, Jill, J. McDermott, Dorothea P. Simon, and H. A. Simon. 1980. Expert and novice performance in solving physics problems. *Science,* 208(4450): 1335–1342.

Lauer, J., and M. Lindauer. 1971. *Genetisch fixierte Lerndispositionen bei der Honigbiene.* Akademie der Wissenschaften und der Literatur, Mainz. 87 pp.

Laughlin, C. D., and E. G. d'Aquili. 1974. *Biogenetic structuralism.* Columbia University Press, New York. x + 211 pp.

Leaf, M. J. 1979. *Man, mind, and science: a history of anthropology*. Columbia University Press, New York. xvi + 376 pp.

Lee, R. B. 1969. Eating Christmas in the Kalahari. *Natural History,* 78(10): 14, 16, 18, 21, 22, 60–63.

Lee, R. B. 1976. !Kung spatial organization: an ecological and historical perspective. In R. B. Lee and I. DeVore, eds., *Kalahari hunter-gatherers: studies of the !Kung San and their neighbors,* pp. 73–97. Harvard University Press, Cambridge, Mass.

Lee, R. B. 1979. *The !Kung San: men, women, and work in a foraging society.* Cambridge University Press, New York. xxvi + 526 pp.

Leibenstein, H. 1976. *Beyond economic man: a new foundation for microeconomics.* Harvard University Press, Cambridge, Mass. xiv + 297 pp.

Leinhardt, S., ed. 1977. *Social networks: a developing paradigm.* Academic Press, New York. xxxiv + 465 pp.

Lenneberg, E. H. 1967. *Biological foundations of language.* John Wiley, New York. xvi + 489 pp.

Lenski, G., and Jean Lenski. 1970. *Human societies: a macrolevel introduction to sociology.* McGraw-Hill Book Co., New York. xvi + 515 pp.

Levarie, S., and N. Rudolph. 1978. Can newborn infants distinguish between tone and noise? *Perceptual and Motor Skills,* 47(3): 1123–1126.

Lévi-Strauss, C. 1969a. *The elementary structures of kinship* (Les structures élémentaires de la parenté), rev. ed., trans. by J. H. Bell; J. R. von Sturmer and R. Needham, eds. Beacon Press, Boston. xlii + 541 pp. (Original edition published in French, 1949.)

Lévi-Strauss, C. 1969b. *The raw and the cooked: introduction to a science of mythology,* vol. 1. Harper & Row, New York. xiv + 387 pp.

LeVine, R. A., and D. T. Campbell. 1972. *Ethnocentrism: theories of conflict, ethnic attitudes, and group behavior.* John Wiley, New York. x + 310 pp.

Levins, R. 1962. Theory of fitness in a heterogeneous environment, I: The fitness set and adaptive function. *American Naturalist,* 96(891): 361–373.

Levins, R. 1968. *Evolution in changing environments: some theoretical explorations.* Princeton University Press, Princeton, N.J. x + 120 pp.

Levinthal, F., E. Macagno, and C. Levinthal. 1975. Anatomy and development of identified cells in isogenic organisms. *Cold Spring Harbor Symposia on Quantitative Biology,* 40: 321–331.

Lewontin, R. C. 1974. *The genetic basis of evolutionary change.* Columbia University Press, New York. xiv + 346 pp.

Leyhausen, P. 1965. The communal organization of solitary mammals. *Symposia of the Zoological Society of London,* no. 14, pp. 249–263.

Li, T.-Y., and J. A. Yorke. 1975. Period three implies chaos. *American Mathematical Monthly,* 82(10): 985–992.

Liberman, A. M., F. S. Cooper, D. P. Shankweiler, and M. Studdert-Kennedy. 1967. Perception of the speech code. *Psychological Review,* 74(6): 431–461.

Liboff, R. L. 1970. Physical laws and the structure of society. *The Cornell Engineer,* 36(2): 3–15.

Lindsay, P. H., and D. A. Norman. 1977. *Human information processing: an introduction to psychology,* 2nd ed. Academic Press, New York. xxiv + 777 pp.

Lindzey, G., C. S. Hall, and R. F. Thompson. 1975. *Psychology*. Worth Publishers, New York. xiv + 802 pp.

Lisker, L., and A. S. Abramson. 1964. A cross-language study of voicing in initial stops: acoustical measurements. *Word*, 20(3): 384–422.

Lockard, Joan S., ed. 1980. *The evolution of human social behavior*. Elsevier, New York. xvi + 336 pp.

Lockard, Joan S., P. C. Daley, and Virginia M. Gunderson. 1979. Maternal and paternal differences in infant carry: U.S. and African data. *American Naturalist*, 113(2): 235–246.

Loehlin, J. C., and R. C. Nichols. 1976. *Heredity, environment, and personality*. University of Texas Press, Austin. xii + 202 pp.

Loftus, G. R., and Elizabeth F. Loftus. 1976. *Human memory: the processing of information*. Lawrence Erlbaum Associates, Hillsdale, N.J. xii + 179 pp.

Logue, A. W. 1979. Taste aversion and the generality of the laws of learning. *Psychological Bulletin*, 86(2): 276–296.

Luce, R. D. 1959. *Individual choice behavior: a theoretical analysis*. John Wiley, New York. xii + 153 pp.

Lumsden, C. J. 1977. On the dynamics of biological ensembles: canonical theory and computer simulation. Ph.D. dissertation, University of Toronto.

Lumsden, C. J., and L. E. H. Trainor. 1976. On the physical content of kinetic Ising models. *Canadian Journal of Physics*, 54(23): 2340–2345.

Lumsden, C. J., and E. O. Wilson. 1980a. Translation of epigenetic rules of individual behavior into ethnographic patterns. *Proceedings of the National Academy of Sciences of the United States of America*, 77(7): 4382–4386.

Lumsden, C. J., and E. O. Wilson. 1980b. Gene-culture translation in the avoidance of sibling incest. *Proceedings of the National Academy of Sciences of the United States of America*, 77(10): 6248–6250.

Macagno, E. R., V. Lopresti, and C. Levinthal. 1973. Structure and development of neuronal connections in isogenic organisms: variations and similarities in the optic system of *Daphnia magna*. *Proceedings of the National Academy of Sciences of the United States of America*, 70(1): 57–61.

MacArthur, R. H., and E. O. Wilson. 1967. *The theory of island biogeography*. Princeton University Press, Princeton, N.J. xii + 203 pp.

McCall, R. B., and J. Kagan. 1967. Attention in the infant: effects of complexity, contour, perimeter, and familiarity. *Child Development*, 38(4): 939–952.

McClearn, G. E., and J. C. DeFries. 1973. *Introduction to behavioral genetics*. W. H. Freeman, San Francisco. 349 pp.

Maccoby, Eleanor E., and Carol N. Jacklin. 1974. *The psychology of sex differences*. Stanford University Press, Stanford, Calif. xvi + 634 pp.

Mach, E. 1942. *The science of mechanics*, 9th ed., trans. from the German by T. J. McCormack. Open Court Publishing Co., LaSalle, Ill. xxxii + 634 pp.

McKusick, V. A., and F. H. Ruddle. 1977. The status of the gene map of the human chromosomes. *Science*, 196(4288): 390–405.

Maller, O., and J. A. Desor. 1974. Effect of taste on ingestion by human newborns. In J. Bosma, ed., *Fourth symposium on oral sensation and perception: development in the fetus and infant*, pp. 279–311. Government Printing Office, Washington, D.C.

Marks, I. M. 1969. *Fears and phobias*. Academic Press, New York. viii + 302 pp.
Marshack, A. 1979. Upper Paleolithic symbol systems of the Russian plain: cognitive and comparative analysis. *Current Anthropology*, 20(2): 271–311.
Martin, N. G., L. J. Eaves, and H. J. Eysenck. 1977. Genetical, environmental and personality factors in influencing the age of first sexual intercourse in twins. *Journal of Biosocial Science*, 9(1): 91–97.
Marx, K. 1971 (1859). *A contribution to the critique of political economy*, trans. from the German by S. W. Ryazanskaya and ed. by M. Dobb. Lawrence & Wishart, London. 264 pp.
Massaro, D. W. 1975. *Experimental psychology and information processing*. Rand-McNally, Chicago. ii + 651 pp.
May, R. M. 1976. Simple mathematical models with very complicated dynamics. *Nature*, 261(5560): 459–467.
May, R. M., and G. F. Oster. 1976. Bifurcations and dynamic complexity in simple ecological models. *American Naturalist*, 110(974): 573–599.
Mayer, J., Margaret M. Dickie, Margaret W. Bates, and J. J. Vitale. 1951. Free selection of nutrients by hereditarily obese mice. *Science*, 113(2948): 745–746.
Maynard Smith, J. 1974. The theory of games and the evolution of animal conflicts. *Journal of Theoretical Biology*, 47(1): 209–221.
Maynard Smith, J. 1976. Evolution and the theory of games. *American Scientist*, 64(1): 41–45.
Maynard Smith, J. 1978. Optimization theory in evolution. *Annual Review of Ecology and Systematics*, 9: 31–56.
Maynard Smith, J., and J. Haigh. 1974. The hitch-hiking effect of a favourable gene. *Genetical Research*, 23(1): 23–35.
Mead, Margaret. 1963. Socialization and enculturation. *Current Anthropology*, 4(1): 184–188.
Milgram, S., L. Bickman, and L. Berkowitz. 1969. Note on the drawing power of crowds of different size. *Journal of Personality and Social Psychology*, 13(2): 79–82.
Milkman, R. 1979. The posterior crossvein in *Drosophila* as a model phenotype. In J. N. Thompson, Jr., and J. M. Thoday, eds., *Quantitative genetic variation*, pp. 157–176. Academic Press, New York.
Miller, G. A. 1956a. The magical number seven, plus or minus two: some limits on our capacity for processing information. *Psychological Review*, 63(2): 81–97.
Miller, G. A. 1956b. Information and memory. *Scientific American*, 195(2): 42–46.
Miller, G. A., and Patricia E. Nicely. 1955. An analysis of perceptual confusions among some English consonants. *Journal of the Acoustical Society of America*, 27(2): 338–352.
Mollon, J. D. 1980. Post-receptoral processes in colour vision. *Nature* 283(5748): 623–624.
Money, J., and A. A. Ehrhardt. 1972. *Man and woman, boy and girl*. Johns Hopkins University Press, Baltimore. xvi + 311 pp.
Morgan, G. A., and H. N. Ricciuti. 1973. Infants' response to strangers during the first year. In L. J. Stone, Henrietta T. Smith, and Lois B. Murphy, eds., *The*

competent infant: research and commentary, pp. 1128–1138. Basic Books, New York.

Morris, D. 1971. *Intimate behavior.* Random House, New York. 253 pp.

Morris, Ramona, and D. Morris. 1965. *Men and snakes.* McGraw-Hill Book Co., New York. 224 pp.

Morsbach, Gisela, and Caroline Bunting. 1979. Maternal recognition of their neonates' cries. *Developmental Medicine and Child Neurology,* 21(2): 178–185.

Mortensen, R. E. 1969. Mathematical problems of modeling stochastic nonlinear dynamic systems. *Journal of Statistical Physics,* 1(2): 271–296.

Moschis, G. P., and R. L. Moore. 1979. Decision making among the young: a socialization perspective. *Journal of Consumer Research,* 6(2): 101–112.

Murdock, G. P. 1949. *Social structure.* Macmillan Co., New York. xx + 387 pp.

Nabokov, V. 1970. *Mary, a novel,* trans. from the Russian by Michael Glenny. McGraw-Hill Book Co., New York. xiv + 114 pp.

Napier, J. R., and P. H. Napier. 1967. *A handbook of living primates.* Academic Press, New York. xiv + 456 pp.

Napier, J. R., and P. H. Napier, eds. 1970. *Old World monkeys: evolution, systematics, and behavior.* Academic Press, New York. xvi + 660 pp.

Navon, D., and D. Gopher. 1979. On the economy of the human-processing system. *Psychological Review,* 86(3): 214–255.

Needham, R. 1979. *Symbolic classification.* Goodyear, Santa Monica, Calif. xii + 78 pp.

Needham, R., ed. 1973. *Right & left.* University of Chicago Press, Chicago. xl + 449 pp.

Neisser, U. 1976. *Cognition and reality: principles and implications of cognitive psychology.* W. H. Freeman, San Francisco. xiv + 230 pp.

Newell, A., and H. A. Simon. 1972. *Human problem solving.* Prentice-Hall, Englewood Cliffs, N.J. xvi + 920 pp.

Nicolis, G., and I. Prigogine. 1977. *Self-organization in nonequilibrium systems.* Wiley Interscience, John Wiley, New York. xii + 491 pp.

Nijhout, H. F. 1978. Wing pattern formation in Lepidoptera: a model. *Journal of Experimental Zoology,* 206(2): 119–136.

Nisbett, R., and L. Ross. 1980. *Human inference: strategies and shortcomings of social judgment.* Prentice-Hall, Englewood Cliffs, N.J. xvi + 325 pp.

Norman, D. A., and D. E. Rumelhart. 1975. *Explorations in cognition.* W. H. Freeman, San Francisco. xvi + 430 pp.

Oades, R. D. 1979. Search and attention: interactions of the hippocampal-septal axis, adrenocortical and gonadal hormones. *Neuroscience and Biobehavioral Reviews,* 3(1): 31–48.

O'Connor, Susan M., P. M. Vietze, J. B. Hopkins, and W. A. Altemeier. 1977. Post-partum extended maternal-infant contact: subsequent mothering and child health. *Pediatric Research,* 11(4): 380.

Oden, G. C. 1977. Integration of fuzzy logical information. *Journal of Experimental Psychology,* 3(4): 565–575.

Oden, G. C., and D. W. Massaro. 1978. Integration of featural information in speech perception. *Psychological Review,* 85(3): 172–191.

O'Laughlin, Bridget. 1975. Marxist approaches in anthropology. *Annual Review of Anthropology,* 4: 341–370.

Oliverio, A. 1979. Uses of recombinant inbred strains. In J. N. Thompson, Jr., and J. M. Thoday, eds., *Quantitative genetic variation,* pp. 197–218. Academic Press, New York.

Orr, D. W. 1979. Catastrophe and social order. *Human Ecology,* 7(1): 41–52.

Ortony, A., R. E. Reynolds, and Judith A. Arter. 1978. Metaphor: theoretical and empirical research. *Psychological Bulletin,* 85(5): 919–943.

Osgood, C. E., G. J. Suci, and P. H. Tannenbaum. 1957. *The measurement of meaning.* University of Illinois Press, Urbana. viii + 342 pp.

Oster, G. F., and E. O. Wilson. 1978. *Caste and ecology in the social insects.* Princeton University Press, Princeton, N.J. xvi + 352 pp.

Patterson, P. H. 1979. Epigenetic influences in neuronal development. In F. O. Schmitt and F. G. Worden, eds., *The neurosciences: fourth study program,* pp. 929–936. MIT Press, Cambridge, Mass.

Pendse, S. G. 1978. Category perception, language and brain hemispheres: an information transmission approach. *Behavioral Science,* 23(6): 421–428.

Penrose, O. 1979. Foundations of statistical mechanics. *Reports on Progress in Physics,* 42(12): 1937–2006.

Peterson, C. R., and L. R. Beach. 1967. Man as an intuitive statistician. *Psychological Bulletin,* 68(1): 29–46.

Petryszak, N. G. 1979. The biosociology of the social self. *Sociological Quarterly,* 20(2): 291–303.

Pfeiffer, J. E. 1969. *The emergence of man.* Harper & Row, New York. xxiv + 477 pp.

Piaget, J. 1952. *The origins of intelligence in children,* trans. by Margaret Cook. International Universities Press, New York. xii + 419 pp.

Pielou, E. C. 1979. *Biogeography.* John Wiley, New York. xii + 351 pp.

Pollack, R. H. 1972. Perceptual development: a progress report. In Sylvia Farnham-Diggory, ed., *Information processing in children,* pp. 25–42. Academic Press, New York.

Posner, M. I. 1973. *Cognition: an introduction.* Scott, Foresman and Co., Glenview, Ill. xii + 208 pp.

Preston, F. W. 1962a. The canonical distribution of commonness and rarity, part I. *Ecology,* 43(2): 185–215.

Preston, F. W. 1962b. The canonical distribution of commonness and rarity, part II. *Ecology,* 43(3): 410–432.

Pribram, K. H. 1971. *Languages of the brain: experimental paradoxes and principles in neuropsychology.* Prentice-Hall, Englewood Cliffs, N.J. xvi + 432 pp.

Price, D. de Solla. 1975. *Science since Babylon,* enlarged ed. Yale University Press, New Haven, Conn. xvi + 215 pp.

Pulliam, H. R., and C. Dunford. 1979. *Programmed to learn: an essay on the evolution of culture.* Columbia University Press, New York. xiv + 144 pp.

Purpura, D. P., A. Hirano, and J. H. French. 1976. Polydendritic Purkinje cells in

X-chromosome linked copper malabsorption: a Golgi study. *Brain Research,* 117(1): 125–129.

Pusey, Anne E. 1980. Inbreeding avoidance in chimpanzees. *Animal Behaviour,* 28(2): 543–552.

Quillian, M. R. 1967. Word concepts: a theory and simulation of some basic semantic capabilities. *Behavioral Science,* 12(5): 410–430.

Rachlin, H. 1976. *Behavior and learning.* W. H. Freeman, San Francisco. xvi + 613 pp.

Rainer, J. D. 1979. Heredity and character disorders. *American Journal of Psychotherapy,* 33(1): 6–16.

Rakic, P. 1975a. Local circuit neurons. *Neurosciences Research Program Bulletin,* 13(3): 291–446.

Rakic, P. 1975b. Synaptic specificity in the cerebellar cortex: study of anomalous circuits induced by single gene mutations in mice. *Cold Spring Harbor Symposia on Quantitative Biology,* 40: 333–346.

Rakic, P. 1979. Genetic and epigenetic determinants of local neuronal circuits in the mammalian central nervous system. In F. O. Schmitt and F. G. Worden, eds., *The neurosciences: fourth study program,* pp. 109–127. MIT Press, Cambridge, Mass.

Ramirez, I., and R. L. Sprott. 1979. Diet/taste and feeding behavior of genetically obese mice (C57BL/6J-*ob/ob*). *Behavioral and Neural Biology,* 25(4): 449–472.

Ramón-Moliner, E., and W. J. H. Nauta. 1966. The isodendritic core of the brain stem. *Journal of Comparative Neurology,* 126(3): 311–335.

Rappaport, R. A. 1971. The sacred in human evolution. *Annual Review of Ecology and Systematics,* 2: 23–44.

Rashevsky, N. 1960. *Mathematical biophysics: physico-mathematical foundations of biology,* 3rd ed., vol. 2. Dover, New York. xiv + 462 pp.

Ratliff, F. 1976. On the psychophysiological basis of universal color terms. *Proceedings of the American Philosophical Society,* 120(5): 311–330.

Reidhead, V. A. 1979. Linear programming models in archaeology. *Annual Review of Anthropology,* 8: 543–578.

Reidhead, V. A. 1980. The economics of subsistence change: test of an optimization model. In T. K. Earle and A. L. Christenson, eds., *Modeling change in prehistoric subsistence economies,* pp. 141–186. Academic Press, New York.

Reif, F. 1965. *Fundamentals of statistical and thermal physics.* McGraw-Hill Book Co., New York. xx + 651 pp.

Reijnders, L. 1978. On the applicability of game theory to evolution. *Journal of Theoretical Biology,* 75(1): 245–247.

Rendel, J. M. 1967. *Canalisation and gene control.* Logos Press, London. 166 pp.

Rendel, J. M. 1979. Canalisation and selection. In J. N. Thompson, Jr., and J. M. Thoday, eds., *Quantitative genetic variation,* pp. 139–156. Academic Press, New York.

Renfrew, C., and K. L. Cooke, eds. 1979. *Transformations: mathematical approaches to culture change*. Academic Press, New York. xxii + 515 pp.

Rice, J., C. R. Cloninger, and T. Reich. 1978. Multifactorial inheritance with cultural transmission and assortative mating: I, description and basic properties of the unitary models. *American Journal of Human Genetics*, 30(6): 618–643.

Richards, Audrey I. 1939. *Land, labour and diet in northern Rhodesia: an economic study of the Bemba tribe*. Oxford University Press, New York. xvi + 415 pp.

Richardson, Jane, and A. L. Kroeber. 1940. Three centuries of women's dress fashions: a quantitative analysis. *University of California Anthropological Records*, 5(2): i–iv, 111–153.

Richelle, J., and A. Ghysen. 1979. Determination of sensory bristles and pattern formation in *Drosophila:* I, a model. *Developmental Biology*, 70(2): 418–437.

Richerson, P. J., and R.Boyd. 1978. A dual inheritance model of the human evolutionary process: I, basic postulates and a simple model. *Journal of Social and Biological Structures*, 1(2): 127–154.

Richter, C. P., and Katherine K. Rice. 1945. Self-selection studies on coprophagy as a source of vitamin B complex. *American Journal of Physiology*, 143(3): 344–354.

Roederer, J. G. 1978. On the relationship between human brain functions and the foundations of physics, science, and technology. *Foundations of Physics*, 8(5/6): 423–438.

Rohner, R. P. 1975. *They love me, they love me not*. HRAF Press, New Haven, Conn. 300 pp.

Rosch, Eleanor. 1973. Natural categories. *Cognitive Psychology*, 4(3): 328–350.

Rosch, Eleanor. 1975. Universals and cultural specifics in human categorization. In R. W. Brislin, S. Bochner, and W. J. Lonner, eds., *Cross-cultural perspectives on learning*, pp. 177–206. Halsted Press, Wiley, New York.

Rosch, Eleanor, and Barbara B. Lloyd, eds. 1978. *Cognition and categorization*. Lawrence Erlbaum Associates, Hillsdale, N.J. viii + 328 pp.

Rosch, Eleanor, Carolyn B. Mervis, W. D. Gray, D. M. Johnson, and Penny Boyes-Braem. 1976. Basic objects in natural categories. *Cognitive Psychology*, 8(3): 382–439.

Rose, Hilary, and S. Rose, eds. 1976. *The radicalisation of science*, vols. 1 and 2. Macmillan Co., London. Vol. 1, xxvi + 218 pp.; vol. 2, xxvi + 205 pp.

Rosenblatt, J. S. 1972. Learning in newborn kittens. *Scientific American*, 227(6): 18–25.

Rosenthal, T. L., and B. J. Zimmerman. 1978. *Social learning and cognition*. Academic Press, New York. xiv + 336 pp.

Roughgarden, J. 1979. *Theory of population genetics and evolutionary ecology: an introduction*. Macmillan Co., New York. x + 634 pp.

Routtenberg, A. 1978. The reward system of the brain. *Scientific American*, 239(5): 154–164.

Rozin, P. 1976. The selection of foods by rats, humans, and other animals. *Advances in the Study of Behavior*, 6: 21–76.

Russell, M. J. 1976. Human olfactory communication. *Nature*, 260(5551): 520–522.

Ryle, G. 1971. *Collected papers, vol. 2: collected essays, 1929–1968*. Hutchinson, London. viii + 496 pp.

Salapatek, P. 1973. The visual investigation of geometric pattern by the one- and two-month-old infant. In L. J. Stone, Henrietta T. Smith, and Lois B. Murphy, eds., *The competent infant: research and commentary*, pp. 631–637. Basic Books, New York.

Salisbury, R. F. 1962. *From stone to steel: economic consequences of a technological change in New Guinea*. Cambridge University Press, Cambridge. xxii + 237 pp.

Salk, L. 1973. The role of the heartbeat in the relations between mother and infant. *Scientific American*, 228(5): 24–29.

Samuelson, P. A. 1976. *Economics*, 10th ed. McGraw-Hill Book Co., New York. xxviii + 917 pp.

Savage-Rumbaugh, E. Sue, and D. M. Rumbaugh. 1978. Symbolization, language, and chimpanzees: a theoretical reevaluation based on initial language acquisition processes in four young *Pan troglodytes*. *Brain and Language*, 6(3): 265–300.

Schank, R. C., and K. M. Colby, eds. 1973. *Computer models of thought and language*. W. H. Freeman, San Francisco. ix + 454 pp.

Schelling, T. C. 1978. *Micromotives and macrobehavior*. W. W. Norton, New York. 252 pp.

Schmitt, F. O., and F. G. Worden, eds. 1979. *The neurosciences: fourth study program*. MIT Press, Cambridge, Mass. xvi + 1185 pp.

Schmitt, F. O., P. Dev, and B. H. Smith. 1976. Electrotonic processing of information by brain cells. *Science*, 193(4248): 114–120.

Schnakenberg, J. 1976. Network theory of microscopic and macroscopic behavior of master equation systems. *Reviews of Modern Physics*, 48(4): 571–585.

Schneider, D. 1969. Insect olfaction: deciphering system for chemical messages. *Science*, 163(3871): 1031–1037.

Schneider, D. M. 1980. *American kinship: a cultural account*, 2nd ed. University of Chicago Press, Chicago. x + 137 pp.

Schroder, H. M., M. J. Driver, and S. Streufert. 1967. *Human information processing*. Holt, Rinehart & Winston, New York. x + 224 pp.

Seemanová, Eva. 1971. A study of children of incestuous matings. *Human Heredity*, 21(1): 108–128.

Seligman, M. E. P. 1972a. Introduction. In M. E. P. Seligman and Joanne L. Hager, eds., *Biological boundaries of learning*, pp. 1–6. Appleton-Century-Crofts, New York.

Seligman, M. E. P. 1972b. Phobias and preparedness. In M. E. P. Seligman and Joanne L. Hager, eds., *Biological boundaries of learning*, pp. 451–460. Appleton-Century-Crofts, New York.

Seligman, M. E. P., and Joanne L. Hager, eds. 1972. *Biological boundaries of learning*. Appleton-Century-Crofts, New York. xiv + 480 pp.

Shepard, R. N. 1958a. Stimulus and response generalization: deduction of the generalization gradient from a trace model. *Psychological Review,* 65(4): 242–256.

Shepard, R. N. 1958b. Stimulus and response generalization: tests of a model relating generalization to distance in psychological space. *Journal of Experimental Psychology,* 55(6): 509–523.

Shepard, R. N. 1978. The circumplex and related topological manifolds in the study of perception. In S. Shye, ed., *Theory construction and data analysis in the behavioral sciences,* pp. 29–80. Jossey-Bass, San Francisco.

Shepard, R. N., and P. Arabie. 1979. Additive clustering: representation of similarities as combinations of discrete overlapping properties. *Psychological Review,* 86(2): 87–123.

Shepher, J. 1971. Mate selection among second-generation kibbutz adolescents and adults: incest avoidance and negative imprinting. *Archives of Sexual Behavior,* 1(4): 293–307.

Shepherd, G. M. 1974. *The synaptic organization of the brain: an introduction.* Oxford University Press, New York. xii + 364 pp.

Shettleworth, Sara J. 1972. Constraints on learning. *Advances in the Study of Behavior,* 4: 1–68.

Simon, H. A. 1957a. *Administrative behavior: a study of decision-making processes in administrative organization,* 2nd ed. Free Press, New York. xlix + 259 pp.

Simon, H. A. 1957b. *Models of man.* John Wiley, New York. xvi + 287 pp.

Simon, H. A. 1979. *Models of thought.* Yale University Press, New Haven, Conn. xviii + 524 pp.

Slatkin, M., and J. Maynard Smith. 1979. Models of coevolution. *Quarterly Review of Biology,* 54(3): 233–263.

Slobin, D. 1971. *Psycholinguistics.* Scott, Foresman, and Co., Glenview, Ill. xii + 148 pp.

Smets, Gerda. 1973. *Aesthetic judgment and arousal: an experimental contribution to psycho-aesthetics.* Leuven University Press, Leuven (Belgium). xviii + 106 pp.

Smith, E. A. 1979. Human adaptation and energetic efficiency. *Human Ecology,* 7(1): 53–74.

Smith, E. A. 1981. The application of optimal foraging theory to the analysis of hunter-gatherer group size. In B. P. Winterhalder and E. A. Smith, eds., *Hunter-gatherer foraging strategies: ethnographic and archaeological analyses.* University of Chicago Press, Chicago, forthcoming.

Sneath, P. H. A., and R. R. Sokal. 1973. *Numerical taxonomy: the principles and practice of numerical classification.* W. H. Freeman, San Francisco. xvi + 573 pp.

Sorokin, P. 1957. *Social and cultural dynamics.* Porter Sargent, Boston. xii + 719 pp.

Spickett, S. G. 1963. Genetic and developmental studies of a quantitative character. *Nature,* 199(4896): 870–873.

Spottswood, P. J., and G. M. Burghardt. 1976. The effects of sex, book weight, and grip strength on book-carrying styles. *Bulletin of the Psychonomic Society,* 8(2): 150–152.

Steele, B. F., and C. B. Pollock. 1968. A psychiatric study of parents who abuse infants and small children. In R. E. Helfer and C. H. Kempe, eds., *The battered child,* pp. 103–147. University of Chicago Press, Chicago.

Stein, M., P. Ottenberg, and N. Roulet. 1958. A study of the development of olfactory preferences. *Archives of Neurology and Psychiatry,* 80: 264–266.

Steiner, J. E. 1979. Oral and facial innate motor responses to gustatory and to some olfactory stimuli. In J. H. A. Kroeze, ed., *Preference behaviour and chemoreception,* pp. 247–261. Informational Retrieval, London.

Stephan, G. E. 1979. Derivation of some social-demographic regularities from the theory of time-minimization. *Social Forces,* 57(3): 812–823.

Stephens, W. R., Jr. 1979. The rise of the Hittite Empire: a comparison of theories on the origins of the state. *Mid-American Review of Sociology,* 4(1): 39–55.

Stern, C. 1973. *Principles of human genetics,* 3rd ed. W. H. Freeman, San Francisco. xii + 891 pp.

Stevens, Janice R. 1979. Schizophrenia and dopamine regulation in the mesolimbic system. *Trends in Neurosciences,* 2(4): 102–105.

Steward, J. H. 1955. *Theory of culture change: the methodology of multilinear evolution.* University of Illinois Press, Urbana. 244 pp.

Strausfeld, N. J. 1976. *Atlas of an insect brain.* Springer-Verlag, New York. xiv + 214 pp.

Strobeck, C. 1975. Selection in a fine-grained environment. *American Naturalist,* 109(968): 419–425.

Swanson, C. P. 1973. *The natural history of man.* Prentice-Hall, Englewood Cliffs, N.J. xiv + 402 pp.

Symons, D. 1979. *The evolution of human sexuality.* Oxford University Press, New York. ix + 358 pp.

Szentágothai, J. 1979. Local neuron circuits of the neocortex. In F. O. Schmitt and F. G. Worden, eds., *The neurosciences: fourth study program,* pp. 399–415. MIT Press, Cambridge, Mass.

Szentágothai, J., and M. Arbib. 1974. Conceptual models of neural organization. *Neurosciences Research Program Bulletin,* 12(3): 307–510.

Tanaka, J. 1976. Subsistence ecology of central Kalahari San. In R. B. Lee and I. DeVore, eds., *Kalahari hunter-gatherers: studies of the !Kung San and their neighbors,* pp. 98–119. Harvard University Press, Cambridge, Mass.

Taylor, M. A., and R. Abrams. 1977. More on genetic transmission in schizophrenia. *American Journal of Psychiatry,* 134(4): 457.

Terrace, H. S., L. A. Petitto, R. J. Sanders, and T. G. Bever. 1979. Can an ape create a sentence? *Science,* 206(4421): 891–902.

Terray, E. 1975. Classes and class consciousness in the Abron Kingdom of Gyaman. In M. Bloch, ed., *Marxist analyses and social anthropology,* pp. 85–135. John Wiley, New York.

Terrell, J. 1974. Comparative study of human and lower animal biogeography in the Solomon Islands. *Solomon Island Studies in Human Biogeography, Field Museum of Natural History, Chicago,* no. 3. ii + 44 pp.

Terrell, J. 1977. Human biogeography in the Solomon Islands. *Fieldiana: Anthropology* (Chicago), 68(1): 1–47.

Thoday, J. M. 1979. Polygene mapping: uses and limitations. In J. N. Thompson, Jr., and J. M. Thoday, eds. *Quantitative genetic variation,* pp. 219–233. Academic Press, New York.

Thomas, H. A., Jr. 1971. Population dynamics of primitive societies. In S. F. Singer, ed., *Is there an optimum level of population?,* pp. 127–155. McGraw-Hill Book Co., New York.

Thompson, J. N., Jr. 1979. Polygenic influences upon development in a model character. In J. N. Thompson, Jr., and J. M. Thoday, eds., *Quantitative genetic variation,* pp. 243–261. Academic Press, New York.

Thompson, J. N., Jr., and T. N. Kaiser. 1979. Computer simulation of the breeding program for polygene action. In J. N. Thompson, Jr., and J. M. Thoday, eds., *Quantitative genetic variation,* pp. 235–242. Academic Press, New York.

Thompson, J. N., Jr., and J. M. Thoday. 1979. Synthesis: polygenic variation in perspective. In J. N. Thompson, Jr., and J. M. Thoday, eds., *Quantitative genetic variation,* pp. 295–301. Academic Press, New York.

Tiger, L., and R. Fox. 1971. *The imperial animal.* Holt, Rinehart & Winston, New York. xii + 308 pp.

Travers, J., and S. Milgram. 1969. An experimental study of the small world problem. *Sociometry,* 32: 425–443.

Trinkhaus, E., and W. W. Howells. 1979. The Neanderthals. *Scientific American,* 241(6): 118–133.

Tulving, E. 1972. Episodic and semantic memory. In E. Tulving and W. Donaldson, eds., *Organization of memory,* pp. 382–403. Academic Press, New York.

Turner, J. R. G. 1970. Changes in mean fitness under natural selection. In K. Kojima, ed., *Mathematical topics in population genetics,* pp. 32–78. Springer-Verlag, New York.

Tversky, A., and D. Kahneman. 1971. Belief in the law of small numbers. *Psychological Bulletin,* 76(2): 105–110.

Tversky, A., and D. Kahneman. 1973. Availability: a heuristic for judging frequency and probability. *Cognitive Psychology,* 5(2): 207–232.

Tversky, A., and D. Kahneman. 1974. Judgment under uncertainty: heuristics and biases. *Science,* 185(4157): 1124–1131.

van den Berghe, P. L. 1979. *Human family systems: an evolutionary view.* Elsevier, New York. xii + 254 pp.

van den Berghe, P. L., and G. M. Mesher. 1980. Royal incest and inclusive fitness. *American Ethnologist,* 7(2): 300–317.

Vandenberg, S. G. 1967. Heredity factors in normal personality traits (as measured by inventions). *Recent Advances in Biological Psychiatry,* 9: 65–104.

Vandenberg, S. G., and K. Wilson. 1979. Failure of the twin situation to influence twin differences in cognition. *Behavior Genetics,* 9(1): 55–60.

VanDeventer, A. D., and D. R. Laws. 1978. Orgasmic reconditioning to redirect sexual arousal in pedophiles. *Behavior Therapy,* 9(5): 748–765.

Waddington, C. H. 1953. Genetic assimilation of an acquired character. *Evolution,* 7(2): 118–126.

Waddington, C. H. 1957. *The strategy of the genes: a discussion of aspects of theoretical biology.* George Allen & Unwin, London. x + 262 pp.

Waddington, C. H. 1960. *The ethical animal.* George Allen & Unwin, London. 230 pp.

Waddington, C. H. 1962. *New patterns in genetics and development.* Columbia University Press, New York. xvi + 271 pp.

Wald, G. 1969. The molecular basis of human vision. In B. R. Straatsma, M. O. Hall, R. A. Allen, and F. Crescitelli, eds., *The retina: morphology, function and clinical characteristics,* pp. 281–295. University of California Press, Berkeley.

Wallace, A. F. C. 1970. *Culture and personality,* 2nd ed. Random House, New York. x + 271 pp.

Walls, D. F. 1976. Non-equilibrium phase transitions in sociology. *Collective Phenomena,* 2: 125–130.

Wang, M. C., and G. E. Uhlenbeck. 1945. On the theory of Brownian motion, II. *Reviews of Modern Physics,* 17(2,3): 323–342.

Wason, P. C., and P. N. Johnson-Laird. 1972. *Psychology of reasoning: structure and content.* Harvard University Press, Cambridge, Mass. viii + 264 pp.

Wasserman, L. 1979. Alienation incident. *The Humanist,* 39(3): 4–10.

Wassermann, G. D. 1978. *Neurobiological theory of psychological phenomena.* Macmillan Co., London. xii + 301 pp.

Wattenwyl, A. von, and H. Zollinger. 1979. Color-term salience and neurophysiology of color vision. *American Anthropologist,* 81(2): 279–288.

Weidlich, W. 1972. The use of statistical models in sociology. *Collective Phenomena,* 1: 51–59.

Weinberg, S. K. 1976. *Incest behavior,* rev. ed. Citadel Press, New York. xxx + 291 pp.

Wessells, N. K. 1977. *Tissue interactions and development.* W. A. Benjamin, Menlo Park, Calif. xii + 246 pp.

West, B. J. 1974. Speculators in a model market. *Collective Phenomena,* 1: 195–217.

Whissell-Buechy, D., and J. E. Amoore. 1973. Odour-blindness to musk: simple recessive inheritance. *Nature,* 242(5395): 271–273.

White, L. A. 1948. Review of *From savagery to civilization,* by G. Clark; and *History,* by V. G. Childe. *Antiquity,* 22(88): 217–218.

White, L. A. 1949a. Ethnological theory. In R. W. Sellars, V. J. McGill, and M. Farber, eds., *Philosophy for the future,* pp. 357–384. Macmillan Co., New York.

White, L. A. 1949b. *The science of culture: a study of man and civilization.* Grove Press, New York. xx + 444 pp.

White, L. A. 1963. Individuality and individualism: a culturological interpretation. *Texas Quarterly,* 6: 111–127.

Whorf, B. L. 1956. *Language, thought, and reality.* MIT Press, Cambridge, Mass. xii + 278 pp.

Wickelgren, W. A. 1979a. *Cognitive psychology.* Prentice-Hall, Englewood Cliffs, N.J. xii + 436 pp.

Wickelgren, W. A. 1979b. Chunking and consolidation: a theoretical synthesis of semantic networks, configuring in conditioning, S-R versus cognitive learning, normal forgetting, the amnesic syndrome, and the hippocampal arousal system. *Psychological Review,* 86(1): 44–60.

Wilbert, J. 1976. To become a maker of canoes: an essay in Warao enculturation. In J. Wilbert, ed., *Enculturation in Latin America: an anthology* (Latin American Studies, vol. 33), pp. 303–358. Latin American Center, University of California, Los Angeles.

Wilbert, J. 1979. Geography and telluric lore of the Orinoco Delta. *Journal of Latin American Lore,* 5(1): 129–250.

Williams, R. S., P. C. Marshall, I. T. Lott, and V. S. Caviness, Jr. 1978. The cellular pathology of Menkes steely hair syndrome. *Neurology,* 28(6): 575–583.

Williams, T. R. 1972a. *Introduction to socialization: human culture transmitted.* C. V. Mosby, St. Louis, Mo. xiv + 308 pp.

Williams, T. R. 1972b. The socialization process: a theoretical perspective. In F. E. Poirier, ed., *Primate socialization,* pp. 207–260. Random House, New York.

Wilson, D. S. 1980. *The natural selection of populations and communities.* Benjamin/Cummings Co., Reading, Mass. xviii + 186 pp.

Wilson, E. O. 1975. *Sociobiology: the new synthesis.* Belknap Press of Harvard University Press, Cambridge, Mass. x + 697 pp.

Wilson, E. O. 1978. *On human nature.* Harvard University Press, Cambridge, Mass. xii + 260 pp.

Wilson, E. O. 1980a. Comparative social theory. In S. M. McMurrin, ed., *The Tanner Lectures on Human Values,* vol. 1, pp. 49–73. University of Utah Press, Salt Lake City.

Wilson, E. O. 1980b. Caste and division of labor in leaf-cutter ants (Hymenoptera: Formicidae: *Atta*): II, the ergonomic optimization of leaf cutting. *Behavioral Ecology and Sociobiology,* 7(2): 157–165.

Wilson, E. O., T. Eisner, W. R. Briggs, R. E. Dickerson, R. L. Metzenberg, R. D. O'Brien, M. Susman, and W. E. Boggs. 1978. *Life on earth,* 2nd ed. Sinauer Associates, Sunderland, Mass. xiv + 846 pp.

Wilson, R. S. 1978. Synchronies in mental development: an epigenetic perspective. *Science,* 202(4371): 939–948.

Winter, S. G. 1971. Satisficing, selection, and the innovating remnant. *Quarterly Journal of Economics,* 85(2): 237–261.

Winterhalder, B. P. 1977. Foraging strategy of the boreal forest Cree: an evaluation of theory and models from evolutionary ecology. Ph.D. dissertation, Cornell University.

Winterhalder, B. P., and E. A. Smith, eds., 1981. *Hunter-gatherer foraging strategies: ethnographic and archaeological analyses.* University of Chicago Press, Chicago, forthcoming.

Wobst, H. M. 1974. Boundary conditions for paleolithic social systems: a simulation approach. *American Antiquity,* 39(2): 147–178.

Wolf, A. P. 1966. Childhood association, sexual attraction, and the incest taboo: a Chinese case. *American Anthropologist,* 68(4): 883–898.

Wolf, A. P. 1968. Adopt a daughter-in-law, marry a sister: a Chinese solution to the problem of the incest taboo. *American Anthropologist,* 70(5): 864–874.

Wolf, A. P. 1970. Childhood association and sexual attraction: a further test of the Westermarck hypothesis. *American Anthropologist,* 72(3): 503–515.

Wolf, A. P., and C. S. Huang. 1980. *Marriage and adoption in China, 1845–1945.* Stanford University Press, Stanford, Calif. xxii + 426 pp.

Wolff, P. H. 1970. "Critical periods" in human cognitive development. *Hospital Practice,* 5(11): 77–87.

Wright, S. 1932. The roles of mutation, inbreeding, crossbreeding and selection in evolution. In D. F. Jones, ed., *Proceedings of the sixth international congress of genetics* (Ithaca, N.Y., 1930), vol. 1, pp. 356–366. Brooklyn Botanic Gardens, Brooklyn, N.Y.

Wright, S. 1970. Random drift and the shifting balance theory of evolution. In K. Kojima, ed., *Mathematical topics in population genetics,* pp. 1–31. Springer-Verlag, New York.

Yakovlev, P. I., and A.-R. Lecours. 1967. The myelogenetic cycles of regional maturation of the brain. In A. Minkowski, ed., *Regional development of the brain in early life,* pp. 3–70. Blackwell, Oxford.

Yerkes, R. M., Ada W. Yerkes. 1936. Nature and conditions of avoidance (fear) response in chimpanzees. *Journal of Comparative Psychology,* 21(1): 53–66.

Young, J. Z. 1978. *Programs of the brain.* Oxford University Press, Oxford. viii + 325 pp.

Zachary, W. W. 1977. An information flow model for conflict and fission in small groups. *Journal of Anthropological Research,* 33(4): 452–473.

Zajonc, R. B. 1968. Cognitive theories in social psychology. In G. Lindzey and A. Aronson, eds., *The handbook of social psychology,* 2nd ed., vol. 1: *Historical introduction, systematic positions,* pp. 320–411. Addison-Wesley, Reading, Mass.

Index

Book design by Marianne Perlak
Composed in VIP Times Roman by Progressive Typographers
Paper: 55 pound Glatfelter
Binding: Joanna Arrestox B
Printed and bound by Halliday Lithograph